Angiogenesis in
Health and Disease

NATO ASI Series

Advanced Science Institutes Series

*A series presenting the results of activities sponsored by the NATO Science Committee,
which aims at the dissemination of advanced scientific and technological knowledge,
with a view to strengthening links between scientific communities.*

The series is published by an international board of publishers in conjunction with the
NATO Scientific Affairs Division

A	Life Sciences	Plenum Publishing Corporation
B	Physics	New York and London
C	Mathematical and Physical Sciences	Kluwer Academic Publishers
D	Behavioral and Social Sciences	Dordrecht, Boston, and London
E	Applied Sciences	
F	Computer and Systems Sciences	Springer-Verlag
G	Ecological Sciences	Berlin, Heidelberg, New York, London,
H	Cell Biology	Paris, Tokyo, Hong Kong, and Barcelona
I	Global Environmental Change	

Recent Volumes in this Series

Series A: Life Sciences

Angiogenesis in Health and Disease

Edited by

Michael E. Maragoudakis

University of Patras Medical School
Patras, Greece

Pietro Gullino

University of Turin Medical School
Turin, Italy

and

Peter I. Lelkes

University of Wisconsin Medical School
Milwaukee, Wisconsin

Plenum Press
New York and London
Published in cooperation with NATO Scientific Affairs Division

Proceedings of a NATO Advanced Study Institute
on Angiogenesis in Health and Diseases,
held June 16–27, 1991,
in Porto Hydra, Greece

Library of Congress Cataloging-in-Publication Data

Angiogenesis in health and disease / edited by Michael E.
  Maragoudakis, Pietro Gullino, and Peter I. Lelkes.
       p.   cm. -- (NATO ASI series. Series A, Life sciences ; v.
  227.)
     "Published in cooperation with NATO Scientific Affairs Division."
     "Proceedings of a NATO Advance Study Institute on Angiogenesis in
  Health and Diseases, held June 16-27, 1991, in Porto Hydra, Greece"-
  -T.p. verso.
     Includes bibliographical references and index.
     ISBN 0-306-44196-9
     1. Neovascularization--Congresses.   I. Maragoudakis, Michael E.
  II. Gullino, Pietro M.   III. Lelkes, Peter I.   IV. North Atlantic
  Treaty Organization. Scientific Affairs Division. V. NATO Advanced
  Stuoy Institute on Angiogenesis in Health and Diseases (1991 :
  Hydra, Greece) VI. Series.
     [DNLM: 1. Neovascularization--congresses.   WG 500 A5855 1991]
  QP106.6A52 1992
  612.1'3--dc20
  DNLM/DLC
  for Library of Congress   92-6789
                                                                CIP

ISBN 0-306-44196-9

© 1992 Plenum Press, New York
A Division of Plenum Publishing Corporation
233 Spring Street, New York, N.Y. 10013

Printed in the United States of America

PREFACE

Angiogenesis describes the formation of new blood vessels, which arise as outgrowths from existing vessels. In many physiological processes such as ovulation and wound healing angiogenesis is involved for a relatively short time. Otherwise under normal physiological conditions in the adult organism angiogenesis is an extremely slow process. By contrast in certain disease states such as diabetic retinopathy, arthritis, chronic inflammation, hemangiomas, etc., angiogenesis persists and contributes to the pathology of these disease states. Some 50 such "angiogenic diseases" have been described where angiogenesis is involved. Also in tumor growth and metastasis angiogenesis is an essential process and precedes neoplastic transformation. Hence, angiogenesis could become an important diagnostic tool and a target for developing therapeutic agents.

This book contains the proceedings of the NATO Advanced Study Institute on "Angiogenesis in Health and Disease" held in Porto Hydra, Greece, from June 16-27, 1991. This meeting was a comprehensive review of endothelial cell biology and endothelial cell phenotypic and functional heterogeneity in relation to angiogenesis under physiological and pathological conditions. Numerous in vitro and in vivo models were presented, which are used to study angiogenesis at the molecular and cellular levels and to evaluate chemical compounds or naturally occuring substances for their effect on angiogenesis. The presentations and discussions at this meeting provided an opportunity for the basic science and the clinical disciplines to meet, exchange information and provide future research directions for many investigators engaged in the study of angiogenesis.

I thank Drs. Pietro Gullino and Peter Lelkes, co-Directors of this meeting and the International Organizing Committee that included Drs. Juliana Denekamp, Olga Hudlicka and Shant Kumar for their invaluable help in the selection of topics, speakers and arranging the Scientific Program. I thank also all the participants for their enthusiastic participation and their complimentary comments on the success of this conference. In addition, I thank the Scientific Affairs Division of NATO who provided the major portion of the funds for the meeting and the publication of this book. The contribution of the following organizations: Abbott (Greece), Bristol-Meyers-Squibb (Greece), Ciba-Geigy (Greece), Daiichi Pharm. (Japan), Dupont Merck Pharm. Co. (U.S.A.), Galenica (Greece), Genentech (U.S.A.), Help (Greece), Jensen (Greece), National Drug Industry (Greece), National Drug Organization (Greece), Smith Kline and French Beecham (Greece), Upjohn (U.S.A.), Vianex (Greece), which was used to support the participation of many young scientists, is gratefully acknowledged. For the daily operation of the conference and travel arrangements I am thankful to the hotel management Mrs. Afroditi Constantinidou and Mrs. Lydia Argyropoulou. I am particularly grateful to Mrs. Anna Marmara for her dedicated and enthusiastic work in the organization of the meeting and the editing of this monograph.

Michael E. Maragoudakis (Greece)

v

CONTENTS

THE DEVELOPEMNT OF VASCULAR SYSTEM

BIOLOGY OF ENDOTHELIAL CELLS AND ANGIOGENESIS

ANGIOGENESIS IN DISEASE STATES

PROMOTERS AND INHIBITORS OF ANGIOGENESIS

THERAPEUTIC POTENTIAL OF PROMOTERS AND INHIBITORS OF ANGIOGENESIS

EMBRYONIC DEVELOPMENT OF BLOOD VESSELS

Françoise   Dieterlen-Lièvre   and   Luc   Pardanaud

Institut d'Embryologie cellulaire et moléculaire du CNRS et
du Collège de France, 49bis, av. de la Belle Gabrielle
94130 Nogent s/Marne, France

The first step in the formation of blood vessels is the emergence
of their inner lining cells, the endothelial cells (EC). These cells then
associate into tubular structures and other cells types, notably
smooth muscle cells or pericytes, organize themselves around them to
make the vessel wall. Structural specializations such as tight junctions
in vessels of the blood-brain barrier or regionalized adhesion
molecules are acquired by differentiating EC. The development of
new blood vessels is an uncommon physiological event in adults. It
occurs during development of the corpus luteum or the placenta,
during wound healing and regeneration or during tumorigenesis. The
latter phenomenon has been elected by many groups for detailed
morphological and biochemical analysis. Various *in vivo* and *in vitro*
models have been devised to study this process. It could be shown
that endothelial cells do not arise *de novo* in the adult but derive
from already established blood vessels. The basement membrane
surrounding the capillaries breaks down; endothelial cells undergo
mitogenesis and migrate into the tumor under the effect of
chemoattractive and mitogenic factors (review in Folkman, 1985).
This process has been designated as angiogenesis and several growth
factors have been identified with angiogenic properties. The
quiescent state normal in most physiological situations is maintained
through an interplay between endothelial cells and pericytes (Sato et
al., 1990).

Angiogenesis in Health and Disease, Edited by M.E. Maragoudakis *et al.*
Plenum Press, New York, 1992

During ontogeny endothelial cells must clearly emerge *de novo*. To study this process, designated as vasculogenesis (Risau and Lemmon, 1988) means of detecting these cells as soon as possible after their determination are crucial. Early investigators explanted the whole chick embryo *in vitro* and observed the development of the vascular tree by following the movements of red cells (Sabin, 1920). In the scanning electron microscope, cords of cells that eventually converge into tubular structures can be diagnosed because of this evolution as endothelial cell forerunners (Hirakow and Hiruma 1981).

Monoclonal antibodies directed against the quail hemangioblastic lineage

But real progress came from monoclonal antibodies that recognize isolated endothelial cells with great sensitivity (Péault et al., 1983; Pardanaud et al., 1987). These antibodies were obtained against immunogens from the quail species. They actually recognize the hemangioblastic lineage, i.e. endothelial and hemopoietic cells (HC). These antibodies gave a boost to research on the embryonic emergence of endothelial cells, making the avian embryo a priviledged model for this study. The association or substitution of cells or rudiments between two different species, the chick and the quail, has been devised on the basis of the presence in quail cell nuclei of a large heterochromatin mass that serves as a marker (Le Douarin, 1973). In most cells, this nucleolar-associated DNA is easily identified, though its aspect may differ depending on the cell type. In endothelial cells where the nucleus is usually elongated and flattened, the nucleolar marker is often less visible. Moreover the close association of EC and pericytes makes them difficult to tell apart (Jotereau and Le Douarin, 1978). Thus the antibodies, that recognize a definite lineage in the quail only, provide a decisive advantage. Two such antibodies have been obtained, MB1 (Péault et al., 1983) and QH1 (Dieterlen-Lièvre, 1984; Pardanaud et al., 1987). They were raised against different antigens, yet their specificities are closely similar. MB1 was produced in mice immunized against the mu chain of quail IgM. It has been extensively characterized (Péault et al., 1983; Péault, 1987; Labastie et al., 1986, 1989). It recognizes several glycoproteins of apparent molecular masses ranging from 80 to 200 kDa. One of them is the immunizing mu chain, another is the plasma protease inhibitor alpha2-macroglobulin, while the other bands are glycoproteins, probably differing by their degree of glycosylation. Quail endothelial cells have been isolated and cultured *in vitro* and could be shown to secrete alpha2-macroglobulin in the medium (Labastie et al., 1986, Péault, 1987). The other antibody, QH1, was obtained in our group from mice immunized with E13 quail embryo bone marrow cells (Dieterlen-Lièvre, 1984, Pardanaud et al., 1987).

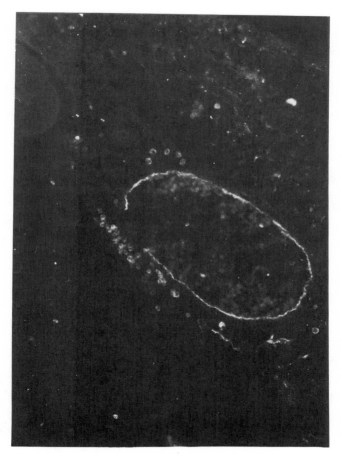

Figure 1.   QH1 antibody applied to an E6 quail/chick chimera (sagittal section). This chimera was obtained by associating complementary parts of blastodiscs from the two species (Martin *et al.*, 1980). Quail endothelial cells have colonized most of the aortic endothelium at this level and also participate to smaller chimeric vessels. Round isolated QH1 positive cells are quail hemopoietic cells.

The antigen first becomes expressed at an early stage of evolution of these lineages (Péault et al., 1988). It is not present on mature erythrocytes. The immunogen appears highly antigenic as this specificity is often obtained. In contrast, no mab recognizing chick endothelial cells has been obtained to date, while several antibodies directed to the hemopoietic lineage are restricted to this species (common leukocyte antigen, CL1, Houssaint et al., 1987) as well as others specific for lymphocyte subpopulations (Chen et al., 1991). Both MB1 and QH1 are reliable and sensitive markers and neither has affinity for chick cells of any type (figure 1). Thus they constitute a powerful probe to trace EC and HC in quail/chick chimeras. QH1 is deposited in the Developmental Studies Hybridoma Bank Baltimore (Md, USA) and has already been extensively used to investigate various problems related or not to vascular development (Sariola et al., 1984; Coffin and Poole, 1988; Schramm and Solursh, 1990; Williams et al., 1989; Noden, 1989).

Mesoderm partitioning

Many different questions may be asked concerning the ontogeny of endothelial cells, notably:
1) how does the angioblastic lineage segregate from the mesoderm ?
2) is there a common precursor of the angioblastic and hemopoietic lineage ?
3) are there regional differences in the mode of vascularization of the embryo ?
4) what is the role of the microenvironment in the differentiation of blood vessels ?
5) what happens to endothelial cells during remodelling of the vascular tree ?

Our investigations addressed mainly the first three of these questions. Older evidence established that endothelial cells and blood cells originate from the mesoderm. This germ layer forms through the convergence of cells towards and the ingression through the primitive streak, followed by their centrifugal migration in the median plane of the blastodisc (figure 2).

Embryonic development proceeds through serial choices during which cell potentialities progressively become restricted. The mesoderm laid down during this process of gastrulation indeed subdivides into structures with defined fates (table 1 and figure 3). The history of many of these derivatives has been precisely established through cell marking or cell transplantation experiments.

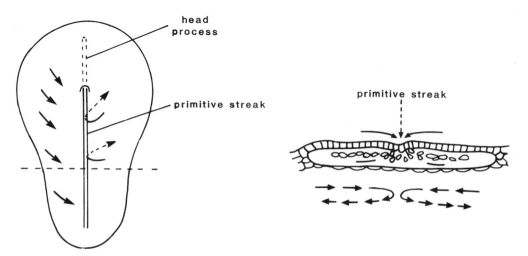

Figure 2. Gastrulation in the chick embryo. Cells from the superficial layer invaginate through the primitive streak and diverge giving rise to the mesoderm. The process is schematized from the top and in section.

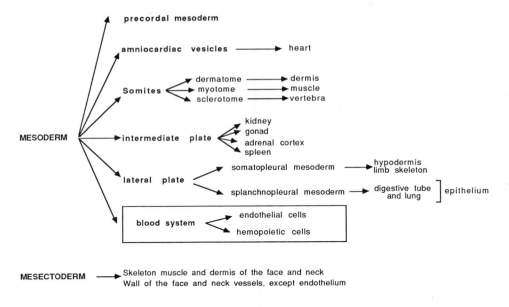

Table 1

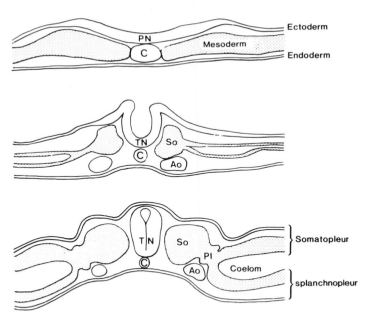

Figure 3. Scheme of truncal mesoderm evolution in the avian embryo. Mesoderm is first a homogeneous cell sheet, then it splits, forming the coelom. Finally it divides into various substructures. The first endothelial cells in the embryonic area appear close to the endoderm and form the aorta. Ao = aorta. PI = intermediate plate. PN = neural plate. So = somite. TN = neural tube

By comparison the history of the blood forming and vascular system still has many gaps. In particular no definite experimental argument attributes unequivocally the origin of vascular and blood cells to one of the mesodermal substructures indicated in table 1. Neither is there experimental evidence in favour of the existence of a common hemangioblastic anlage, though this is suggested by the tight topographic and functional links between the two lineages.

Though mobility is a prominent property of embryonic endothelial cells, migration in the plane of the blastodisc does not account for the development of the endothelial network. Endothelial cells emerge in all the surface of the blastodisc, as was shown in "yolk sac chimeras" associating the extraembryonic area of a chick and the embryonic body of a quail. The frontiers between the peripheral chick territory and the central quail territory were always neat and endothelia did not encroach upon them (Beaupain et al., 1979).

The case of the mesectoderm

In the head of the embryo, where there are no somites, most supporting tissues have a special origin; they derive from a subpopulation of neural crest cells called the mesectoderm. The mesectoderm gives rise to the bone, cartilage, connective tissue and to most of the muscle in the head of the embryo. It also gives rise to the wall of blood vessels in the head and upper trunk. Notably, however, endothelial cells in these blood vessels is never derived from the mesectoderm, as demonstrated in mapping experiments performed by transplanting the mesencephalic and rhombencephalic neural tube from the quail into the chick (Le Lièvre and Le Douarin, 1975). In this region, the vessel walls build up through the anatomical and physiological association of lineages originating from different embryonic germ layers. In particular, EC and pericytes from different origins cooperate.

Development of blood vessels in the germ layers of the embryo

When the early blastodisc is stained *in toto* with QH1 at various stages, endothelial cells are seen to appear progressively and shortly after to associate until a complete network invests the whole surface (Pardanaud et al., 1987). On sections vascularisation appears to proceed asymmetrically. The positive cells (MB1 in this study) always appear among mesodermal cells in close vicinity to the endoderm

(Péault et al., 1988). In the embryonic area they appear at the level of the first pair of somites and the emerge in parallel with somitogenesis (figure 4). These cells canalize and become the two dorsal aortae in the primitive embryo. The dorsalmost mesodermal cells remains devoid of MB1+ cells for a couple of hours more. Then the dorsal aortae emit branches that penetrate between the somites. These aspects strongly suggest a phenomenon of angiogenesis from the aortic endothelium. In the vascular area it is unclear whether the dorsalmost MB1+ cells differentiate in situ from mesodermal precursors or whether they immigrate in this dorsal site. Hence the notion that splanchnopleural and somatopleural mesoderm have different angioblastic potentialities, that was submitted to experimental analysis.

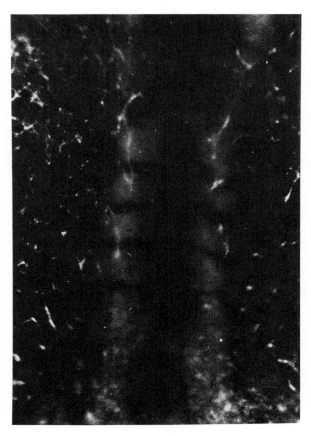

Figure 4. Somitic level of a quail blastodisc stained in toto with QH1. Forerunners of the aortae are seen along the somites.

Development of blood vessels during organogenesis

We have shown that, at the organogenesis period, distinct mechanisms are responsible for the formation of the endothelial network, depending on the germ layer makeup of the rudiment (Pardanaud et al., 1989). The experimental scheme is to transplant various rudiments from one species into the coelom of a host of the other species. When the transplant is quail into a chick host, any QH1+ cell is intrinsic. When the transplant is from the chick into a quail host, positive cells are immigrants from the host. The rudiments of internal organs were found to develop their endothelial network from intrinsic precursors, while external organ rudiments (limb bud) acquired their endothelial network from host immigrating cells. Internal organs on the other hand were populated by host hemopoietic stem cells. Thus at this period of development the two lineages clearly have distinct origins. In contrast, in the limb bud and bone marrow, cells of the two lineages develop from colonizing extrinsic precursors. This pattern of development has been disclosed in the bone marrow many years ago in quail/chick combinations analyzed by means of the Feulgen reaction (Jotereau and Le Douarin, 1978). Our experiments using the QH1+ mab fully confirmed these early results and completed them by disclosing the distinct origins of the two lineages in internal organs.

Our recent experiments (unpublished) about the potentialities of the early germ layers (somatopleura versus splanchnopleura) extend the conclusion to earlier stages: endothelial cells emerge from the mesoderm only when this germ layer is tightly associated to the endoderm. Extrinsic colonization by endothelial precursors has been demonstrated by similar experiments in the case of the mouse kidney (a mesodermal rudiment) (Sariola et al., 1988) and the brain (neural ectoderm) (Stewart and Wiley, 1981; Risau et al., 1988). Thus we conclude that endoderm exerts a obligatory influence probably of an inductive nature on the associated mesoderm. A similar influence had been disclosed by Wilt (1965) concerning the differentiation of blood cells in the extraembryonic area. It is fair to mention that this role of the endoderm is not accepted by everybody. According to Noden (1989), transplantation of any mesodermal tissue into the head region of a host (quail/chick system + polyclonal ab) yields endothelial cells in all cases. More work is necessary to resolve the discrepancy between Noden's results and ours. However we would like to argue the following: when somites are retrieved, endothelial cells of the future aortae are closely associated to their ventral side and are probably taken along. In this context interesting data obtained by Stern et al. (1988) are relevant. These authors have marked with a fluorescent dye single cells of either the last somite or the segmental plate. Two days later, clones were found distributed exclusively in the different components of the somite, when the cell was marked in the somite or the cephalicmost side of the segmental plate. In contrast when a cell was marked in the caudalmost side of the unsegmented plate, the clone became distributed in the somite but also in the blood system: unexpectedly endothelial cells were marked exclusively on the ventral aspect of the aorta. Furthermore in some embryos a large number of hemopoietic cells in the aorta were fluorescent. The latter cells are part of the early intraembryonic process of hemopoiesis that we have documented earlier (Dieterlen-Lièvre and Martin, 1981). Thus it appears that some progenitors in the unsegmented plate have pluripotentialities including somitic and blood system forming capacities. Why these cells never gave rise to the dorsal aortic endothelium remains to be understood. It should be emphasized that this region where endothelial cells emerge from the segmental plate is precisely the one where these cells have a chance for close contact with the endoderm (see figure 3). Further mapping experiments should disclose where the precursors for the dorsal endothelial cells arise, and whether their localisation involves a migratory process.

Our hypothesis that a contact with the endoderm is an obligatory requirement for the emergence of endothelial cells is supported by an experiment of nature; among the appendages of the avian embryo, the amnios never becomes vascularized; indeed it is constituted by mesoderm and ectoderm, while the yolk sac and allantois that develop a rich vascular network and produce hemopoietic cells, are both made of mesoderm and endoderm.

Is there a common hemangioblastic rudiment ?

Descriptive aspects suggest its existence at two phases of early development. The blood islands of the extraembryonic area appear as homogeneous aggregates of cells. Eventually the outer cells will become endothelial and the inner ones hemopoietic. The second phase concerns the budding of hemopoietic cells by the wall of the aorta at E3. Several morphological and developmental properties of the early erythropoietic cells set them apart from the latter series. Among them, we want to review here recent findings about the expression of some nuclear oncogenes, because they may turn out useful markers for the divergence of the endothelial and hemopoietic lineage. These protooncogenes, c-ets1 and c-myb, were known to be expressed in the adult's immature lymphoid cells. In the chick embryo, c-ets1 protein is present in the nuclei of immature endothelial cells and c-myb is expressed in the nuclei of immature hemopoietic cells (Vandenbunder et al., 1989). In the E3 aortic wall, complementary patterns of expression are found: the c-ets1 molecular probe gives a positive hybridization signal on the aortic endothelium, while the c-myb probe lits up the ventral intraluminal aggregates of hemopoietic cells. However in the E2 chick embryo extraembryonic blood islands, no signal is obtained with the c-myb probe, while an ubiquitous signal appears with the ets probe, both on the presumptive endothelial and hemopoietic cells. When the blood islands mature, the ets signal disappears from the inside hemopoietic cells.

These differences is expression patterns between early and later hemopoiesis processes are also the rule in mice embryo, as evinced in mice with a targeted deletion of the c-myb gene (Mucenski et al., 1991). These mice develop normally until E13 and die thereafter: they produce perfectly normal yolk sac blood cells but fail to undergo the hematopoietic switch from primitive to definitive erythropoiesis. In other words, the c-myb protein is not necessary during the yolk sac phase but becomes indispensable for the liver phase of hemopoiesis.

The complementary gene expression should turn out interesting at the least to probe lineage determination.

Conclusion

Many aspects remain to be understood about the emergence and differentiation of endothelial cells in the embryo, not to mention other processes about which no information is available. For instance how do endothelial cells and pericytes associate, how do organ specific differentiation of blood vessels occur ? If the necessity of endodermal contact for endothelial emergence becomes confirmed, then it will be interesting to approach other problems: nature of the inductive stimulus in rudiments with intrinsic endothelial precursors, nature of attractive factors in rudiments colonized by precursors. The already abundant knowledge about angiogenesis in the adult will certainly guide the approaches to these problems. Concerning the relationship between hemopoietic and angioblastic lineages, future research will be aimed at finding whether there exists a common progenitor, the hemangioblast, and how its early commitment to one or the other lineages fits in with the experimental data, i.e. the colonization of some organ rudiments by distinct progenitors.

REFERENCES

Beaupain, D., Martin, C. and Dieterlen-Lièvre, F., 1979, Are developmental hemoglobin changes related to the origin of stem cells and site of erythropoiesis ? Blood, 53: 212.

Chen, C.H., Bucy, R.P. and Cooper, M.D., 1991, T cell differentiation in birds, Sem. Immunol.

Coffin, J.D. and Poole, T.J., 1988, Embryonic vascular development: immunohistochemical identification of the origin and subsequent morphogenesis of the major vessel primordia in quail embryos, Development, 102: 735.

Dieterlen-Lièvre, F., 1984, Emergence of intraembryonic blood stem cells studied in avian chimeras by means of monoclonal antibodies, Dev. Comp. Immunol., supplt.3: 75.

Dieterlen-Lièvre, F. and Le Douarin, N.M., 1991, The use of avian chimeras in developmental biology, In: "Avian Immunology, Basis and Practice", A. and P. Toivanen eds, CRC Press, sous presse.

Dieterlen-Lièvre, F. and Martin, C., 1981, Diffuse intraembryonic hemopoiesis in normal and chimeric avian development, Dev. Biol., 88: 180.

Folkman, J., 1985, Toward an understanding of angiogenesis: search and discovery, Perspectives in Biology and Medicine, 29: 10.

Hirakow, R. and Hiruma, T., 1981, Scanning electron microscopy study on the development of primitve blood vessels in chick embryos at the early somite-stage, Anat. Embryol., 163: 299.

Houssaint, E., Tobin, E., Cihak, J. and Lösch, U., 1987, A chicken leukocyte common antigen: biochemical characterization and ontogenetic study, Eur. J. Immunol., 17: 287.

Jotereau, F. and Le Douarin, N.M., 1978, The developmental relationship between osteocytes and osteoclasts: a study using the quail-chick nuclear marker in endochondral ossification, Dev. Biol,. 63: 253.

Kessel, J. and Fabian, B., 1987, Inhibitory and stimulatory influences on mesodermal erythropoiesis in the early chick blastoderm, Development, 101: 45.

Labastie, M.C., Poole, T.J., Péault, B.M. and Le Douarin, N.M., 1986, MB1, a quail leukocyte-endothelium antigen: partial characterization of the cell surface and secreted forms in cultured endothelial cells. Proc. Natl. Acad. Sci. USA, 83: 9016.

Labastie, M.C., 1989, MB1, a quail leukocyte-endothelium antigen : further characterization of soluble and cell-associated forms. Cell Diff. and Dev., 27: 151.

Le Douarin, N.M., 1973, A biological cell labeling technique and its use in experimental embryology, Dev. Biol., 30: 217.

Le Lièvre, C. and Le Douarin, N., 1975, Mesenchymal derivatives of the neural crest: analysis of chimaeric quail and chick embryos, J. Embryol. exp. Morphol., 34: 125.

Martin, C., Beaupain, D. and Dieterlen-Lièvre, F., 1980, A study of development of the hemopoietic system using Quail-Chick chimeras obtained by blastoderm recombination, Dev. Biol., 75: 303.

Mucenski, M.L., McLain, K., Kier, A.B., Swerdlow, S.H., Schreiner, C.M., Miller, T.A., Pietryga, D.W., Scott, W.J. and Potter, S.S., 1991, A functional c-myb gene is required for normal murine fetal hepatic hematopoiesis, Cell, 65: 677.

Noden, D.M., 1989, Embryonic origins and assembly of blood vessels, Am. Rev. Resp. Dis., 140: 1097.

Pardanaud, L., Altmann, C., Kitos, P., Dieterlen-Lièvre, F. and Buck,

C.A., 1987, Vasculogenesis in the early quail blastodisc as studied with a monoclonal antibody recognizing endothelial cells, Development, 100: 339.

Pardanaud L., Yassine F. and Dieterlen-Lièvre F., 1989, Relationship between vasculogenesis, angiogenesis and hemopoiesis during avian ontogeny, Development. 105, 473.

Péault, B., Thiery, J.P., and Le Douarin, N.M., 1983, Surface marker for the hemopoietic and endothelial cell lineages in the quail that is defined by a monoclonal antibody, Proc. Natl. Acad. Sci. USA, 80: 2976.

Péault, B., 1987, MB1, a quail leukocyte/vascular endothelium antigen: characterization of the lymphocyte-surface form and identification of its secreted counterpart as alpha2-microglobulin, Cell Differentiation, 21: 175.

Péault, B., Coltey, M. and Le Douarin, N.M., 1988, Ontogenic emergence of a quail leukocyte/endothelium cell surface. Cell Differentiation, 23: 165.

Risau, W. and Lemmon, V., 1988, Changes in the vascular extracellular matrix during embryonic vasculogenesis and angiogenesis, Dev. Biol., 125: 441.

Sabin, F.R., 1920, "Studies on the origin of the blood-vessels and of red-blood corpuscles as seen in the living blastoderm of chicks during the second day of incubation," In: Contributions to Embryology, vol.9, Carnegie Inst. Wash., pp.214-262.

Sariola, H., Péault, B., Le Douarin, N.M., Buck, C., Dieterlen-Lièvre, F. and Saxen, L., 1984, Extracellular matrix and capillary ingrowth in interspecies chimeric kidneys, Cell Differentiation, 15: 43.

Sato, Y., Tsuboi, R., Lyons, R., Moses, H. and Rifkin, D.B., 1990, Characterization of the activation of latent TGF-b by co-cultures of endothelial cells and pericytes or smooth muscle cells: a self-regulation system, J. Cell Biol., 111: 757.

Schramm, C. and Solursh, M., 1990, The formation of premuscle masses during chick wing bud development. <u>Anat. Embryol.</u>, 182: 235.

Stern, C.D., Fraser, S.E., Keynes, R.J. and Primmett, D.R.N., 1988, A cell lineage analysis of segmentation in the chick embryo, <u>Development</u>, 104: supplt., 231.

Stewart, P.A. and Wiley, M.J., 1981, Developing nervous tissue induces formation of blood-brain barrier characteristics in invading endothelial cellls: a study using quail-chick transplantation chimeras, <u>Dev. Biol.</u>, 84, 183.

Vandenbunder, B., Pardanaud, L., Jaffredo, T., Mirabel, M.A. and Stéhelin, D., 1989, Complementary patterns of expression of c-ets1, c-myb and c-myc in the blood-forming system of the chick embryo, <u>Development</u>, 106: 265.

Williams, R.L., Risau, W., Zerwes, H.G., Drexler, H., Aguzzi, A. and Wagner, E.F., 1989, Endothelioma cells expressing the polyoma middle T oncogene induce hemangiomas by host cell recruitment, <u>Cell</u>, 57: 1053.

Wilt, F.H., 1965, Erythropoiesis in the chick embryo: the role of endoderm. <u>Science</u>, 147, 1588.

REGULATION OF EMBRYONIC BLOOD VESSEL FORMATION

Hannes Drexler, Harald Schnürch,
Georg Breier, Werner Risau

Max-Planck-Institut für Psychiatrie
Abt. Neurochemie
Am Klopferspitz 18A
8033 Martinsried
FRG

INTRODUCTION

Endothelial cell turnover within normal adult tissues
generally appears to be very low and is in the range of months
or even years (Denekamp, 1984). In contrast, the
proliferation rate of endothelial cells lining the blood
vessels that are involved in tumor vascularization is very
high and in the range of a few days (Engermann et al., 1967).
The massive formation of new blood vessels is not an exclusive
characteristic of tumor development, but also occurs in other
pathological conditions like diabetic retinopathies or
rheumatoid arthritis. It can also be observed during wound
healing as well as during ovulation and corpus luteum
formation in the female adult, but in these instances the
angiogenic process seems to be regulated by control mechanisms
that prevent an overshooting reaction. This controlled
angiogenic process may be termed physiological angiogenesis to
distinguish it from the aberrant formation of blood vessels
under various pathological conditions.
    Physiological angiogenesis also occurs during embryonic
development as the demands of the embryo for supplies of
oxygen and nutrients and for the removal of waste products are
steadily increasing.

MODEL SYSTEMS TO STUDY EMBRYONIC BLOOD VESSEL FORMATION

In the very early embryo the blood islands (also called
hemangioblasts) differentiate from the splanchnopleuric
mesoderm of the area vasculosa. In the center of the blood
island blood cell precursor cells differentiate whereas in the
periphery the blood islands are surrounded by differentiating
endothelial cell precursors (Sabin, 1917; Clark and Clark,
1939; Haar and Ackermann, 1971).

Angiogenesis in Health and Disease, Edited by M.E. Maragoudakis *et al.*
Plenum Press, New York, 1992

One system to study the early vascular development are embryonic stem cells. Embryonic stem cells are directly established from the inner cell mass of mouse blastocysts. When cultured on a feeder layer of embryonic fibroblasts they proliferate but do not differentiate. Upon withdrawal from the feeder layer, they spontaneously differentiate in suspension culture through a series of embryoid bodies of increasing complexity. Finally they form so called cystic embryoid bodies.In the yolk sac like structure of these bodies, blood islands begin to differentiate spontaneously in a simple cell culture medium (Doetschman et al., 1985). They can be recognized on the surface of the cystic embryoid bodies as red dots.The experiments using embryonic stem cells show that blood island differentiation is an intrinsic program of the development of the yolk sac and that embryonic stem cells are able to supply themselves with all factors necessary for angioblast induction. Initial experiments with purified growth and hemopoietic factors to augment this formation of blood islands were negative. However, treatment of the cystic embryoid bodies with human cord serum resulted in a slight but significant increase of blood island formation. The nature of the active component(s) within this serum remain elusive.

When blood island induction has occured the first event that follows is the formation of a primary capillary plexus from growing and fusing blood islands. This process could not be observed in the embryonic stem cell derived cystic embryoid bodies when cultured in vitro. However, when they were implanted into the peritoneal cavity of syngeneic mice a capillary plexus in the cystic embryoid bodies was formed (Risau et al., 1988). This result suggests that additional factors are needed for the development of a primary capillary plexus from blood islands. At the moment the nature of this inducing signal in this in vivo system is unknown.

Based on morphological studies of developing chick embryos it has been hypothesized that it is the interaction between endoderm and mesoderm which is necessary for the differentiation of blood islands (Flamme, 1989) and the formation of a primary capillary plexus, but direct evidence for such an interaction is still missing. So in contrast to the induction of blood islands which is an intrinsic program of the yolk sac development, the subsequent formation of a primary capillary plexus by the fusion of the growing blood islands does not seem to be a intrinsic step, but requires some kind of regulatory interaction or signalling in the embryo.

A distinction can be made between vasculogenesis, the development of blood vessels from in situ differentiating angioblasts and angiogenesis, the formation of new blood vessels by sprouting from preexisting ones (Risau and Lemmon, 1988; Risau et al, 1988; Pardanaud et al., 1989; Poole and Coffin, 1989). Vasculogenesis is not exclusively restricted to the development of the yolk sac but is also operative during vascularization of endodermal organs like the lung (Pardanaud et al., 1989). On the other hand, vascularization of the brain which is of ectodermal origin or the kidney which is of mesenchymal origin occurs through an angiogenic process which was demonstrated by transplantation experiments using the chick-quail chimera system (Stewart and Wiley, 1981; Ekblom et al., 1982). During organogenesis endothelial cells

sprout from already existing vessels that are located outside
the organ anlagen and invade the organ rudiment.

We have focused our interest on the vascularization of
the brain and the subsequent differentiation of the brain
endothelium. At day 3 of embryonic development a primary
capillary plexus surrounds the neural tube of the chick. The
endothelial cells lining the vessels of this capillary plexus
are derived from migratory angioblasts that have invaded the
head region early on (Bär, 1980). From this so called
perineural plexus, capillary sprouts begin to radially invade
the neuroectoderm. The endothelial cells of this perineural
vascular plexus at first locally degrade the perineural
basement membrane, then migrate into the neuroectoderm and
finally branch in the subependymal layer in a highly
reproducible pattern. The molecular signal which triggers the
coordinated invasion of the neuroectoderm at this stage of
development is unknown.

Angiogenesis generally involves basement membrane
degradation, which is a prerequisite for endothelial cells to
be able to migrate out in to the perivascular tissue. It also
involves directional endothelial migration and endothelial
cell proliferation. Subsequently there is a final maturation
process with the formation of a new basement membrane and the
appearance of pericytes and smooth muscle cells.

As far as the degradation of the basement membrane is
concerned, experiments with endothelioma cells expressing the
middle T oncogen of the polyoma virus, provided interesting
cues on how the structure of the blood vessel wall might be
regulated (Williams et al., 1988).

When these endothelioma cells were injected into chick or
quail embryos multiple hemangiomas formed rapidly in the yolk
sac as well as within the embryo.The formation of hemangiomas
was not abrogated when thymidin labeled and mitomycin C
treated endothelioma cells were injected, which were unable to
proliferate and which could be detected in sections of
hemangiomas by autoradiography (Williams et al., 1989). Only
1-5% of the cells enclosing the blood filled sac comprising
the hemangiomas contained thymidine labelled nuclei. These
labelled nuclei were distributed throughout the inner wall of
the hemangiomas within a discontinous layer of host
endothelial cells (Williams et al., 1989). Taken together,
these results indicate that 1) endothelioma cells had
integrated into the vessel wall, 2) that the wall of the
hemangiomas are not exclusively constructed by the injected
endothelioma cells but that host cells participate in the
establishment of the hemangiomas and 3) that the integrated
endothelioma cells had disrupted the normal vessel morphology.

When endothelioma cells are grown within fibrin gels they
proliferate rapidly and form large cyst-like structures, which
are strikingly reminiscent of the morphological appearance of
the cavernous hemangiomas formed in vivo. This is in contrast
to the behaviour of normal endothelial cells that formed tube
like structures within the fibrin gel (Montesano, 1990).

Interestingly, by the addition of serin protease
inhibitors like Trasylol to the culture medium, the
endothelioma cells regained the phenotype of normal
endothelial cells indicating an involvement of protease
activity in the observed malformations (Montesano, 1990).

It is known that activated endothelial cells are able to secrete a variety of degrading enzymes like the serine protease Plasmingenactivator (u-PA), as well as metalloproteases like type IV-specific collagenase, interstitial collagenase and stromelysin. The activity of these enzymes eventually results in a fragmentation of the basement membrane and the extracellular matrix. The plasminogen-plasmin system is one of the enzyme systems that has been extensively studied in cellular invasion processes. The active component of this system is plasmin, a tryptic protease capable of degrading components of the ECM and of activating other matrix degrading enzymes e.g. the metalloproteinase zymogens.

In normal endothelial cells u-PA activity remains cell associated or is secreted into the culture medium where it is complexed with PAI.When the proteolytic properties of the endothelioma cells were analyzed, it was found that these cells displayed a dramatically increased PA activity than their normal counterparts which is due to the transcription of a high level of u-PA mRNA. In addition, the endothelioma cells synthesize much lower amounts, if any PAI-1 (Montesano et al., 1990). This emphasizes the importance of a finely tuned balance between proteolytic and antiproteolytic activities in normal vessel development and points at the possibility that any disturbance of this balance has a severe impact on the morphology of the blood vessel wall.

FGF AND THE VASCULARIZATION OF THE BRAIN

Endothelial cell migration and endothelial cell proliferation are two other processes are as well important during angiogenesis.

Using chemotaxis and proliferation assays we have investigated whether the embryonic chick brain produces factor(s) that could be relevant for the angiogenic response. By using heparin sepharose chromatography and RP-HPLC we were able to characterize two growth factors that stimulated endothelial cell proliferation in vitro and that were also angiogenic in vivo and were identified as acidic and basic fibroblast growth factor (aFGF, bFGF; Risau, 1986; Risau et al., 1988b).

The mitogenic activity purified from chick brain extracts increased several hundredfold during chick brain development reaching a maximum at around day 14, which is consistent with the expression of the aFGF gene during chick brain development. Expression of aFGF is already detectable on embryonic day 3 and increases prominently on day 9 until day 14 (Schnürch et al., 1991). In addition, endothelial cell proliferation in the chick brain also reaches a maximum on embryonic day 14, suggesting that the aFGF-like activity from chick brain might be involved in the regulation of embryonic brain vascularization.

However, northern blot analysis showed that the aFGF gene transcription is not down regulated at the time when endothelial cell proliferation stops in the postnatal chick brain (Schnürch et al., 1991) and that comparatively large amounts of aFGF protein can be purified from adult brain extracts (Böhlen et al., 1985; Gospodarowicz et al., 1984; Lobb et al., 1984; Risau et al., 1988b).

An important question is whether aFGF, a polypeptide that is lacking a signal sequence, get out of the cells that are synthesizing it and what cells in the brain are synthesizing the polypeptide? At least the classical pathway of protein secretion obviously cannot be used in the case of aFGF and so far, the mechanism of aFGF secretion, if any, remains elusive.

In situ hybridization studies using a chick specific single stranded DNA probe for aFGF indicate that the pattern of aFGF expression during chick brain development can not be correlated with endothelial cell proliferation and angiogenesis. According to our results aFGF is expressed in neurons at a time when neurons are already postmitotic which happens between embryonic day 4 - day 9 in the central nervous system of the chick. Furthermore, endothelial cells and smooth muscle cells of blood vessels (cortical capillaries, veins and pial vessels) as well as ependymal cells and choroid plexus epithelial and endothelial cells did not express aFGF mRNA at a detectable level.

These results suggest that aFGF has a function as differentiation or maintenance factor for postmitotic neurons rather than being the primary angiogenic factor in embryonic chick brain (Schnürch et al., 1991).

What do we know about the expression of receptors for FGF during the time when brain angiogenesis occurs? One possibility might be the down-regulation of receptor levels during embryonic brain development. This hypothesis is supported by recent experiments of Olwin and Hauschka (Olwin et al., 1990).

However, in situ hybridization data demonstrate that the receptors for FGF do not seem to be present on embryonic endothelial cells at the time of brain vascularization (Heuer et al., 1990; Wanaka et al., 1991), also indicating that it might not be the aFGF polypeptide which is the crucial angiogenic factor during brain vascularization.

Are there other polypeptides that might function as angiogenic factors during embryonic brain vascularization? Several factors have been identified up to now that belong to the growing family of heparin binding growth factors. An important structural difference between acidic and basic FGF and other members of this family is the presence of a hydrophobic signal peptide in all other polypeptides indicating that they are secreted via the known pathways.

It is not yet clear whether the other members of the FGF family have an influence on embryonic vascular development. At least FGF-5 is a potent mitogen for endothelial cells and is expressed in neonatal rat brain (Zhan et al., 1988). Virtually nothing is known about the role of the other members of the FGF family with respect to brain angiogenesis or angiogenesis in general.

SPECIFIC ENDOTHELIAL CELL GROWTH FACTORS

Recently several mitogens have been purified and characterized that are specific for endothelial cells. Platelet-derived endothelial cell growth factor (Pd-ECGF) specifically stimulates endothelial cell proliferation and chemotaxis in vitro and angiogenesis in vivo (Ishikawa et al., 1989). Pd-ECGF, like acidic and basic FGF, lacks a hydrophobic signal sequence. Since Pd-ECGF is a platelet product, it may promote the repair of the endothelial cell layer over denuded areas of the vessel wall. Pd-ECGF may also be involved in embryonic angiogenesis because is also expressed in human placenta (Usuki et al., 1990), a tissue that is highly vascularized and where the proliferation rate of endothelial cells is comparatively higher than in normal tissues.

More recently, another factor, called vascular endothelial growth factor (VEGF) with distinct target cell specificity for vascular endothelial cells has been characterized from the culture medium of different cell types (Ferrara et al., 1989; Gospodarowicz et al., 1989; Keck et al., 1989). The cDNA cloning of the mouse gene for VEGF revealed the existance of three molecular forms of VEGF which differ by the insertion or deletion of short domains near the carboxy-terminal end of the protein (Breier et al, in press). When transfected into COS cells the two lower molecular weight forms were secreted into the culture medium whereas the high molecular weight form of VEGF remained cell associated. The secretion pattern of the different VEGF forms was reflected by the mitogenic activity for endothelial cells that could be identified within the culture supernatants of the transfected COS cells.

In situ hybrization studies using an antisense RNA probe detected VEGF transcripts in various organs of the developing mouse embryo. In the brain of 15 and 17 day old mouse embryos VEGF mRNA was present in cells of the ependymal layer that is invaded by capillary sprouts during brain vascularization and also in choroid plexus epithelium. When adult brain was analyzed, only background labeling was observed in the ependymal layer (Breier et al, in press). These results support our hypothesis that factors which act specifically on endothelial cells probably are involved in the regulation of brain vascularization. It is tempting to think of VEGF in this context as it seems to be expressed at the right time and in the appropriate locations.

REGRESSION OF BLOOD VESSELS

During embryonic development the formation of new blood vessel occurs in parallel with the regression of blood vessels in certain regions of the embryo, a phenomenon that has been ignored for quite a long time.

The regression or the death of part of the vasculature is observed under various developmental conditions:
1) The sinus marginalis within the developing limb bud regresses probably due to changes in blood flow during the course of limb outgrowth.
2) During limb development when the digits and toes are freed from their interdigital tissue, a simultaneous regression of tissue concomitantly with its vasculature can be observed (Hurle et al., 1985).
3) Regression of blood vessels without regression of the surrounding tissue and without prior changes in blood flow occurs during the development of the skeletal elements in the limb bud (Feinberg et al., 1986; Hallmann et al., 1987).

The last example is the most interesting one because only the endothelium in the respective regions of the limb bud seem to be affected and appear to be specifically eliminated.

When a limb bud of a HH stage 28 chick embryo is injected with ink to reveal the blood vessel system and stained for cartilage one can recognize the prospective areas of the digits which are already avascular. Morphological analysis of the development of these avascular areas within the limb bud demonstrated that the blood capillaries in the prechondrogenic regions of the limb bud regress before the onset of cartilage differentiation (Hallmann et al., 1987). Our current working hypothesis is that factors specific for endothelial cells are present locally in the limb that may control endothelial cell proliferation proliferation or even induce endothelial cell death in certain areas thereby permitting chondrocytes to differentiate. It is as well conceivable that chondrocyte precursor cells produce such factors.

Attempts to extract an endothelial cell death inducing activity from day 4 - day 10 chick embryos resulted in the purification of a low molecular weight compound with an molecular weight of 2.5 kDa as judged by SDS-PAGE. This compound induced the death of most endothelial cells in vitro within several hours and is able to induce the fragmentation of the DNA into discrete fragments of 180 bp and multiples thereof which is a typical characteristic for apoptotically dying cells (Drexler et al, unpunblished). Direct N-terminal gasphase sequencing of this low molecular weight compound was impossible probably due to a blocked N-terminus. At the moment we are carrying out further studies using mass spectroscopy methods to elucidate the primary structure of the inhibitor as well as studies on the cell specifity. We are also preparing antibodies which may help us to learn more about its tissue distribution in the embryo.

REFERENCES

Bär, T. (1980). The Vascular Sytem of the Cerebral Cortex. Springer Verlag. Heidelberg.

Böhlen, P., Esch, F., Baird, A. and Gospodarowicz, D. (1985). Acidic fibroblast growth factor (FGF) from bovine brain: amino-terminal sequence and comparison with basic FGF. EMBO J. **4**: 1951-1956.

Clark, E.R.and Clark, E.L. (1939).Microscopic observations on the growth of blood capillaries in the living mammal. Am. J. Anat. **64**: 251-301.

Denekamp, J. (1984).Vascular endothelium as the vulnerable element in tumors. Acta Radiol. Oncol. **23**: 217-217

Doetschman, T.C., Eistetter, H., Katz, M., Schmidt, W. and Kemöer, R. (1985). The in vivo development of blastocyst-derived embryonic stem cell lines: formation of visceral yolk sac, Blood islands and myocardium. J. Embryol. exp. Morphol. **87**: 27-45.

Ekblom, P., Sariola, H., Karkinen, M. and Saxen, L. (1982). The origin of the glomerular endothelium. Cell Differ. **11**: 35-39.

Engermann, R. L., Pfaffenbach, D. and Davis, M. D. (1967). Cell turnover of capillaries. Lab. Invest. **17**: 738-743.

Feinberg, R. N., Latker, C. H. and Beebe, D. C. (1986). Localized vascular regression during limb morphogenesis in the chicken embryo. I. Spatial and temporal changes in the vascular pattern. Anat. Rec. **214**: 405-409.

Ferrara, N. and Henzel, W. J. (1989). Pituitary follicular cells secrete a novel heparin-binding growth factor specific for vascular endothelial cells. Biochem. Biophys. Res. Comm. **161**: 851-858.

Flamme, I. (1989). Is extraembryonic angiogenesis in the chick embryo controlled by the endoderm. Anat. Embryol. **180**: 259-272.

Gospodarowicz, D., Cheng, J., Lui, G., Baird, A. and Böhlen, P. (1984). Isolation of brain fibroblast growth factor by heparin-sepharose affinty chromatography: identity with pituitary fibroblast growth factor. Proc. Natl. Acad. Sci. USA **81**: 6963-6967.

Gospodarowicz, D. and Lau, K. (1989). Pituitary follicular cells secrete both vascular endothelial growth factor and follistatin. Biochem. Biophys. Res. Comm. **165**: 292-298.

Haar, J.L. and Ackermann, G.A. (1971). A phase and electron microscopic study of vasculogenesis and erythropoiesis in the yolk sac of the mouse. Anat. Rec. **170**: 199-224.

Hallmann, R., Feinberg, R. N., Latker, C. H., Sasse, J. and Risau, W. (1987). Regression of bloodvessels precedes cartilage differentiation during chick limb development. Differentiation **34**: 98-105.

Heuer, J. G., von Bartheld, C. S., Kinoshita, Y., Evers, P. C. and Bothwell, M. (1990). Alternating phases of FGF receptor and NGF receptor expression in the developing chicken nervous system. Neuron **5**: 283-296.

Hurle, J. M., Colvee, E. and Fernendez-Teran, M. A. (1985). Vascular regression during the formation of the free digits in the avian limb bud: a comparativ study in chick and duck embryos. J. Embryol. exp. Morph. **85**: 239-250.

Ishikawa, F., Miyazonz, K., Hellman, U., Drexler, H., Wernstedt, C., Hagiwara, K., Usuki, K., Takaku, F., Risau, W. and Heldin, C.-H. (1989). Identification of angiogenic activity and the cloning and expression of platelet-derived endothelial growth factor. Nature **338**: 557-562.

Keck, P. J., Hauser, S. D., Krivi, G., Sanzo, K., Warren, T., Feder, J. and Connolly, D. T. (1989). Vascular permeability factor, an endothelial cell mitogen related to PDGF. Science **246**: 1309-1312.

Lobb, R. R. and Fett, J. W. (1984). Purification of two distinct growth factors from bovine neural tissue by heparin affinty chromatography. Biochem. **23**: 6265-6299.

Montesano, R., Pepper, M. S., Möhle-Steinlein, U., Risau, W., Wagner, E. F. and Orci, L. (1990). Increased proteolytic activity is responsible for the aberrant morphogenetic behaviour of endothelial cells expressing the middle T oncogen. Cell **62**: 435-445.

Olwin, B. B. and Hauschka, S. D. (1990). Fibroblast growth factor receptor level decrease during chick embryogenesis. J. Cell Biol. **110**: 503-509.

Pardanaud, L., Yassine, F. and Dieterlen-Lievre, F. (1989). Relationship between vasculogenesis, angiogenesis and haemopoesis during avian ontogeny. Development **105**: 473-485.

Poole, T.J. and Coffin, J.D. (1989). Vasculogenesis and angiogenesis - two different mechanisms establish embryonic vascular pattern. J. Exp. Zool. **251**: 224-231.

Risau, W. (1986). Developing brain produces an angiogenesis factor. Proc. Natl. Acad. Sci. USA **83**: 3855-3859.

Risau W. and Lemmon, V. (1988). Changes in the vascular extracellular matrix during embryonic vasculogenesis and angiogenesis. Devl. Biol. **125**: 441-450.

Risau, W., Sariola, H., Zerwes, H.-G., Sasse, J., Ekblom, P., Kemler, R. and Doetschman, T. (1988a). Vasculogenesis and angiogenesis in embryonic-stem-cell-derived embryoid bodies. Development **102**: 471-478.

Risau, W., Gautschi-Sova, P., Böhlen, P. (1988b). Endothelial cell growth factors in embryonic and adult chick brain are related to human acidic fibroblast growth factors. EMBO J. **7**: 959-962.

Sabin, F.R. (1917). Origin and development of the primitive vessels of the chick and of the pig. Carnegie Contrib. Ebryol. **6**: 61-124.

Schnürch, H. and Risau, W. (1991). Differentiating and mature neurons express the acidic fibroblast growth factor gene during chick neural development. Development **111**: 1143-1154.

Stewart, P.A. and Wiley, M.J. (1981). Developing nervous tissue induces the formation of blood brain barrier characteristics in invading endothelial cells: A study using quail-chick translpantation chimeras. Devl. Biol. **84**: 183-192.

Usuki, K., Norberg, L., Larsson, E., Miyazono, K., Hellman, U., Wernstedt, C., Rubin, K. and Heldin, C.-H. (1990). Localisation of platelet-derived endothelial cell growth factor in human placenta and purification of an alternatively processed form. Cell Regulation **1**: 577-596.

Wanaka, A., Milbrandt, J. and Johnson, E.M. (1991). Expression of FGF receptor gene in rat development. Development **111**: 455-468.

Williams, R. L., Courneidge, S. A. and Wagner, E. F. (1988). Embryonic lethalities and endothelial tumors in chimeric mice expressing polyoma middle T oncogene. Cell **52**: 121-131.

Williams, R. L., Risau, W., Zerwes, H.-G., Drexler, H., Aguzzi, A. and Wagner, E. F. (1989). Endothelioma cells expressing the polyoma middle T oncogene induce hemangiomas by host cell recruitment. Cell **57**: 1053-1063.

Zhan, X., Bates, B., Hu, X. and Goldfarb, M. (1988). The human FGF-5 oncogene encodes a novel protein related to fibroblast growth factors. Mol. Cell. Biol. **8**: 3487-3495.

ROLE OF HETEROTYPIC INTERACTIONS BETWEEN ENDOTHELIAL CELLS AND PARENCHYMAL CELLS IN ORGANSPECIFIC DIFFERENTIATION: A POSSIBLE TRIGGER FOR VASCULOGENESIS

Peter I. Lelkes and Brian R. Unsworth[*]

Lab. Cell Biology, Dept. of Medicine, Univ. Wisconsin Medical School, and
[*]Dept. of Biology, Marquette University, Milwaukee, WI

Endothelial cell organ-specificity, diversity and heterogeneity. The endothelial cell (EC) lining of the vascular wall is presumably the largest complex functional organ in the body, in terms of exposed surface area. Among the most striking features of ECs are phenotypic diversity within the vascular tree, functional heterogeneity, and organ-specificity[1-3]. Details of the cellular and molecular basis for organ-specificity of ECs are not understood[4,5]. Early embryonic differentiation of ECs occurs via two distinct mechanisms, namely angiogenesis and vasculogenesis[6]. *Angiogenesis* is defined as the formation of new blood vessels by endothelial cell-sprouting from existing blood conduits. In some organs, for example the brain and the kidneys, angiogenesis is the primary form of vascularization. Angiogenic mechanisms are also responsible for the vascularization of tumors[7]. *Vasculogenesis* indicates the in situ transdifferentiation of endothelial cells from local mesenchymal precursor cells. For example, the initial formation of the dorsal aortae and the vascularization of several organs, such as the liver, lung, and pancreas occur, predominantly, via vasculogenesis[8].

Organ-specific EC differentiation requires a number of divergent local stimuli, such as humoral factors, e.g., growth factors or cytokines, contacts with other cells of the vascular wall, and interactions with extracellular matrix proteins[9,10]. Organ-specificity is illustrated primarily by the distinct morphology of ECs: large vessels are lined by a continuous endothelium, while in spleen liver and bone marrow the predominant morphology of ECs is sinusoidal. However, even within a given blood vessel there seems to be a certain degree of EC morphological heterogeneity[11].

Beyond phenotypic heterogeneity, recent studies have revealed significant functional diversity amongst ECs within the vascular tree [11,12]. Arterial and venous ECs exhibit a number of well characterized functional differences, e.g., fibrinolytic activity [13], and stimulus-response-coupling [14,15]. Furthermore the location of ECs within the vascular tree, e.g., in either a large- or a micro-vessel, determines their functional heterogeneity[16,17]. Even ECs of similar morphology show biological differences, depending upon organ derivation. The reason for this functional diversity is not clear, but might be related to differential gene expression/activation, induced by organ-specific cues. Alternately, this heterogeneity might also reflect distinct cellular responses to differences in local hemodynamics.

In vivo, EC heterogeneity is illustrated by distinct cell surface antigens on the luminal side. For example, the adressins or selectins provide for the selective, organ-specific homing of blood borne cells[18]. Similarly, tumor metastasis involves the recognition of organ-specific glycoconjugates on the surface of ECs[19-21]. Several laboratories have described anti-EC antibodies[11,22,23] which recognize selected, organ-specific EC surface antigens.

Transplantation experiments confirm the pivotal role of the local microenvironment in establishing organ-specificity of the vascular lining. For example, implants, of either PC12 cells into the rat spinal cord[24], or of pancreatic ß-cells onto the chick allantoic membrane[25], become neovascularized. The ECs within the transplanted tissues are derived from the host, but acquire the morphology of an attenuated, fenestrated endothelium, characteristic of endocrine glands, in striking contrast to the endothelial phenotype of the surrounding host tissue.

In vitro, isolated ECs rapidly de-differentiate. However, some degree of organ-specificity is maintained, or can be (re)introduced, in culture. For example, tumor cells preferentially adhere to cultured capillary ECs derived from organs which, in vivo, are the known targets for metastasis[26,27]. Organ-specificity in vitro is inducible[28]: for example, large vessel ECs develop blood-brain-barrier-specific morphology and functions, upon coculture with astrocytes. This phenotypic switch is caused by humoral factors contained in the astrocyte conditioned medium, acting in conjunction with certain extracellular matrix proteins[29]. Similarly, the particular mix of extracellular matrix proteins, in conjunction with humoral factors, confers organ-specificity on cultured capillary ECs[10].

These data stress the importance of microenvironmental cues in conferring EC differentiation and organ-specificity. The opposite scenario, viz. the importance of EC-derived cues in modulating parenchymal cell differentiation, has received little attention, and until recently, has been mentioned en passant only. Yet, close scrutiny of the literature discloses that heterotypic cell contacts between ECs and parenchymal cells occur, e.g., during the development of mouse spinal cord[30], chick embryonic adrenal gland [31], and neoanal opossum adrenal medulla[32]. During capillary growth in the mesentery of young rats, cell contacts between fibroblasts and ECs result in fibroblasts transdifferentiating into pericytes[33].

Taken together, this body of evidence supports our hypothesis that EC-derived cues will modulate the differentiation of parenchymal cells in a given organ. In particular, we hypothesize that organ-specific differentiation of neural crest-derived cells of the sympathoadrenal lineage is modulated by heterotypic contacts with vascular ECs[34-37]. This hypothesis, based on our work using isolated adrenal medullary cells as an in vitro model, is corroborated by ultrastructural studies of the developing rat adrenal gland.

On the possible role of endothelial cells in organ-specific sympathoadrenal differentiation in vitro. The final fate and differentiation of neural crest-derived cells in situ, is determined by the microenvironment, viz., by a combination of humoral (trophic) factors, membrane-associated signals (induced e.g., by cell contacts), and/or interactions with extracellular matrix proteins[38,39]. Neural-crest-derived cells of the sympathoadrenal lineage differentiate into sympathetic neurons, or endocrine secretory cells, such as the chromaffin cells in the adrenal medulla[40,41]. Foremost amongst humoral factors, that modulate the differentiation of the sympathoadrenal progenitor cells in situ, are corticosteroids and growth factors, such as nerve growth factor[42,43]. Maturation of chromaffin cells has traditionally been associated with their exposure to glucocorticoids, secreted from the neighboring cortical cells[41,44]. Other differentiative cues in chromaffin cell differentiation are less well documented, however, it is reasonable to assume that such cues exist: for example, fibroblast growth factor, might be locally deposited into the extracellular matrix (ECM) by the surrounding mesenchymal cells[45]. ECM proteins define migratory pathways and neural development[46,47]. Membrane-associated cell adhesion molecules regulate axonal guidance and histogenesis[48]. Heterotypic cell contacts, e.g. via specialized junctions, play a crucial role in embryonic development[49]. Some of the new endothelial (junctional) adhesion molecules, such as PECAM [50] or EndoCAM [51] mediate homotypic and also heterotypic interactions, and might be consequential in intercellular signalling during organogenesis.

The development of the adrenal medulla is pretty well understood, from the perspective of the parenchymal chromaffin cells, but vascularization, in particular the mechanisms leading to the formation of fenestrated capillaries, is virtually unknown[52,53]. In situ differentiation of sympathoadrenal precursors depends on an interplay between a number of genetic and epigenetic factors, which might include both intercellular signalling, such as heterotypic cell contacts, and humoral cues from endothelial (precursor) cells. The literature contains some anecdotal suggestions that ECs might contribute to chromaffin cell differentiation [31,32]. However, to the best of our knowledge, organogenesis of the adrenal medulla has not been addressed from the vantage point of, a) development of the organotypic vasculature and, b) the role of intercellular signalling in the differentiation of both the parenchymal cells and the ECs.

We initially proposed that heterotypic, differentiative, interactions occur between neural crest derived cells of the sympathoadrenal lineage and vascular ECs, during organogenesis of the adrenal medulla [34,36,37]. To test our hypothesis, and to study the role of ECs in sympathoadrenal differentiation in vitro, we have developed a coculture system combining isolated vascular ECs and PC12 cells, a clonal, catecholamine secreting tumor cell line derived from rat adrenal medullary pheochromocytoma [54]. PC12 cells are transdifferentiated by environmental cues, such as nerve growth factor or glucocorticoids, into, respectively, either the neuronal or the endocrine phenotype[55]. Undifferentiated PC12 cells express unique morphological, biochemical, and genetic markers, which change in a characteristic fashion, depending on the response of PC12 cells to environmental cues. Thus, alterations in the expression of these markers provide sensitive indicators for the differentiated state of the cells. Many of the data on PC12 cell transdifferentiation have been confirmed, using immature/neonatal chromaffin cells [55].

Plasticity and availability in pure culture are two reasons for using PC12 cells to study EC-induced modulation of sympathoadrenal differentiation. Specific interactions in vitro between adrenal medullary ECs (AMECs) and PC12 cells have been previously described in detail [36,37]. Suffice it here to summarize some of the pertinent earlier findings and to introduce some newer results.

In coculture with AMECs, PC12 cells preferentially adhere to the ECs and not to AMEC-derived extracellular matrix. The time course and the specificity of this adhesion are similar to those for organ-specific attachment of tumor cells to ECs in vitro, which occurs via putative, organotypic cell surface antigens[26]. Specific PC12 cell/AMEC adhesion was inhibited by proteinaceous cell surface extracts from both these cell types, but not by extracts from unrelated cells. This finding suggests the existence of specific heterotypic cell adhesions molecules, as yet to be identified. In direct coculture with AMECs, PC12 cells became growth arrested in acini-like clusters, reminiscent of in vitro nests of adrenomedullary chromaffin cells. Extending these findings to general mechanisms governing morphogenesis[6], we might conclude that heterotypic intracellular signals involving ECs are also consequential for spatial arrangement (morphogenesis) within a given organ.

In direct coculture with AMECs, PC12 cells failed to extend neurites in response to nerve growth factor (NGF). Significantly, we ascertained that coculture did not change the level of expression of NGF receptors on the PC12 cells [35]. Since NGF modulates the expression of select proteins in PC12 cells cocultured with AMECs (see below), we suggest that the response to coculture conditions is not to completely shut off NGF-signal perception, but rather to modify the intracellular pathways of NGF-induced stimulus-response coupling. One of the consequences then is the abrogation of neurite extension.

Direct coculture with AMECs lead to an increase in PC12 cell met-enkephalin content. Augmentation of met-enkephalin content is believed to be one of the characteristic features of PC12 cell transdifferentiation towards the endocrine phenotype [56]. Direct coculture also resulted in a transient increase (within 30 minutes) of protooncogene c-fos expression in both cell types. The fact that this early marker of subsequent changes in functional/and or differentiative properties, was induced in both cell types, may be indicative of the bi-directionality of intercellular signalling. However, the increase in both met-enkephalin and c-fos are ambiguous indicators for the direction of the differentiation: an increase in these parameters has also been observed upon chemical differentiation of PC12 cells into the sympathetic pathway using NGF [55].

All of these PC12 cell responses specifically occurred in cocultures with AMECs and were not observed with either large vessel-derived ECs or with unrelated cells, e.g. fibroblasts. Most of the original observations, made with *bovine* AMECs, originally termed BAME (bovine adrenal medullary endothelial) cells, have now been repeated with *rat* AMECs, yielding essentially identical results, suggesting that organ-specificity might take precedence over species-specificity.

Except for the c-fos induction, which was also initiated by appropriately conditioned media, the above mentioned effects were all dependent upon direct cell-cell contact between the PC12 cells and the ECs. Taken together, the data indicate that these contacts mediate the transdifferentiation of the PC12 cells towards the endocrine, chromaffin cell phenotype.

Organ-specific, differentiative signaling between ECs and parenchymal cells in vitro, is not restricted to our model system in which we coculture cells derived from the adrenal medulla. For example, similar specific, inter-species interactions exist between cloned (rat) parathyroid epithelial cells and ECs isolated from the (bovine) parathyroid gland [57]. More recently, we observed a long term stabilization of insulin secretion from isolated rat ß-cells, in coculture with ECs isolated from rat pancreatic islets (Mathias and Lelkes, unpublished observations).

In the following, we present some new in vitro data, which strengthen our hypothesis regarding the organ-specificity of the differentiative intercellular signals between ECs and PC12 cells.

Our initial observations demonstrated heterotypic cell contacts, without an intervening basement membrane, in 3 day old PC12/AMEC cocultures [36]. Since it is known that the establishment of a basement membrane is a time-dependent process, we now have investigated the ultrastructure of long term cocultures using light and transmission electron microscopy. The micrographs in **Figure 1** were taken after coculturing PC12 cells and bovine AMEC's for 11 days. The inset (toluidine blue stained thick section) clearly shows the nest-like assembly of the PC12 cells on the surface of the AMEC monolayer. In panel 1, the close (focal) contacts between the PC12 cells and the ECs

29

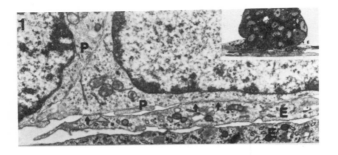

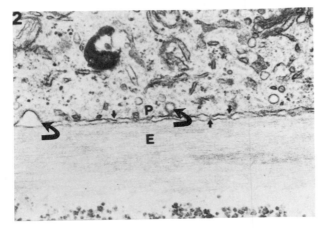

Figure 1. Ultrastructure of PC12 cells and bovine adrenal medullary endothelial cells (BAME) cocultured for 11 days. Panel 1: Arrows indicate the heterotypic contact areas between PC12 cells (P) and endothelial cells (E). (original magnification 12,000 x). Inset: toluidine blue-stained semi-thick section (original magnification 230 x).

Panel 2: Close up of heterotypic contacts between PC12 cell and endothelial cell membranes at higher magnification (32,000 x). Small arrows: putative areas of (gap) junctional contacts. Exocytotic/ endocytotic profiles in the PC12 cell are highlighted by curved arrows.

is evident. There is no intervening basement membrane between the different cell types. At a higher magnification (panel 2), morphology of the focal, heterotypic cell contacts resembles the transmission electron micrograph equivalents of (gap)junctional contacts. Several features on the PC12 cell side are reminiscent of exocytosis/endocytosis, e.g., the Ω-shaped, exocytotic profile and the coated pit. This morphology strongly suggests the existence of at least two independent pathways of intercellular, differentiative signalling in vitro: a) transfer of small molecules (and/or electrical coupling ?) via (gap) junctional contacts, and b) mass transfer of both small and large molecules contained in storage granules. In the latter case, secretion from one cell type into the intercellular space will occur by exocytosis, and the contents of the granules can then be taken up by other cell types, via specific receptors or by receptor mediated endocytosis. Similar interactions might also occur during organogenesis in vivo (see below).

Direct coculture resulted in heterotypic cell contacts between PC12 cells and BAME cells. Under these conditions the proliferation of both cell types was inhibited[37]. We have now further explored the effects of various coculture conditions on cellular proliferation. Coculturing the cells in the same dish, but physically separated using a Millicell system, yielded enhanced growth of both PC12 and BAME cells (**Figure 2**). However, modulation of cellular proliferation by humoral factors is complex, involving labile factors and/or unremittingly secreting ECs: in the continued presence of ECs, these humoral factors have an overall mitogenic effect on PC12 cells growing on the other side of the Millicell system. By contrast, in the presence of 10% BAME conditioned medium, the rate of PC12 cell proliferation was decreased, while the rate of AMEC proliferation was increased by 10% PC12 cell-conditioned medium. Similar data were obtained in cocultures of PC12 cells and rat adrenal medullary ECs (not shown).

Sympathoadrenal differentiation towards either the neuronal or the endocrine phenotype is reflected in qualitative and quantitative differences in catecholamine synthesis and secretion. To assess the effects of coculture conditions on stimulus-secretion coupling, we tested the secretory competence of PC12 cells labelled with [³H]-norepinephrine and seeded onto various substrates. After only 1 hour of cell-contact between PC12 cells and the ECs, the secretory response of the PC12 cells increased significantly (**Figure 3**). By contrast, the amount of radioactivity released

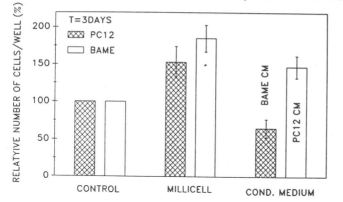

EFFECT OF COCULTURE CONDITIONS ON PC12/BAME PROLIFERATION

Figure 2. Effects of both conditioned medium and physical separation on PC12/BAME proliferation. 1×10^5 PC12 and BAME cells, each, were plated and grown for 3 days under various culture conditions. Hatched histogram - PC12 cells; open histogram - BAME cells. Control - cells growing alone. Millicell - cells grown in the same dish, but physically separated by a Millicell filter. Conditioned medium (CM) - cells grown in the presence of medium conditioned by the other cell type, e.g., PC12 cells grown in BAME-CM.

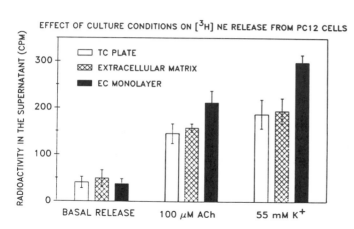

EFFECT OF CULTURE CONDITIONS ON [^{3}H] NE RELEASE FROM PC12 CELLS

Figure 3. Effect of culture conditions on [^{3}H]-Norepinephrine release. PC12 cells were loaded for 60 minutes with [^{3}H]-norepinephrine (NE), and plated, either onto tissue culture plates, on endothelial cell-derived extracellular matrix (ECM), or on a confluent monolayer of BAME cells. After 1 hour, the medium was replaced either with fresh medium, or medium containing 100 μM acetylcholine or 55 mM potassium. Ten minutes later, the supernatants were collected. [^{3}H]-NE release quantitated by liquid scintillation counting.

from PC12 cells, plated onto EC-derived extracellular matrix (ECM), was virtually identical to that of control cells plated onto tissue culture plastic. This short-term augmentation of catecholamine secretion may indicate enhanced efficiency of stimulus-secretion-coupling, rather than a differentiative switch. However, in view of the concomitant activation of *c-fos* (see above), an increase in the secretory competence might be one of the functional changes which herald long-term transdifferentiation of PC12 cells. Indeed, one could consider a novel concept of stimulus-response-synthesis-differentiation coupling.

To make sure that the observed effects are organ-specific, we cocultured PC12 cells with a zoo of ECs, derived from various organs and from different species. Coculture of PC12 cells with large vessel ECs did <u>not</u> induce differentiative signals (e.g. in terms of changes in PC12 morphology, assembly into nest-like structures, or growth patterns). By contrast, AMECs from humans, rats or cows, induced PC12 cell transdifferentiation towards the endocrine phenotype.

As research invariably goes, we did one control experiment too many, namely, we cocultured PC12 cells with ECs isolated from sheep endocardium (SHEC). In contrast to our previous

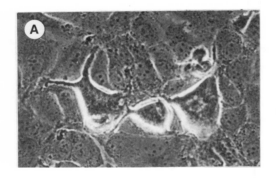
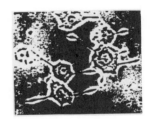

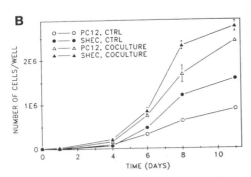
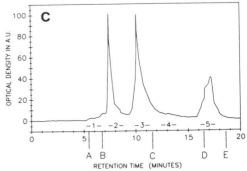

Figure 4. Neurite sprouting and proliferation in PC12 cells cocultured with sheep heart endothelial cells (SHECs). Panel A: PC12 cells plated atop a confluent SHEC monolayer (Orig. mag. 400 x). **Panel B**: Rate of proliferation of both PC12 cells and SHEC under various culture conditions. **Panel C**: Fractionation of the SHEC conditioned medium by size exclusion, HPLC. The numbers 1-5 denote the different fractions that were pooled for biochemical analyses, the retention of the different molecular weight markers is indicated by the letters A-E, ranging from 400 kD (A) to 1.2 kD (E). Inset (top): Video-processed image of the neurite formation in PC12 cells cultured for 48 hours in regular medium supplemented with 10 μg protein from fraction 5.

experience with other ECs, SHECs induced reversible PC12 cell flattening and neurite outgrowth, characteristic of neuronal differentiation (**Figure 4a**) and also stimulated PC12 cell proliferation (**Figure 4b**). In characterizing the factors (it is not clear at present whether it is one compound only which promotes both effects, or several different ones), we determined their humoral nature. Neurite sprouting and PC12 cell proliferation were also induced by SHEC conditioned medium, or by growing the two cell types in the same culture dish but physically separated, e.g. in a Millicell system (not shown). Fractionation of the conditioned medium by gel permeation HPLC (**Figure 4c**) yielded the neurotropically most active fraction in the molecular weight range of 17-44 kD. This molecular weight range and the PC12 cell responses suggest that the active molecule(s) might be a member of the FGF family[55]. However, neutralizing anbtibodies to both aFGF and bFGF failed to inhibit these reponses (not shown).

These data strenghten our hypothesis that differentiative interactions occur between PC12 cells and ECs in coculture. The nature of the EC-derived signals which direct PC12 cell differentiation seem to depend on the provenance of the ECs. Assessment of PC12 cell transdifferentiation is straightforward, since there are markers (e.g. neurite extension, proliferation, morphology, etc.), which uniquely delineate a given differentiative pathway. But what about the reciprocal case, namely, PC12 cell-induced EC differentiation? Light microscopy indicates morphological changes in bovine AMECs in coculture with PC12 cells[37]. Eventually, we would like to employ unique, organspecific EC differentiation markers, for both in vitro experiments and for in vivo studies. Therefore, we have embarked on identifying such markers, by analyzing 2-dimensional gels for the differential expression of total cell proteins following different (co)culture conditions[35]. In a typical silver stained 2-D gel, we can identify in excess of 600 individual proteins. Sophisticated computer-aided image processing techniques yield quantitative analysis of differential protein expression. The partial result of such an analysis for the expression of BAME proteins as a function of the culture

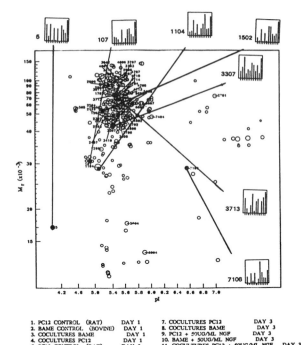

Figure 5. Protein expression in BAME cells cocultured with PC12 cells: Analysis by 2-dimensional polyacrylamide gel. Each protein spot is assigned a unique identifying number. The relative expression of several of these proteins is plotted in the bar graphs for the different culture conditions stated in the Figure.

1. PC12 CONTROL (RAT)	DAY 1	7. COCULTURES PC12	DAY 3	
2. BAME CONTROL (BOVINE)	DAY 1	8. COCULTURES BAME	DAY 3	
3. COCULTURES BAME	DAY 1	9. PC12 + 50UG/ML NGF	DAY 3	
4. COCULTURES PC12	DAY 1	10. BAME + 50UG/ML NGF	DAY 3	
5. PC12 CONTROL (RAT)	DAY 3	11. COCULTURES PC12 + 50UG/ML NGF	DAY 3	
6. BAME CONTROL (BOVINE)	DAY 3	12. COCULTURES BAME + 50UG/ML NGF	DAY 3	

conditions is shown on **Figure 5**. From this and similar experiments, eventually a few proteins of interest will emerge, e.g., those that are turned on in coculture with PC12 cells, but not with other cells. These, putatively organ-specific, proteins will then be identified by microsequencing, etc., thus generating a palette of cellular and molecular tools associated with the term "organ-specific "differentiation" antigens.

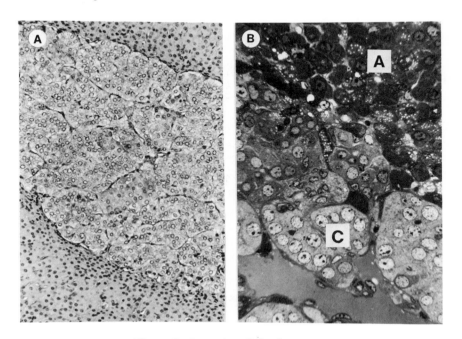

Figure 6. Legend on following page.

33

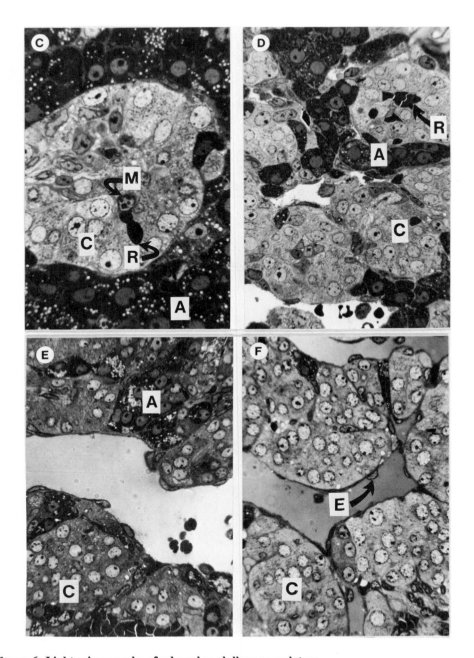

Figure 6. Light micrographs of adrenal medullary vasculature
A: mature organ, B: postnatal day 10, C : birth: avascular medullary island surrounded by cortical cells, note the presence of red blood cells in the vicinity of a "mesenchymal " cell. D: postnatal day 3, E: postnatal day 10, F: postnatal day 21, mature gland.
Panel A: H&E stained paraffin section, orig. mag. 63 x; Panels B-F: Toluidine blue stained semi-thick sections, orig. mag.630 x. Nomenclature used in all labelling: **A** - Adrenocortical cells; **C** - Chromaffin cells; **E** - Endothelial cells; **M** - Mesenchymal cells; **R** - Red blood cells.

Do Heterotypic Cell-Cell Contacts Between Chromaffin Cells and Endothelial Cell Precursors Induce Vasculogenesis in the Developing Rat Adrenal Medulla? To assess the in vivo relevance of our in vitro model, we are re-examining the ultrastructure of the developing rat adrenal from the vantage point of chromaffin cell/endothelial cell interactions. In the following, we will describe vascular development in the neonatal adrenal medulla, leading to the formation of fenestrated capillaries.

Shown in **Figure 6a** is a low power light micrograph of a mature adrenal gland, following conventional Hematoxylin and Eosin staining. The high degree of vascularization throughout the (inner) medulla is characterized by nests of chromaffin cells, wrapped by microvessels. In the mature organ, the cortex is clearly separated from the medulla; blood vessels (capillaries) seem to be well developed with flattened ECs lining the vascular space (**Figure 6b**). Toluidine-blue staining of semi-thick sections clearly differentiates adrenocortical (A) cells (dark, often filled with large translucent lipid droplets), from chromaffin (C) cells and other cells, e.g., endothelial (E) cells, or mesenchymal (M) cells. Chromaffin cells are easily recognized by the circular appearance of their nuclei. By contrast, both vascular ECs lining the blood vessels and mesenchymal cells have irregularly shaped nuclei. Red blood cells (R) appear dark and amorphous.

At birth, there is no clear cut distinction between cortex and medulla; nests of chromaffin cells are intermingled with adrenocortical cells[32]. During the first few days of neonatal development, we frequently find red blood cells (RBCs) within seemingly avascular nests of chromaffin cells (**Figure 6c and 6d**). These extravasated RBCs abut onto chromaffin cells, and onto one or more "indistinct" mesenchymal cell. Serial sections, revealed the lack of a "true" vascular space surrounding these RBCs: we could not detect a contiguous link between established blood vessels and these "quasi blood islands". The presence of RBCs amidst islets of chromaffin cells has previously been observed, but the significance of this observation was unclear at that time[53]. We speculate that RBCs extravasate from immature vasculature[1], and migrate into the "nests" of chromaffin cells, along pathways determined by the least tight intercellular contacts, such as those provided by heterotypic junctions.

Separation between medulla and cortex becomes more evident as the gland matures (**Figure 6e**). By day 10, the organotypic vascular features, characteristic of the adult organ, are prominent (**Figure 6f**).

Ultrastructural analysis, by electron microscopy, confirms the close apposition of extravascular

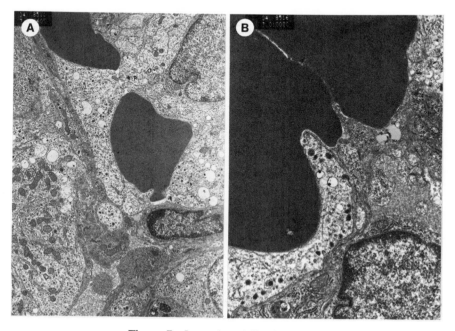

Figure 7. Legend on following page.

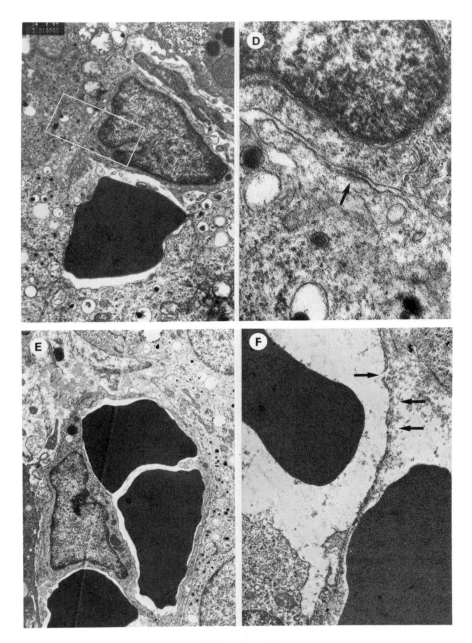

Figure 7. Ultrastructure of adrenal medullary vasculature.
A : neonatal day 1, orig. mag. 4,600 x), **B**: neonatal day 1, orig. mag. 10,000 x, **C**: neonatal day 1, orig. mag 10,000 x. Note the close, bipolar contact of the mesenchymal cell with chromaffin cells. **D**: Close up (orig. mag. 50,000 x) heterotypic (junctional) contact between the mesenchymal cell and the chromaffin cell in C; **E**: neonatal day 3, orig. mag 5,400 x, **F**: neonatal day 5, orig. mag. 42,500 generation of fenestrated capillaries

RBCs and individual chromaffin cells, during early stages of neonatal development of the adrenal gland (**Figure 7a and b**). Higher magnification reveals close, heterotypic contacts between chromaffin cells and mesenchymal cells, without intervening basement membrane (**Figure 7c**). Moreover, at this early post natal stage of differentiation, chromaffin cells and mesenchymal cells seem to communicate via (gap)junctional contacts (**Figure 7d**). The nature of these intimate contacts in situ is reminiscent of the close heterotypic membrane apposition, previously described for the in vitro coculture system between PC12 cells and AMECs (**Figure 1**).

At postnatal day three, some mesenchymal cells, still in contact with, or in close apposition to, chromaffin cells, extend processes which engulf the RBCs (**Figure 7e**). Eventually, these processes develop fenestrae (**Figure 7f**), suggesting transdifferentiation of "indistinct" mesenchymal cells into organotypic endothelial cells. These "mesenchymal" cells now line genuine vascular spaces and develop other EC-like features, such as intercellular junctions of the occluding type (not shown).

Using fluorescent immunomicroscopy on frozen sections, we identified these "indistinct" mesenchymal cells. Although it is difficult to discriminate medullary and cortical regions in the newborn rat adrenal (**Figure 8a**), nests of (tyrosine hydroxylase-positive) chromaffin cells were located in the center of the the gland (**Figure 8c**). Anti-van Willebrand factor antibodies stained ECs lining blood vessels, as well as cells within seemingly avascular spaces(**Figure 8b**), suggesting that these vWF-positive cells might indeed be mesenchymal EC precursors. Similar staining patterns were obtained with other specific markers for chromaffin cells, and ECs, respectively (not shown). Whether these immunological probes will also be useful markers at earlier stages of embryonic organogenesis, remains to be tested. These findings need to be complemented by independent techniques, e.g., by in situ hybridization and immuno-electron microscopy.

These ultrastructural observations add a fascinating new twist to our original hypothesis on the differentiative role of heterotypic cell contacts in chromaffin cell maturation. We propose that these interactions will also trigger neonatal vasculogenesis. Thus, heterotypic contacts between parenchymal and mesenchymal cells will guide the formation of organ-specific capillaries by

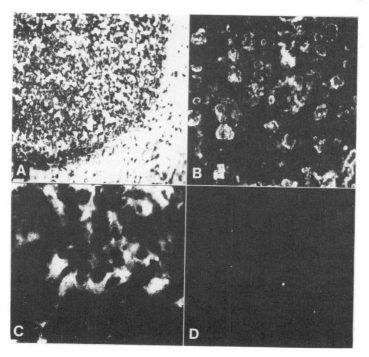

Figure 8. Immunocytochemistry of newborn rat adrenal gland (frozen sections)
A: (Hematoxylin & Eosin stain; original magnification 200 x, **B,C,D:** Indirect immunofluorescence, original magnification 630 x): **B:** Endothelial cell-specific staining (van Willebrand factor), central region. **C:** Chromaffin cell-specific staining (tyrosine hydroxylase), central region. **D:** Control staining, non-immune antibody.

inducing the in situ transdifferentiation of "indistinct" mesenchymal cells into fenestrated ECs. This hypothesis is consistent with recent views that the connective tissue-forming mesenchyme of the embryo exerts control over vascular assembly[58]. If confirmed, this would be the first example of true in situ vasculogenesis during postnatal development, which until now has only been observed in early stages of embryogenesis[59].

Concluding remarks

The cellular and molecular basis of endothelial cell differentiation is an emerging, exciting new field, which involves a variety of disciplines, such as cell biology, biochemistry, physiology, and molecular biology. We believe that the manifestation of EC organspecificity has both a genetic and an epigenetic component. Epigenetic factors comprise microenvironmental cues, including heterotypic interactions with parenchymal cells in the surrounding tissue. Organspecific EC heterogeneity and functional diversity might also be fine tuned by local hemodynamic parameters and modulated by pathological disturbances. Progress in understanding EC organspecificity has been excruciatingly slow, probably due the lack of (organ) specific EC differentiation markers. Such markers are needed to identify EC precursors during embryonic development. Organ-specific endothelial cell differentiation markers could yield the answer to the question why "an endothelial cell is not an endothelial cell is not an endothelial cell"[60]. For example, a newly cloned immediate-early gene, which encodes a product related to the family of GTP-binding proteins, is expressed in cultured human EC upon exposure to the tumor promotor phorbol ester [61]. It remains to be proven, whether the same or similar genes are also induced during organogenesis. Organ-specific differentiation markers might become useful for the early detection of EC transdifferentiation and dysfunction caused by pathophysiological disturbances, such as atherosclerosis, ischemia, and hypertension. We therefore direly need to develop such organ-specific EC developmental markers for studying differential gene activation and protein expression.

Acknowledgements: The early stages of the in vitro experiments were conducted, while one of us (P.I.L) worked in the laboratory of Harvey. B. Pollard (NIDDK,NIH). We would like to acknowledge Harvey's encouragement and the significant contributions of Drs. Philip Lazarovici, Benzion Levi, Yaffa Mizrachi, Alma Munoz, Jose Naranjo and Richard Ornberg. We appreciate the excellent secretarial skills of Barbara Haber. Our recent work has been supported by grants in aid from the Mount Sinai Research Foundation, Milwaukee, and from the American Heart Association National Grant (to P.I.L.), and (to B.R.U) from the Marquette University Committee on Research.

References

1. Majno,G. Ultrastructure of the vascular membrane. In: *Handbook of physiology: circulation*, edited by Hamilton, W.F.& Dow, P. Washington,D.C.: Am. Phys. Soc. p. 2293-2543,(1965).
2. Shepro,D. & D'Amore,P.A. Physiology and biochemistry of the vascular wall endothelium. In: *Handbook of physiology: section 2: the cardiovascular system: volume IV: microcirculation, part I*, ed. by Renkin,E.M. & Michel,C.C. Bethesda, MD: Am. Phys. Soc., p. 103-164, (1984).
3. Simionescu,N. & Simionescu,M. The cardiovascular system. In: *Cell and tissue biology: a textbook of histology*, ed. by Weiss, L. Baltimore: Urban & Schwarzenberg, p. 353-400, (1988).
4. Nugent,J. & O'Connor,M. (eds.) *Development of the vascular system, vol. 100*, London:Ciba Foundation, Pitman, (1983).
5. Feinberg,R.N., Sherer,G.K. & Auerbach,R.(eds.) *Development of the vascular system*, Basel:Karger, (1991).
6. Gilbert,S.F. *Developmental Biology*, Sunderland, MA: Sinauer Asso., 3rd ed., (1991).
7. Folkman,J., Watson,K., Ingber,D. & Hanahan,D. Induction of angiogenesis during the transition from hyperplasia to neoplasia. *Nature* 339:58-61, (1989).
8. Sherer,G.K. Vasculogenic mechanisms and epithelio-mesenchymal specificity in endodermal organs. In: *The development of the vascular system*, ed. by Feinberg,R.N., Sherer,G.K. & Auerbach, R. Basel: Karger, p. 37-57., (1991).
9. Orlidge,A. & D'Amore,P.A. Inhibition of capillary endothelial cell growth by pericytes and smooth muscle cells. *J Cell Biol* 105:1455-1462, (1987).
10. Madri,J.A., Bell,L., Marx,M., Merwin,J.R., Basson,C. & Prinz,C. Effects of soluble factors and extracellular matrix components on vascular cell behavior in vitro and in vivo: models of de-endothelialization and repair. *J.Cell.Biochem.* 45:123-130, (1991).

11. Zetter,B.R. Endothelial heterogeneity: influence of vessel size, organ localization, and species specificity on the properties of cultured endothelial cells. In: *Endothelial cells: vol. II*, ed. by Ryan, U.S. Boca Raton, FL: CRC Press, p. 63-79, (1988).

12. Gerritsen,M.E. Functional heterogeneity of vascular endothelial cells. *Biochem Pharmacol* 36:2701-2711, (1987).

13. Kamio,A., Shiraishi,M., Shitama,K., Sugimoto,Y., Honka,K. & Takebayashi,S. Measurement of endothelial fibrinolytic activity in aorta and vena cava in rabbits: a fibin-agar plate method. *Artery* 5:76-89, (1979).

14. Bartha,K., Muller-Peddinghaus,R. & Van Rooijen,L.A.A. Bradykinin and thrombin effects on polyphosphoinositide hydrolysis and prostacyclin production in endothelial cells. *Biochem.J.* 263:149-155, (1989).

15. Altura,B.M. Pharmacology of arterioles versus venules: an overview. *Microcirc. Endoth. Lymphatics* 4:91-111, (1988).

16. Kumar,S., West,D.C. & Ager,A. Heterogeneity in endothelial cells from large vessels and microvessels. *Differentiation* 36:57-70, (1987).

17. Oestreich,R., Zink,S. & Rosen,P. Eicosanoid production by endothelial cells: a comparison between cells from micro- and macrovascular sources. *Prog Clin Biol Res* 301:371-375, (1989).

18. Berg,E.L., Goldstein,L.A., Jutila,M.A., Nakache,M., Picker,L.J., Streeter,P.R., Wu,N.W., Zhou,D. & Butcher,E.C. Homing receptors and vascular addressins: cell adhesion molecules that direct lymphocyte traffic. *Immunol Rev* 108:5-18, (1989).

19. Pauli,B.U., Augustin-Voss,H.G., el-Sabbah,M.E., Johnson,R.C. & Hammer,D.A. Organ-preference of metastasis: the role of endothelial cell adhesion molecules. *Cancer Metas Rev* 9:175-189, (1990).

20. Auerbach,R. Patterns of tumor metastasis: organ selectivity in the spread of cancer cells. *Lab Invest* 53:361-364, (1988).

21. Nicolson,G.L. Organ specificity of tumor metastasis: role of preferential adhesion, invasion and growth of malignant cells at specific secondary sites. *Cancer Metas Rev* 7:143-188, (1988).

22. Risau,W., Hallmann,R., Albrecht,U. & Henke-Fahle,S. Brain induces the expression of an early cell surface marker for blood-brain barrier-specific endothelium. *EMBO J* 1:3179-3183, (1986).

23. Fattorossi,A., Aurbach,G.D., Sakaguchi,K., Cama,A., Marx,S.J., Streeten,E.A., Fitzpatrick, L.A. & Brandi,M.L. Anti-endothelial cell antibodies: detection and characterization in sera from patients with autoimmune hypoparathyroidism. *P.N.A.S. USA* 85:4015-4019, (1988).

24. Pappas,G.D. & Sagen,J. Fine structure of PC12 cell implants in the rat spinal cord. *Neurosci. Lett.* 70:59-64, (1986).

25. Hart,T.K. & Pino,R.M. Pseudoislet vascularization: induction of diaphragm-fenestrated endothelia from the hepatic sinusoids. *Lab Invest* 54:304-313, (1986).

26. Alby,L. & Auerbach,R. Differential adhesion of tumor cells to capillary endothelial cells in vitro. *P.N.A.S. USA* 81:5739-5743, (1984).

27. Gumkowski,F., Kaminska,G., Kaminski,M., Morrissey,L.W. & Auerbach,R. Heterogeneity of mouse vascular endothelium: in vitro studies of lymphatic, large blood vessel and microvascular endothelial cells. *Blood Vessels* 24:11-23, (1987).

28. Tao-Cheng,J.-H. & Brightman,M.W. Development of membrane interactions between brain endothelial cells and astrocytes in vitro. *Int.J.Devl.Neuroscience* 6:25-37, (1988).

29. Arthur,F.E., Shivers,R.R. & Bowman,P.D. Astrocyte-mediated induction of tight junctions in brain capillary endothelium: an efficient in vitro model. *Brain Research* 433:155-159, (1987).

30. Nakao,T., Ishikawa,A. & Ogawa,R. Observations of vascularization in the spinal cord of mouse embryos, with special reference to development of boundary membranes and perivascular spaces. *Anat.Rec.* 221:663-677, (1988).

31. Bertossi,M., Ribatti,D. & Roncali,L. Ultrastructural features of the vasculogenetic processes in the thyroid and suprarenal glands of the chick embryo. *J.Submicrosc.Cytol.* 19:119-128, (1987).

32. Carmichael,S.W., Spagnoli,D.B., Frederickson,R.G., Krause,W.J. & Culberson,J.L. Opossum adrenal medulla: I. postnatal development and normal anatomy. *Am J Anat* 179:211-219, (1987).

33. Rhodin,J.A.G. & Fujita,H. Capillary growth in the mesentery of normal young rats: intravital video and electron microscope analyses. *J.Submicrosc.Cytol.Pathol.* 21:1-34, (1989).

34. Lelkes,P.I., Mizrachi,Y., Ornberg,R., Munoz,A., Levi,B., Narranjo,J. & Pollard,H.B. Specific interactions between PC12 cells and adrenal medullary endothelial cells in co-culture: an in vitro model for early stages of developmental differentiation in neurosecretion. *Neurosci* 22:273S,(1987).

35. Lelkes, P.I., Munoz, A. and Lazarovici, P. Endothelial cells and secretory cells in coculture: modulation of growth hormone receptors and protein expression. *J Cell Biol* 107:553a, (1989).

36. Mizrachi,Y., Lelkes,P.I., Ornberg,R.L., Goping,G. & Pollard,H.B. Specific adhesion between pheochromocytoma (PC12) cells and adrenal medullary endothelial cells in co-culture. *Cell Tiss Res* 256:365-372, (1989).

37. Mizrachi,Y., Narranjo,J., Levi,B.-Z., Pollard,H.B. & Lelkes,P.I. PC12 cells differentiate into chromaffin cell like phenotype in co-culture with adrenal medullary endothelial cells. *P.N.A.S. USA* 87:6161-6165, (1990).

38. Doupe,A.J., Landis,S.C. & Patterson,P.H. Environmental influences in the development of neural crest derivatives: glucocorticoids, growth factors, and chromaffin cell plasticity. *J.Neurosci.* 5:2119-2142, (1985).

39. Patterson,P.H. The molecular basis of phenotypic choices in the sympathoadrenal lineage. *Ann N Y Acad Sci* 496:20-26, (1986).

40. Patterson,P.H. Control of cell fate in a vertebrate neurogenic lineage. *Cell* 62:1035-1038, (1990).

41. Anderson,D.J. & Michelsohn,A. Role of glucocorticoids in the chromaffin-neuron developmental decision. *Int.J.Devl.Neurosci.* 7:475-487, (1989).

42. Unsicker,K., Krisch,B., Otten,U. & Thoenen,H. Nerve growth factor-induced fiber outgrowth from isolated rat adrenal chromaffin cells-impairment by glucocorticoids. *P.N.A.S. USA* 75:3498-3502, (1978).

43. Doupe,A.J., Patterson,P.H. & Landis,S.C. Small intensely fluorescent cell in culture: role of glucocorticoids and growth factors in their development and interconversions with other neural crest derivatives. *J.Neurosci.* 5:2143-2160, (1985).

44. Seidl,K. & Unsicker,K. The determination of the adrenal medullary cell fate during embryogenesis. *Dev Biol* 136:481-490, (1989).

45. Rogelj,S., Klagsburn,M., Atzmon,R., Kurokawa,M., Haimovitz,A., Fuks,Z. & Vlodavsky,I. Basic fibroblast growth factor is an extracellular matrix component required for supporting the proliferation of vascular endothelial cells and the differentiation of PC12 cells. *J Cell Biol* 109:823-831, (1989).

46. Sieber-Blum,M., Kumar,S.R. & Riley,D.A. In vitro differentiation of quail neural crest cells into sensory-like neuroblasts. *Dev.Brain Res.* 39:69-83, (1988).

47. Perris,R., Paulsson,M. & Bronner-Fraser,M. Molecular mechanisms of avian neural crest cell migration on fibronectin and laminin. *Dev Biol* 136:222-238, (1989).

48. Takeichi,M., Hatta,K., Nose,A., Nagafuchi,A. & Matsunaga,M. Cadherin-mediated cell sorting and axonal guidance. In: *The assembly of the nervous system*, ed. by Landmesser, L.T. Alan R. Liss, p. 129-136, (1989).

49. Edelman,G.M. & Thiery,J. *The cell in contact: adhesions and junctions as morphogenetic determinants*, N.Y.:John Wiley & Sons, (1985).

50. Newman,P.J., Berndt,M.C., Gorski,J., White,G.C., Lyman,S., Paddock,C. & Muller,W.A. PECAM-1 (CD31) cloning and relation to adhesion molecules of the immunoglobulin gene superfamily. *Science* 247:1219-1222, (1990).

51. Albelda,S.M., Oliver,P.D., Romer,L.H. & Buck,C.A. EndoCAM: a novel endothelial cell-cell adhesion molecule. *J Cell Biol* 110:1227-1237, (1990).

52. Coupland,R.E. Blood supply of the adrenal gland. In: *Handbook of physiology: a critical, comprehensive presentation of physiological knowledge and concepts: section 7: endocrinology: volume VI. adrenal gland*, ed. by Greep,R.O. & Astwood,E.B. Washington, D.C.: Am. Phys. Soc. p. 283-294, (1975).

53. Grynszpan-Winograd,O. Ultrastructure of the chromaffin cell. In: *Handbook of physiology: a critical, comprehensive presentation of physiologicall knowledge and concepts: section 7: endocrinology: volume VI. adrenal gland*, ed. by Greep,R.O. & Astwood,E.B. Washington, D.C.: Am. Phys. Soc., p. 295-308, (1975).

54. Tischler,A.S. & Greene,L.A. Morphologic and cytochemical properties of a clonal line of rat adrenal pheochromocytoma cells which respond to nerve growth factor. *Lab Invest* 39:77-89, (1978).

55. Fujita,K., Lazarovici,P. & Guroff,G. Regulation of the differentiation of PC12 pheochromocytoma cells. *Environmental Health Perspectives* 80:127-142, (1989).

56. Byrd,J.C., Naranjo,J.R. & Lindberg,I. Proenkephalin gene expression in the PC12 pheochromocytoma cell line: stimulation by sodium butyrate. *Endocrinology* 121:1299-1305,(1987).

57. Brandi,M.L., Sakaguchi,K., Ornberg,R.L., Lelkes,P.I. & Aurbach,G.D. On the interactions

between clonal parathyroid epithelial (PT-r) and endothelial (BPE) cells: evidence for structural and humoral correlates. *J Cell Biol* 107:553a, (1989).

58. Noden,D.M. Embryonic origins and assembly of blood vessels. *Am Rev Respir Dis* 140:1097-1103, (1989).

59. Risau,W. & Lemmon,V. Changes in the vascular extracellular matrix during embryonic vasculogenesis and angiogenesis. *Dev Biol* 125:441-450, (1988).

60. Lelkes,P.I. New aspects of endothelial cell biology. *J.Cell.Biochem.* 45:242-244, (1991).

61. Hla,T. & Maciag,T. An abundant transcript induced in differentiating human endothelial cells encodes a polypeptide with structural similarities to G-protein-coupled receptors. *J.Biol.Chem.* 265:9308-9313, (1990).

ROLE OF THE MACROPHAGE IN THE REGULATION OF PHYSIOLOGICAL AND

PATHOLOGICAL ANGIOGENESIS

Peter J. Polverini and Luisa A. DiPietro

Department of Pathology
Northwestern University Medical and Dental Schools
303 East Chicago Avenue
Chicago, Illinois 60611-3008

INTRODUCTION

Angiogenesis, the growth of new capillary blood vessels, is one of the most pervasive and essential biological processes encountered in the mammalian organism (Auerbach, 1981; Folkman and Cotran, 1978; Folkman, 1985; Folkman and Klagsbrun, 1987). A wide array of physiological and pathological conditions such as embryonic development, the formation of inflammatory granulation tissue during wound repair, and the growth of malignant solid tumors are strictly dependent upon the recruitment of new capillaries. In the adult organism, angiogenesis occurs infrequently yet can be rapidly induced by a number of physiologic stimuli. In recent years the macrophage (Mø) has been shown to be one of the key players in the regulation of physiologic and pathologic angiogenesis (Polverini et al., 1977b; Polverini, 1989). These cells perform multiple diverse functions within the immune system and influence the course of certain primitive yet fundamentally important processes such as nonspecific inflammation and wound healing (Leibovich and Ross, 1975). This functional versatility is due in large part to their ability to respond rapidly to environmental signals that induce them to acquire new or enhanced properties (Nathan, 1987). One of these functions is the capacity to induce and subsequently down-regulate the formation of new capillary blood vessels.

MØ AND PHYSIOLOGICAL ANGIOGENESIS

The central role of the Mø in tissue reorganization was initially demonstrated by Leibovich and Ross (1975 and 1976), who observed that wound healing was greatly delayed in animals depleted of monocytes and macrophages. While this study implicated the Mø in wound angiogenesis, one of the first indications that Mø did indeed mediate neovascularization came from our laboratory. Using tritiated thymidine as measure of DNA synthesis, the magnitude of endothelial cell replication in delayed-type hypersensitivity reactions in the skin of guinea pigs was shown to coincide with the time course and magnitude of monocyte/Mø infiltration into the tissues (Polverini et al. 1977a). In a more direct evaluation of this relationship, activated Mø or their culture supernatants were shown to potently stimulated neovascularization when

Angiogenesis in Health and Disease, Edited by M.E. Maragoudakis *et al.*
Plenum Press, New York, 1992

introduced into the avascular cornea of the guinea pig eye (Polverini et al., 1977b). Activated Mø derived from several sources are potently angiogenic, and appear to mediate blood vessel growth through the secretion of a number of growth factors and cytokines (Thakral et al., 1979; Hunt et al., 1984; Polverini, 1989). Several mediators produced by Mø have been implicated in the induction of angiogenesis and include among others basic fibroblast growth factor (bFGF), platelet derived growth factor (PDGF), a platelet derived growth factor-like endothelial mitogen, transforming growth factor-alpha (TGF-α), tumor necrosis factor alpha (TNF-α) and several uncharacterized low molecular weight mitogens (Polverini, 1989). While multiple angiogenic factors have been shown to be produced by macrophages, the precise factors produced by these cells and the manner in which they orchestrate the various components of angiogenic response *in vivo* are not well understood. RNA phenotyping studies of adherent cells from healing wounds at day 6 indicate that mRNAs for TGF-α, TGF-β, epidermal growth factor, PDGF-A, and insulin-like growth factor-1 are produced in these cells (Rappolee et al., 1988). However, the time course of production of these factors has not yet been determined.

One of the characteristic features of angiogenesis in wound repair is that the response is tightly regulated temporally and spatially . The rich supply of nutrients provided by the newly formed vessels is essential to meet the enhanced metabolic demands of healing tissues. When tissue integrity is restored the vessels comprising granulation tissue regress. The complement of vessels present in the newly healed tissues is eventually restored to its pre-injury levels. This example of physiologic angiogenesis demonstrates two key aspects of this process: the formation of new capillary blood vessels is rapid and controlled yet it is transient and characterized by regression to a physiologic steady state level.

The abrupt termination of angiogenesis characteristic of normal wound repair suggests two possible mechanisms of control neither of which are mutually exclusive. First, under circumstances not well understood, there is a controlled termination or marked reduction in the synthesis and elaboration of angiogenic mediators. Second, a parallel increase in the levels of substance which inhibit new vessel growth may occur (Bouck, 1990). While most investigations of the angiogenic capabilities of the Mø have focused on the production of positive regulators, growing evidence suggests that Mø may also secrete substances which inhibit blood vessel growth. Thus, by virtue of their ability to secret both positive and negative mediators in a temporal sequence, Mø may dictate the course of neovascular responses as in wound healing.

MØ AND PATHOLOGICAL ANGIOGENESIS

The potential for expression of angiogenic activity is inherent in all cell types. Normally, only a few cell types participate transiently in the induction of angiogenesis. However, other cell types may exhibit aberrant angiogenic activity during pathologic processes. The consequences of a disruption in the regulatory mechanisms controlling angiogenesis are best exemplified by the growth of malignant solid tumors. Early neoplasms are able to survive undetected for an extended period of time as small nonvascularized spheroids, acquiring nutrients by simple diffusion (Gimbrone et al., 1972 and 1974; Folkman, 1990). With the onset of neovascularization the tumor is able to grow exponentially and go on to spread and ultimately overwhelm host defense mechanisms. Presumably the unregulated production of angiogenic mediators is the principal reason for the continued growth of tumors.

While tumor cells often have the capacity to mediate neovascularization, Mø may also promote new blood vessel formation during tumorigenesis. Mø are a frequent if not invariable component of the inflammatory infiltrate that is associated with solid tumors (Evans, 1976; Evans, 1979; Mantovani et al., 1980). Although these cells have been shown to have potent cytostatic/cytotoxic activity for tumors, there is equally compelling evidence that Mø can function to enhance tumor growth (Evans, 1977a; Evans, 1977b; Mostafa et al., 1980; Stenzinger et al., 1982). Mø isolated from solid tumors have been shown to be potently angiogenic in rat corneas and to stimulate the growth of endothelial cells *in vitro* (Polverini and Leibovich, 1984). Furthermore, tumors depleted of Mø exhibit a reduced capacity to induce neovascularization. When Mø are added back to these tumors their growth and angiogenic activity was markedly enhanced. It would therefore appear that tumor associated Mø may contribute to tumor growth and progression by virtue of their ability to enhance tumor neovascularization.

Imbalances in the regulation of angiogenesis are also implicated in the pathogenesis of a variety of angiogenesis-dependent inflammatory diseases, such as adult periodontitis and rheumatoid arthritis (Koch et al., 1986). In rheumatoid arthritis the unrestrained proliferation of fibroblasts and capillary blood vessels leads to prolonged and persistent granulation tissue formation. The granulation tissue which forms is endowed with a rich complement of degradative enzymes that contributes to profound destruction of joint spaces (Harris, 1976). Koch et al. (1986) has shown that a subpopulation of Mø isolated from rheumatoid synovium are potently angiogenic *in vivo* and chemotactic for capillary endothelial cells *in vitro* . Keratinocytes from prelesional and psoriatic skin are potently angiogenic and appear to have a deficiency in thrombospondin, a matrix glycoprotein recently reported to have angiogenesis inhibitory activity (Nickoloff, Mitra, and Polverini, unpublished data). As has been reported for solid tumor growth it may be that a combination of overproduction of mediators of angiogenesis coupled with a lack of naturally occurring inhibitors of this process accounts for the prolonged neovascularization characteristic of this and possibly other chronic inflammatory diseases.

As described above, the expression of aberrant angiogenic activity by Mø and other cell types has been implicated in augmenting angiogenesis in certain chronic inflammatory disorders. Recent evidence suggests that the inability of Mø to express appropriate angiogenic activity may also contribute the pathogenesis of certain diseases associated with defective angiogenesis. Koch and her colleagues (unpublished data) have recently found that blood monocyte - derived Mø from patients with scleroderma fail to stimulate angiogenesis when exposed to the activating agent lipopolysaccharide (LPS) and exhibited decreased activation - associated surface markers. These results suggest that a defect in Mø responsiveness to activating signals and their subsequent inability to stimulate neovascularization may contribute to the aberrant vascularization that is encountered in scleroderma.

GENETIC CONTROL OF EXPRESSION OF THE ANGIOGENIC PHENOTYPE

Several pieces of evidence suggest that the development of angiogenic activity by normal cells such as the macrophage is due to loss of an active factor produced by non-angiogenic cells that prevents angiogenesis (Rastinejad et al. 1989; Polverini, 1989). Hybrids between nonangiogenic monocytes or monocyte cell lines and angiogenic activated, macrophages are uniformly suppressed for the angiogenic phenotype (Table 1, Figures 1 and 2). In addition, conditioned media from these cells was able to suppress the angiogenic activity of known mediators of angiogenesis (Table 2, Figures 3 through 6).

These studies strongly suggest that expression of Mø angiogenic activity is under the negative control of a suppressor element. This finding also implies that deregulated expression of the angiogenic activity as in solid tumors, may involve the loss or inactivation of

Table 1. Angiogenic Activity of *Hybrid Mø and Monocytes

Conditioned media:	§Neovascularization Proportion of Positive Responses (%)		Angiogenic Status
P388D1 6TGr NEOr	10/10	(100)	+
x Balb/c Res. Mø	0/8	(0)	−
Hybrid 1	0/5	(0)	−
Hybrid 2	0/5	(0)	−
†LPS treated revertant	5/5	(100)	+
Hybrid 3	1/5	(20)	−
Spontaneous segregant	4/5	(80)	+
x Human PB monocytes			
Hybrid 1	0/6	(0)	−
Hybrid 2	0/6	(0)	−
LPS treated revertant	7/8	(88)	+
Hybrid 3	1/5	(20)	−
Spontaneous segregant	7/7	(100)	+
Controls			
Balb/c LPS activated peritoneal Mø	5/5	(100)	+
Cell − free CM	0/6	(0)	−
Sham Hydron implant	0/5	(0)	−

*Cells were fused either in surface culture or suspension with polyethylene glycol 1000 and selected in HAT and G418. Hybrids were cloned, serum free conditioned media was prepared and concentrated 20 − fold using a YM10 amicon membrane.
†Hybrid cultures were treated for 24hrs with 10µg/ml. Cells were washed extensively and incubated for another 48hrs without serum.
§Media was incorporated into hydroxyethylmethracrylate (Hydron) and 10µl pellets were implanted into the avascular cornea of rat eyes. Responses were rescored after 7 days and scored as positive when sustained directional ingrowth of capillaries was detected. Responses were recorded as negative when no sustained capillary ingrowth was observed or when no growth was detected. Permanent records of vascularized corneas were prepared following colloidal carbon perfusion (Figures 1 and 2).

suppressor gene function. Evidence from two different tumor systems support this hypothesis. First, fusions between immortalized, non-angiogenic hamster mucosal keratinocytes and fully transformed, angiogenic keratinocytes results in hybrids which were suppressed for the angiogenic phenotype (Moroco et al., 1990; Polverini et al., 1988). In a second system, fusions between non-angiogenic, immortalized BHK hamster kidney fibroblasts and its chemically transformed, potently angiogenic, counterparts yield hybrids that were suppressed for the angiogenic phenotype (Bouck et al. 1987). In the BHK system, the dominant nature of the non-angiogenic phenotype has been shown to be due to the presence of an inhibitor of angiogenesis in the conditioned media of BHK cells and their suppressed hybrids (Rastinejad et al., 1989). Biochemical and immunologic analysis revealed the inhibitory activity was mediated by a molecule which is identical to a portion of the multifunctional matrix glycoprotein thrombospondin (TSP)(Good et al., 1990). In this system, a clear relationship between the secretion of an inhibitor of vascular growth and the maintenance of the non-angiogenic phenotype has been documented. However, genetic studies indicate that the angiogenesis suppressor gene is not TSP itself, as the suppressor appears to lie on

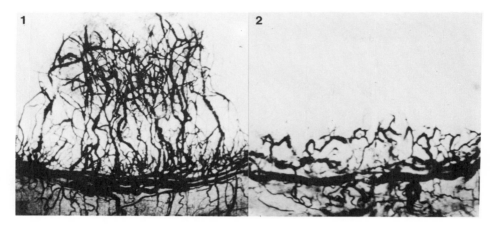

Figure 1. A positive neovascular response induced after 7 days by conditioned media from P388D1 macrophage - like cells.
Figure 2. A representative response from an angiogenesis suppressed P388D1 x human monocyte hybrid.

Table 2. Inhibition of Growth Factor Mediated Angiogenesis by Hybrid Mø Conditioned Media

*Conditioned media or ¶growth factor mixed with:	Neovascularization Proportion of Positive Responses (%)	
	bFGF	TNFα
bFGF (50 ng)	5/5 (100)	–
TNFα (50 ng)	–	6/6 (100)
LPS Act. Balb/c Mø	5/5 (100)	4/4 (100)
P388D1 6TGr NEOr		
x Balb/c Res. Mø		
Hybrid 1	0/3 (0)	1/3 (33)
Hybrid 2	0/4 (0)	0/3 (0)
x Human PB monocytes		
Hybrid 1	1/3 (33)	1/4 (25)
Hybrid 2	0/3 (0)	0/3 (0)
x WEHI-3 monocytes		
Hybrid 1	1/3 (33)	1/4 (25)
Hybrid 2	0/3 (0)	1/5 (20)
Spontaneous segregant	3/3 (100)	4/4 (100)

*Conditioned media was concentrated 40 - fold, mixed with growth factors and incorporated into Hydron pellets.

¶Human recombinant basic fibroblast growth factor and tumor necrosis factor alpha.

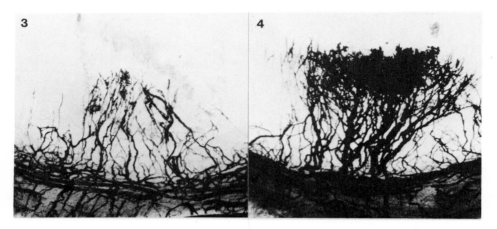

Figure 3. Colloidal carbon perfused corneas demonstrating positive neovascular response induced by 50ng of bFGF.
Figure 4. Colloidal carbon perfused corneas demonstrating positive neovascular response induced by 50ng of TNFα.

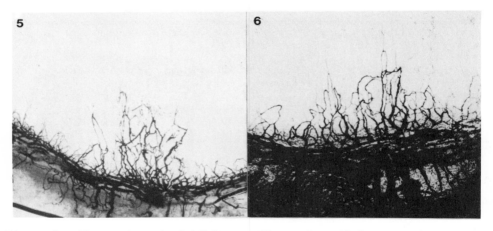

Figure 5. The angiogenic inhibitory effect of conditioned media from an angiogenesis suppressed P388D1 x WEHI-3 Mø hybrid when mixed with bFGF.
Figure 6. The angiogenic inhibitory effect of conditioned media from an angiogenesis suppressed P388D1 x WEHI-3 Mø hybrid when mixed with TNFα.

another chromosome (Good et al., 1990). Therefore, it appears that the angiogenic suppressor acts to modulate TSP production in this system.

Investigations of the transient activation of angiogenesis during embryogenesis also suggest a role for a angiogenesis suppressor during development. We have reported recently that embryonic stem cells induced to differentiate by allowing them to aggregate in culture and clones of mesenchymal cells derived from the secondary palates of 11 through 13 day mouse embryos are potently angiogenic *in vivo* and *in vitro*. Cells isolated after this time fail to express this phenotype. Hybrids produced between angiogenic stem cells or palatal fibroblasts and nonangiogenic mesenchymal cells or fibroblasts are uniformly suppressed for this phenotype. Interestingly, when the angiogenic parents are exposed briefly to retinoic acid, the cells lose their ability to induce angiogenesis. This suggests that transient inactivation of a negative regulatory element, i.e. an "angiogenesis suppressor gene", may account for the temporal expression of angiogenesis that occurs during development. These studies would also suggest that retinoic acid, a potent endogenous morphogen that has previously been reported to be antiangiogenic, may function to down-regulate angiogenesis at defined periods in development by reactivating a differentiation-linked suppressor element (McGill et al., 1991; Moroco et al. 1991). The physiological end product that accounts for suppression in this system is unknown. Preliminary evidence suggests however that it is not thrombospondin.

It is becoming increasing clear that a fundamental mechanism of genetic control may be responsible for regulating the expression of angiogenesis in mulitple systems. Negative control of the angiogenic phenotype has now been described for many cell types, as listed in Table 3. In most cases, the precise molecules responsible for the control of angiogenesis have not been determined. Many inducers and inhibitors of angiogenesis are likely present together in tissues; their temporal and spacial distribution may be critical in determining the neovascular response. The particular inducers and inhibitors may even be unique to a particular tissue and stage of development. The number of angiogenic mediators, as well as the mechanisms responsible for the production of such mediators remains to be elucidated.

Table 3. Cells where Expression of the Angiogenic Phenotype is Under Negative Genetic Control

Cell type:	Conditions Leading to Inactivation or Loss of Negative Control of the Angiogenic Phenotype
Hamster mucosal keratinocytes	Initiation with chemical carcinogens *in vivo*
Baby hamster kidney fibroblasts	Mutation with chemical carcinogens *in vitro*
Embryonic stem cells	Normal and retinoic acid - induced differentiation
Embryonic mesenchymal cells	Normal and retinoic acid - induced differentiation
Human and murine Mø	Activation with lipopolysaccharide or phorbol esters

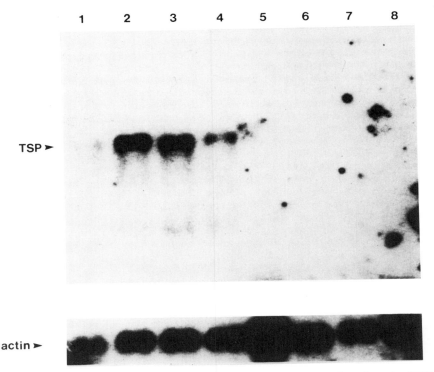

TSP ➤

actin ➤

Figure 7.   Northern analysis of thrombospondin mRNA production in WEHI-3
cells. Lanes 1-4, Cells were incubated with 0, 10, 1, or 0.1ug/ml LPS
respectively for 24 hours and total cytoplasmic RNA was isolated
immediately.   Lanes 5-8, cells were incubated with 0, 10, 1, or 0.1 ug/ml
LPS respectively for 24 hours, washed, and placed in conditioning media
(DMEM + 0.075% BSA) for 48 hours at which time total cytoplasmic RNA was
harvested.   10 ug of total RNA was run in each lane; the probes used were
a 1.5kb Pst fragment of the human TSP cDNA (upper panel) and, as a
positive control, a rat actin cDNA probe(lower panel).

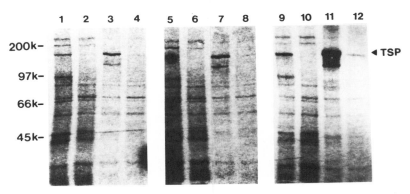

Figure 8.   Immunoprecipitation of thrombospondin from biosynthetically
labelled WEHI-3 cells and their culture supernatants. Lanes 1-4, control
WEHI-3 cells. Lanes 5-8, WEHI-3 cells incubated for 4 hours with 1 ug/ml
LPS. Lanes 9-12, WEHI-3 cells incubated for 8 hours with 1ug/ml LPS.
Lanes 1, 2, 5, 6, 9, 10, cell lysates. Lanes 3, 4, 7, 8, 11, 12,
culture supernatants. Immunoprecipitation performed with either
specifically purified rabbit anti-TSP (odd-numbered lanes) or pre-immune
serum alone (even numbered lanes).

The identification of TSP as an inhibitor of angiogenesis in non-transformed BHK cells leads to the obvious question of whether this molecule plays a role in the control of the angiogenic phenotype in macrophages. In investigations of peritoneal monocytes and Mø, TSP has been described to be secreted by monocytes, resident peritoneal macrophages, and to a lesser degree by illicited and activated peritoneal macrophages (Jaffe et al., 1985). Interestingly, we have observed that TSP transcription and protein production increase in the monocyte/macrophage cell line WEHI-3 when activated with LPS (Figures 7, 8). While this increase in production of an inhibitor of angiogenesis appears to be at odds with the angiogenic capacity of the activated macrophage, active transcription of TSP appears to drop dramatically when the activated WEHI-3 cells are placed in conditioning media (Figure 7). It is unclear how this observation correlates with *in vivo* production of TSP by activated macrophages. A careful analysis of TSP production by normal Mø involved in physiologic angiogenesis will be necessary to answer this question.

The accumulated evidence regarding the development of the angiogenic state in activated Mø suggests that the angiogenic phenotype is under negative control. Thus in Mø, activation presumably leads to a down-regulation of the angiogenic suppressor gene with a concomitant up-regulation of the genes encoding key growth factors. Recently we have found that the development of the angiogenic phenotype *in vitro* can be abrogated when the cells are treated with retinoic acid. Although we do not yet understand the mechanism of suppression, it is tempting to speculate that retinoic acid serves to maintain the expression of the angiogenic suppressor. If this is so, retinoic acid may provide a tool by which suppressor gene function can be up-regulated. Such a tool would be very helpful in devising strategies by which the angiogenic suppressor gene might be cloned. As has been seen in other systems, the cloning of such a suppressor gene is not easily accomplished. Thee ability to up-regulate transcription of the suppressor might allow for the successful use of traditional strategies such as subtraction libraries and subtractive hybridization. The use of novel strategies such as technical knockout (TKO), in which genes are inactivated through the transfection of anti-sense cDNA, hold considerable promise for isolating these elusive suppressor elements (Deiss and Kimchi, 1991).

THERAPEUTIC IMPLICATIONS

The existence of both positive and negative mediators capable of regulating angiogenesis has been known for some time. The complexity of control of angiogenesis is evident through the description of growing numbers of both inducers and inhibitors of capillary growth. Once the mechanisms orchestrating this duel control are clarified it may be possible to customize therapy for those disorders where neovascular-ization has been shown to contribute to disease. Endothelial cells share a number of features that make attractive as a target for such therapy. These homogeneous, karyotypically stable cells are normally primed to respond to both positive and negative regulators of angiogenesis. Normally primed to regress these cells would be ideal targets for stimulatory or inhibitory factors which might function to up- or down-regulate angiogenesis. The ability to take advantage of this "internal' control mechanism may permit the effective development of strategies for the control of a number of pathophysiologic disorders that can be characterized as a deficiency in either the "up-regulation or "down-regulation" of angiogenesis. Mø might fit into this picture as a potential delivery system of angiogenic mediators. For example, during tumor development Mø progenitors that have been genetically engineered to produce specific inhibitors of angiogenesis upon entering the tumor could be programmed to release factors that target the responsive endothelial cells. A similar approach might be envisioned with chronic inflammatory disorders such as rheumatoid arthritis that are characterized by prolonged angiogenesis and significant Mø infiltration. In conditions when angiogenic activity is inadequate, such as the aberrant wound healing of diabetes and certain ocular diseases where blindness is often the consequence of inappropriate angiogenesis, Mø mediated angiogenic stimulatory therapy might be effective. A better understanding of the angiogenic response will lead to successful therapeutic strategies.

REFERENCES

Auerbach, R., Angiogenesis-inducing factors: in: "Lymphokines," E. Pick. ed., Academic Press, New York (1981).

Bouck, N.P., Stoler, A., and Polverini, P.J., 1986, Coordinate control of anchorage independence, actin cytoskeleton, and angiogenesis by human chromosome 1 in hamster-human hybrids, Cancer Res. 46:5101.

Bouck, N., 1990, Tumor angiogenesis: the role of oncogenes and tumor suppressor genes, Cancer Cells 2:179.

Deiss, L.P., and Kimchi, A., 1991, A genetic tool used to identify thioredoxin as a mediator of a growth inhibitory signal, Science, 252:117.

Evans, R., Tumor macrophages in host immunity to malignancy: in "The Macrophage in Neoplasia," M.A.Fink ed., Academic Press, New York (1976).

Evans, R., 1979, Host cells in transplanted murine tumors and their possible relevance to tumor growth, J. Reticuloendothelial. Soc. 26:427.

Evans, R., 1977a, The effect of azothioprine on host-cell infiltration and growth of a murine fibrosarcoma, Int. J. Cancer 20:120.

Evans, R., 1977b, Effects of x-irradiation on host-cell infiltration and growth of a murine fibrosarcoma, Br. J. Cancer 35:557.

Folkman, J., and Cotran, R., 1976, Relation of vascular proliferation to tumor growth, Int. Rev. Exp. Pathol. 16:207.

Folkman, J., 1985, Tumor angiogenesis, Adv. in Cancer Res. 43:175.

Folkman, J, and Klagsbrun M., 1987, Angiogenic factors, Science 235:442.

Folkman, J., 1990, What is the evidence that tumors are angiogenesis dependent, J. Natl Cancer Inst. 82:4.

Gimbrone, M.A.Jr., Leapman, S.B., Cotran, R.S., and Folkman, J., 1972, Tumor dormancy in vivo by prevention of neovascularization, J. Exp. Med. 136:261.

Gimbrone, M.A.Jr., Cotran, R.S., Leapman, S.B., and Folkman, J., 1974, Tumor growth and neovascularization: an experimental model using the rabbit cornea, J. Natl. Cancer Inst. 52:413.

Good, D.J., Polverini, P.J., Rastinejad, F., LeBeau, M.M., Lemons, R.S., Frazier, W.A., and Bouck, N.P., 1990. A tumor suppressor-dependent inhibitor of angiogenesis is immunologically and functionally indistinguishable from a fragment of thrombospondin, Proc. Natl. Acad. Sci. U.S.A. 87:6624.

Harris, E.D.Jr., 1976, Recent insights into the pathogenesis of the proliferative lesion in rheumatoid arthritis, Arthritis Rheum. 19:68.

Hunt, T.K., Knighton, D.R., Thakral, K.K., Goodson, W.H., and Andrews, W.S., 1984, Studies on inflammation and wound healing; angiogenesis and collagen synthesis stimulated in vivo by resident and activated wound macrophages, Surgery 96:48.

Jaffe, E.A., Ruggiero, J.T., and Falcone, D.J., 1985, Monocytes and macrophages synthesize and secrete thrombospondin, Blood 65:79.

Koch, A.E., Polverini, P.J., and Leibovich, S.J., 1986, Stimulation of neovascularization by human rheumatoid synovial tissue macrophages, Arthritis Rheum. 29:471.

Leibovich, S.J., and Ross, R., 1975, The role of the macrophage in wound repair: a study with hydrocortisone and antimacrophage serum, Am. J. Pathol. 78:71.

Leibovich, S.J., and Ross, R., 1976, A macrophage-dependent factor that stimulates the proliferation of fibroblasts *in vitro* , Am. J. Pathol. 84:501.

Mantovani, A., Giavazzi, R., Polentarutti, N., Spreafico, F., and Garattini, S., 1980, Divergent effects of macrophage toxins on growth of primary tumor and lung metastases in mice, Int. J. Cancer 25:617.

McGill, J.S., Moroco, J.R., and Polverini, P.J., 1991, Transient inactivation of an angiogenesis suppressor gene during palatogenesis. J. Dent. Res. 70:515.

Moroco, J.R., Solt, D.B., and Polverini, P.J., 1990, Sequential loss of suppressor genes for three specific functions during *in vivo* carcinogenesis, Lab. Invest. 63:298.

Moroco, J.R., McGill, J.S., and Polverini, P.J., 1991, Angiogenesis suppressor gene inactivation in differentiated stem cells and teratocarcinomas. J. Dent. Res. 70:595.

Mostafa, L.K., Jones, D.B., and Wright, D.H., 1980, Mechanisms of the induction of angiogenesis by human neoplastic lymphoid tissue: studies on the chorioallantoic membrane (CAM) of the chick embryo, J. Pathol. 132:191.

Nathan, C., 1987, Secretory products of macrophages, J. Clin. Invest. 79:319.

Polverini, P.J., Macrophage-induced angiogenesis: a review: in: "Cytokines," C. Sorg. ed., S. Karger, Basel (1989).

Polverini, P.J., Cotran, R.S., and Sholley, M.M., 1977a, Endothelial proliferation in the delayed hypersensitivity response: an autoradiographic study, J. Immunol. 118:529.

Polverini, P.J., Cotran, R.S., Gimbrone, M.A.Jr., and Unanue, E.R., 1977b, Activated macrophages induce vascular proliferation, Nature 269:804.

Polverini, P.J., and Leibovich, S.J., 1984, Induction of neovascularization in vivo and endothelial proliferation in vitro by tumor associated macrophages, Lab. Invest. 51:635.

Rappolee, D.A., Mark, D., Banda, M.J., and Werb, Z., 1988, Wound macrophages express TGF-α and other growth factors *in vivo*: analysis by mRNA phenotyping, Science 241:708.

Polverini, P.J., Shimizu, K., and Solt, D.B., 1988, Control of angiogenic activity in carcinogen-initiated and neoplastic hamster pouch keratinocytes and their hybrid cells, J. Oral Pathol. 18:522.

Rastinejad, F., Polverini, P.J., and Bouck, N.P., 1989, Regulation of the activity of a new inhibitor of angiogenesis by a cancer suppressor gene, Cell 56:345.

Stenzinger, W., Bruggen, J., Macher, E., and Sorg, C., 1983, Tumor angiogenic activity (TAA) production *in vitro* and growth in the nude mouse by human malignant melanoma, Eur. J. Cancer Clin. Oncol. 19:649.

Thakral, K.K., Goodson, W.H., and Hunt, T.K., 1979, Stimulation of wound blood vessel growth by wound macrophages, J. Surg. Res. 26:430.

ENDOTHELIAL CELL HETEROGENEITY AND DIFFERENTIATION[1]

Robert Auerbach, Johanna Plendl, and Barmak Kusha

Laboratory of Developmental Biology
University of Wisconsin
Madison, WI 53706 USA

INTRODUCTION

The biology of the endothelial cell is a burgeoning area of scientific interest. Once considered relatively inactive and homogenous, endothelial cells in recent years have been shown to be functionally interactive with other cells in many complex ways and to manifest extensive heterogeneity. Early studies of their heterogeneity focussed on such differences as fenestrated vs. non-fenestrated, large vessel vs. capillary, lymphatic vs. vascular. More current studies have investigated the heterogeneity of the cell surface of endothelial cells, particularly microvascular cells, as demonstrated by organ and tissue specificity. Selectivity in the seeding of metastatic cells and site-specificity of lymphocyte homing are examples of preferential adhesion to the endothelial cell surface. Organ specific enzymes such as glutamyl transpeptidase, structural heterogeneity such as the absence or presence of Weibel-Palade bodies, differences in the synthesis of specific cytokines such as IL-6, and variability in the production of vascular regulatory products such as prostaglandins are testimony to the complexity of the endothelium. Several other papers in this volume will stress the internal and structural variability among endothelial cells obtained from different sites, and from different developmental stages.

Improvement in the methods and tools available has greatly aided recent progress in this area. Among these tools were the generation of monoclonal antibodies, the increased sophistication of flow cytometers, the ability to grow microvascular endothelial cells in cell, tissue and organ culture, and the identification and isolation of growth factors that could be used to maintain and modulate endothelial cells in vitro (Simionescu and Simionescu, 1988; 1991; Ryan, 1989; Feinberg, Sherer and Auerbach, 1991; Orr and Weiss, 1991).

Our own interest in endothelial cell heterogeneity came from two separate questions: (i) What determinants are responsible for the exquisite

---

[1] The original studies reported in this paper have been supported by grants EY 3243 from the National Institutes of Health and a gift from Nova Pharmaceuticals. JP has been supported by a grant from the Deutsche Forschungsgesellschaft.

selectivity demonstrated by tumor cells, by migrating stem cells, by lymphocytes, by microorganisms and by parasites all of which are disseminated through the vascular or lymphatic system and all of which predictably recognize their destination at specific secondary sites? (ii) What are the developmental origins and heterotypic interactions that lead to the differentiation of endothelial cells at these different sites? Since both of these questions clearly involve cell surface phenomena, our studies have been directed at characterizing the cell surface of endothelial cells from different organ sites and different developmental ages (Auerbach, 1991 a,b).

EMBRYOLOGICAL CONSIDERATIONS

There have been two opposing schools of thought with respect to endothelial cell origins: an extrinsic origin as originally proposed by His and an intrinsic one argued convincingly by several workers such as Sabin and the Clarks in the 1920s (see Wagner, 1980; Auerbach and Joseph, 1983, for references to the extensive early studies in this field). As has so often been true in science, the resolution of the controversy has come through the recognition that both processes occur: organs such as the brain and limb bud appear to acquire blood vessels by the extension of existing vascular endothelial cells from preexisting vessels (angiogenesis) whereas organs such as the spleen, pancreas, liver and heart develop blood vessels from endothelial cell precursors in situ (vasculogenesis).[2]

We have previously pointed out that the argument for organ-specific differentiation of endothelial cells does not depend on the developmental origins of the endothelium (Auerbach and Joseph, 1983). On the one hand, precursors to endothelial cells that are resident in an organ rudiment during its early development are certainly influenced by the tissue/organ environment in which they have been developing. On the other hand, even where endothelial cells arise by extension of existing vessels extrinsic to the organ (e.g. brain, tumors) they are subject to extensive modulation once they have entered their new environment. However, one might well assume that the extent of influence of the microenvironment on endothelial cell differentiation would be greater in the former instance.

This does not mean that the impact of the organ environment is not different for the two modes of endothelial cell origins. More than a half-century ago, Waddington[3] argued for the distinction between "induction" and "individuation". To the former he assigned the ability of a heterotypic tissue (inducer) to cause a fundamental developmental change

---

[2] The terms "angiogenesis" vs "vasculogenesis" have been proposed by several workers in the area of vascular developmental biology. Angiogenesis has been equated with the development of blood vessels from pre-existing blood vessels, while vasculogenesis has been reserved for the formation of blood vessels from preendothelial or stem cells. Unfortunately, the terminology is based on the underlying assumption that these are two discrete processes, an assumption which has been put in question by the observations of Noden (1989), who has presented convincing evidence that both processes may occur during the formation of blood vessels even within a single organ or site. Sherer (1991), using his own terminology of "angiotrophic" and "angioblastic" has recently summarized the conflicting views, emphasizing the dual origin of blood vessels in several developing organ rudiments.

(determination) in the responding cells. The latter (individuation) yielded a less profound influence (modulation). This distinction, overly rigid even when first made, has become muddied as the complex sequence of regulatory events governing cell differentiation has begun to unfold, yet it serves to emphasize the differences between influencing a cell early in its differentiative sequence, i.e. at a stem cell or pre-stem cell stage, modifying it at the point that it has already acquired many of the definitive characteristics of a differentiated endothelial cell.[3] Moreover, if, in fact, a given organ rudiment is vascularized in part by entry of endothelial cells from preexisting vessels and in part by in situ differentiation of non-committed cells or pre-endothelial stem cells then one can almost presume that there will exist, even within a given organ, a fundamental dichotomy which may be reflected in distinct endothelial cell subpopulations with differing functional capacities.

There are several ways of characterizing the cell surface of endothelial cells: 1) expression of type-specific molecules. These include such markers as the receptor for acetylated low density lipoprotein and surface-associated angiotensin converting enzyme. These markers are constitutive, although they may be modulated in vitro depending on the state of confluence of the culture; 2) expression of function-associated molecules such as Class II antigens of the major histocompatibility complex. These molecules may be up-regulated by cytokines, leading to increased capacity for antigen presentation to lymphocytes; 3) presence of cell-surface associated glycoproteins important in cell-cell interactions such as lymphocyte and tumor cell adhesion; and 4) expression of other cell surface associated antigens as yet only identified on the basis of their being detected by various monoclonal antibodies.

ORGAN-SPECIFIC ANTIBODIES

Our own studies were designed to identify organ-specific antigens by means of a cross-species immunization/screening protocol that favored the detection of conserved epitopes (Auerbach et al, 1985). This rationale was based on the long-established fact that inductive tissue interactions (e.g. lung and kidney epithelio-mesenchymal interactions) and reorganization within cell aggregates (e.g. heart vs. liver) operated with tissue specificity but not species specificity.

Following this line of reasoning we immunized mice with rat ovary cells, generated antibody-forming clones by standard hybridoma technology, screened these using hamster ovary fibroblasts, and selected among the positive clones those that were also positive on mouse ovary endothelial cells. By subsequent screening we were able to identify three clones that produced antibodies that labelled mouse ovary endothelial cells and mouse ovary fibroblasts but failed to label either fibroblasts or endothelial cells from other murine organs. In a similar way we were able to isolate antibodies with a high degree of specificity for lung or brain endothelial cells.

Using these antibodies as well as standard reagents such as antibodies to Thy 1 and Class II (I[a]) antigens we were able to distinguish among different endothelial cell lines derived from brain, ovary and lung, as shown in Table I.

---

[3]For a more detailed discussion the reader is referred to the excellent textbook of Gilbert (1990) who provides both a historical perspective and a contemporary interpretation of these concepts.

Table 1.   Organ-specific antibodies against murine microvascular
            endothelial cells

Reagent	Brain	Lung	Ovary
P5E3 (anti-brain)	+++	±	-
Thy 1	++	-	-
Ia^d (Class II MHC)	-	+	ND
RO15 (anti-ovary)	-	-	+++
RO21 (anti-ovary)	-	-	+
α-ACE 3.1.1	+	++	ND

Additional antibodies such as the series of reagents generated against mouse endothelial cells by E. Butcher and his colleagues at Stanford (MECA antibodies) provided further evidence of endothelial cell surface antigen heterogeneity (Oliver and Auerbach, unpublished observations).   Quantitative differences in the expression of such antigens as Class I histocompatibility antigens (Auerbach, unpublished) and transferrin (Friden et al, 1991) provided yet more evidence of such heterogeneity.

LECTIN-BINDING PROPERTIES

Lectins, with their general property of recognizing specific terminal sugar residues, have proven to be useful in delineating cell surface-associated carbohydrates and glycoproteins.   Some terminal sugars are present on too many of these molecules to prove useful for characterizing specific endothelial cell subsets, but others are expressed on only few or a single molecule(s), and thereby become suitable as cell surface markers.

For murine endothelial cells the Dolichus biflorus lectin (DBA) has proven most interesting.   DBA is not found on endothelial cells from most mouse strains (e.g. BALB/c, C57BL/6), but is present on many endothelial cells from RIII and NMRI strain mice (Ponder and Wilkinson, 1983).   Expression of molecules that are recognized by DBA is not uniform, however.   For example, embryonic brain endothelial cells are DBA positive but postnatal brain endothelial cells are not; DBA binding is seen in all cardiac endothelial cells but never in liver-derived endothelial cells (Plendl and Schmahl, 1986).   Moreover, in several organs, microvascular endothelial cells are bimodal, with two distinct subsets differing in the number of DBA binding sites/cell. Thus DBA binding becomes suitable in the "DBA+" strains of mice for distinguishing among endothelial cells from different organs, among endothelial cells within a single organ, and among endothelial cells obtained at different times during development.

Although the studies with DBA lectin are the most dramatic, similar results have been seen using the peanut-derived lectin, PNA.   This lectin generally does not bind to endothelial cells from BALB/c mice, but there is one notable exception: endothelial cells from the neonatal or adult brain (Gumkowski, Kaminska, Kaminski et al, 1987).   Indeed, this lectin-binding has been exploited to identify a brain-derived cell surface protein by lectin affinity chromatography (Hua and Auerbach, unpublished observations).

Flow cytometric analysis, because of its capacity to quantitate the amount of lectin bound, has provided more information on endothelial cell heterogeneity: even when a particular lectin binds to many different endothelial cells, the extent of binding as defined by fluorescence intensity varies for different lectins on different endothelial cells (Gumkowski, Kaminska, Kaminski et al, 1987; Belloni and Nicolson, 1988). For example, although wheat germ agglutinin (WGA) binds to virtually all endothelial cells, the degree of binding may be as much as 100X for some endothelial cells as for others (Table 2).

Table 2. Mouse strain differences in lectin-binding properties of endothelial cells

Mice	Brain			Liver			Aorta			Heart		
	DBA	WGA	PNA	DBA	WGA	PNA	DBA	WGA	PNA	DBA	WGA	PNA
NMRI	++	+++	+++	++	+++	++	++	+++	++	++++	+++	++
BALB/c	-	++	++	-	++	-	-	++	-	-	+++	+/-

DIFFERENTIAL IN VITRO RESPONSES TO CELL MATRIX COMPONENTS

One of the striking examples of organ differentiation is seen when endothelial cells grown to confluence in vitro begin to organize into three-dimensional structures. As first described by Folkman, Haudenschild and Zetter (1979), microvascular endothelial cells but not large vessel endothelial cells grown on gelatin substrata were able to form such structures spontaneously once growth was arrested by contact-mediated inhibition, especially in the absence of extraneous growth factors. Subsequently, it was found that embryonic large vessel endothelial cells also were able to form such three-dimensional tubes (Madri and Pratt, 1986).

Addition of various extracellular matrix components such as collagen and laminin enhanced tube formation for endothelial cells. Differences are seen, however, between tube formation by microvascular endothelial cells and endothelial cells from large vessels. It seems likely, moreover, that endothelial cells derived from different organ sites may respond selectively to various extracellular matrix components, but systematic studies of different microvascular endothelial cells have not yet been reported.

DIFFERENCES IN CELL SURFACE-ASSOCIATED ADHESION MOLECULES

Cells disseminated through the vascular system extravasate by attaching to the luminal surface of endothelial cells following which they pass through the endothelial cell barrier by forcing a transient retraction of endothelial cell-cell attachment sites. The heterogeneity of endothelial cells with respect to permitting cell attachment is seen most vividly when considering the specialized functional role played by the high endothelial cells of the postcapillary venules. It is here that lymphocytes attach and extravasate, and it is now well documented that even among these special endothelial cells there are subsets that express adhesion molecules recognized by different lymphocyte subsets (Butcher, 1990).

A similar selective preference has been shown for tumor cells which preferentially adhere to microvascular endothelial cells of one or another organ bed (Alby and Auerbach, 1984; Auerbach et al, 1987; Nicolson, 1988a,b; Schirrmacher, 1986). Hepatoma cells, for example, preferentially adhere to hepatic endothelial cells, glioma cells to endothelial cells of the brain, teratoma cells to endothelial cells of the ovary, and mammary tumor cells to lymphatic endothelial cells.

Recently, several adhesion-mediating molecules important in endothelial cell recognition of circulating blood cells have been identified, molecules specific for neutrophil or lymphocyte attachment and extravasation. As yet, however, the molecules mediating tumor cell adhesion have not been defined. It is likely, however, that they will be part of the emerging family of adhesion-mediating glycoproteins that show structural homology for different domains but are distinct in their carbohydrate-rich ligand-binding properties.

DIFFERENTIAL RESPONSES TO ANGIOGENESIS-INDUCING FACTORS

Heterogeneity can be demonstrated among endothelial cells responding to growth- and migration-inducing stimuli. For example, endothelial cells obtained from different sites were found to vary in their chemokinetic response to complex factors contained in the bronchoalveolar lavage fluids obtained from normal individuals or from patients with various pulmonary pathologies (Weber, Meyer, Calhoun et al, 1989). Other studies have shown that endothelial cells obtained from large vessels differ from microvascular endothelial cells in their response to low-molecular weight angiogenesis factors isolated from tumor cells (Odedra and Weiss, 1991).

Almost unexplored, however, is the question of heterogeneity among microvascular endothelial cells in terms of their response to angiogenic factors released by different types of tumor cells. The assumption that there is a single angiogenic factor (TAF) unique to tumors has long ago been replaced by the recognition that there are many angiogenesis-inducing factors (see Blood and Zetter, 1990 for an excellent overview). Given the selective nature of tumor cell adhesion to endothelial cells and the precision with which tumor vascularization occurs at sites of secondary colonization it would not be surprising to find that among the angiogenesis-stimulating factors released by tumor cells one will find specificity as well.

CONCLUDING REMARKS

Much progress has been made in the identification of endothelial cell growth factors and in the development of methods for the study of endothelial cell behavior in vitro. In contrast, as yet our knowledge concerning endothelial cell heterogeneity is limited and our understanding of the regulatory events leading to endothelial cell differentiation is rudimentary. Given in the central position occupied by vascular endothelial cells in regulating the cellular and molecular interchange between the circulation and each of the organs in which they are located, a greater emphasis needs now to be placed on gaining a better understanding of the developmental biology of the vascular endothelium.

ACKNOWLEDGMENTS

We want to thank Laura Hartwell for help with endothelial cell cultures and flow cytometry, Jane Bielich for cell sorting, and Wanda Auerbach for editorial and bibliographical assistance.

REFERENCES

Auerbach, R., Vascular endothelial cell differentiation: organ specificity and selective affinities as the basis for developing anti-cancer strategies. Int. J. Rad. Biol. (in press, 1991)

Auerbach, R., Endothelial cell heterogeneity: Its role as a determinant of selective metastasis. in "Endothelial Cell Dysfunction," Orr, F.W. and Weiss, L., eds. CRC Press, Boca Raton, FL, p.169, (1991)

Auerbach, R., Alby, L., Morrissey, L.W., Tu, M., and Joseph, J., Expression of organ-specific antigens on capillary endothelial cells. Microvascular Res., 29:401 (1985).

Auerbach, R., and Joseph, J., Cell surface markers on endothelial cells: A developmental perspective. in "The Biology of Endothelial Cells", E.A.Jaffe, ed., Martinus Nijhoff, The Hague, p.393 (1983).

Belloni, P.N., and Nicolson, G.L., Differential expression of cell surface glycoproteins on various organ-derived microvascular endothelia and endothelial cell cultures. J. Cell. Physiol., 136:398 (1988).

Blood, C.H., and Zetter, B.R., Tumor interactions with the vasculature: angiogenesis and tumor metastasis. Bioch. Biophys. Acta, 1032:89 (1990).

Butcher, E.C., Cellular and molecular mechanisms that direct leukocyte traffic. Amer. J. Pathol., 136:3 (1990).

Feinberg, R., Sherer, G.K, and Auerbach, R., eds., "Development of the Vascular System", (Issues in Medicine, vol. 14), Karger, Basel

Folkman, J., Haudenschild, C.C., and Zetter, B.R., Long-term culture of capillary endothelial cells. Proc. Natl. Acad. Sci. U.S.A., 76:5217 (1979).

Friden, P.M., Walus, L.R., Musso, G.F., Taylor, M.A., Malfroy, B., and Starzyk, R.M., Anti-transferrin receptor antibody and antibody-drug conjugates cross the blood-brain barrier. Proc. Natl. Acad. Sci. U.S.A., 88:4771 (1991).

Gumkowski, F., Kaminska, G., Kaminski, M., Morrissey, L.W., and Auerbach, R., Heterogeneity of mouse vascular endothelium. In vitro studies of lymphatic, large blood vessel and microvascular endothelial cells. Blood Vessels, 24:11 (1987).

Madri, J.A., and Pratt, B.M., Endothelial cell-matrix interactions: in vitro models of angiogenesis. J. Histochem. Cytochem. 34:85 (1986).

Nicolson, G.L., Organ specificity of tumor metastasis: role of preferential adhesion, invasion and growth of malignant cells at specific secondary sites. Cancer Metastasis Rev., 7:143 (1988a)

Nicolson, G.L., Cancer metastasis: tumor cell and host organ properties important in metastasis to specific secondary sites. Bioch. Biophys. Acta, 948:175 (1988b).

Noden, D.M., The formation of avian embryonic blood vessels. Amer. Rev. Respir. Dis., 140:1097 (1989).

Odedra, R., and Weiss, J.B. Low molecular weight angiogenesis factors. Pharm. Therap., 49:111 (1991).

Orr, F.W., and Weiss, L., eds., "The Microcirculation in Cancer Metastasis," CRC Press, Boca Raton, FL (in press).

Plendl, J., and Schmahl, W., Binding of dolichos biflorus agglutinin (DBA) to endothelial cells during the embryonal period is restricted to the NMRI-mouse. in "Lectins", vol.V, T.C. Beg-Hansen and E. van Driessche, eds. Walter de Gruyler, Berlin (1986).

Ponder, A.J., and Wilkinson, M.M., Organ-related differences in binding of Dolichos biflorus agglutinin to vascular endothelium. Dev. Biol., 96:535 (1983).

Ryan, U., ed., "Endothelial Cells," vol. I-III, CRC Press, Boca Raton, FL, 1989.

Sherer, G.K., Vasculogenic mechanisms and epithelio-mesenchymal specificity in endodermal organs. in "The Development of the Vascular System", Feinberg, R.N., Sherer, G.K., and Auerbach, R., eds. Karger, Basel, 37.

Schirrmacher, V., Mechanisms of metastatic spread to liver. Recent Res. Cancer Res., 100:23 (1986).

Simionescu, N., and Simionescu, M., eds., "Endothelial Cell Biology in Health and Disease." Plenum Press, New York, 1988.

Simionescu, N., and Simionescu, M., eds., "Endothelial Cell Dysfunction", Plenum Press, New York (in press).

Wagner, R.C., Endothelial cell embryology and growth. Microcirculation 9:45 (1980).

HETEROGENEITY IN ENDOTHELIAL CELLS WITH SPECIAL REFERENCE TO THEIR GROWTH RELATED PROTEINS

S Kumar, P Kumar, D Pye, A Sattar, M Wang and J Ponting

Christie Hospital, Manchester Polytechnic, Manchester M20 9BX
England

The endothelium has been calculated to occupy an area of 7000 m² ($\sim 6 \times 10^{13}$ endothelial cells) and to weigh nearly 1 kg (Wolinsky, 1980 and Simionescu, 1988). The use of a variety of techniques has shown that the simple looking endothelial membrane lining blood and lymphatic vessels is dynamic, complex and heterogeneous with regard to its histogenesis, morphology, metabolism, biochemical and immunological characteristics (Table 1). Furthermore, the *in vitro* culture conditions can profoundly modulate cell behaviour.

The formation of new blood vessels occurs by one of two processes. The term angiogenesis is used to describe the emergence of new blood vessels from pre-existing ones (Herting, 1935), while vasculogenesis denotes the *de novo* occurrence of blood vessels. The acquisition of a vasculature by solid tumours is the result of angiogenesis. In the developing embryo the vasculature is formed by both angiogenesis and vasculogenesis. Whether or not endothelial cells (EC) that arise intrinsically (from graft) or extrinsically (from host) differ from each other is not known.

A specialised type of endothelium lines the marginal sinus capillaries of the spleen and post-capillary venules in the lymph node. These post-capillary venules or high endothelial venules (HEV) are lined by rather plump looking high endothelial cells (HEC). HEC differ from other vascular EC in many respects: (i) HEC have been noted to be significantly more activated than the other EC within or outside lymph nodes, (ii) Both synthetic and secretory pathways are highly upregulated and HEC possess a unique biosynthetic pathway resulting in continuous secretion of sulphated glycolipids, (iii) They possess greatly increased levels of cytoplasmic esterase, (iv) HEC lack von Willebrand Factor (vWF). (v) The use of monoclonal antibodies has identified several distinct antigens on HEC and (vi) lymphocyte binding to HEC is nearly 50 times greater than to flat EC (Ager 1990 a & b).

Until the early 1970's no reliable method for the isolation of EC was available. Jaffe et al (1973) succeeded in obtaining EC from human umbilical vein using collagenase instead of trypsin digestion. Over the years the availability of special media, growth supplements and better quality sera has enormously facilitated the isolation and culture of EC from large and small vessels. As the number of endothelial cell markers that can be used to identify EC has mushroomed, so it has become apparent that there is no standard EC and no standard EC marker. In this communication a practical account of some of the methods available to culture and purify EC is given and

Angiogenesis in Health and Disease, Edited by M.E. Maragoudakis *et al.*
Plenum Press, New York, 1992

TABLE 1. A selective list of endothelial cell characteristics

Characteristic	Frequency	Examples
1. Embryonic origins	By angiogenesis By vasculogenesis	Kidney, Brain Stomach, Pancreas
2. Morphology	Typical (Cobblestone) Atypical (Fibroblastic)	Large Vessels BBEC, HEC
3. W-P bodies	Common Rare	Large Vessels Microvessels
4. vWF	Present Absent	Most EC Liver & Spleen Sinusoids; Glomerular Capillaries
5. HLA DR	Positive Negative	Most EC Umbilical Vein, Brain
6. Mab IF 10	Positive Negative	Most EC Spleen & Liver Sinusoids
7. Mab PAL-E	Positive Negative	Most EC Lymphatic & Large Vessel EC
8. γ - glutamyl transpeptidase	Positive Negative	Brain EC Most other EC
9. Cell Proliferation	Greatly increased	Tumour EC
10. Radiation Sensitivity	More Less	Bovine EC Rabbit EC
11. Dil-Ac-LDL	Positive Negative	Most EC HEC, BBEC
12. Albumin binding protein	Present Absent	Continuous Capillaries Arterial EC

Key references: Auerbach, 1991; Blood and Zetter, 1990; Bouck, 1990; Denekemp, 1991; Fajardo, 1989; Ingber et al, 1990; Kumar et al, 1987; Le Lievre et al, 1975; Paweletz and Knierim, 1989

HEC - High endothelial cells; BBEC - Bovine brain endothelial cells; W-P body - Weibel-Palade body we show that angiogenesis promoters and inhibitors differ in their action on EC. Endothelial cell heterogeneity was also investigated by comparing the protein-profiles of proliferating and quiescent (non-dividing) endothelial cells.

MATERIALS AND METHODS

All tissues were obtained taking aseptic precautions. A number of good reference books on tissue culture techniques are available (Freshney, 1987).

A₁   ISOLATION OF ENDOTHELIAL CELLS

(a)   Large vessels

Human umbilical vein endothelial cells (HUVEC)

Umbilical cord is one of the most widely used and reliable sources of human EC. Briefly, a vein in an umbilical cord is cannulated, washed with phosphate buffered saline (PBS) and incubated with 0.1% collagenase solution. The cell pellet is washed free of collagenase and resuspended in fresh medium. We have tried various tissue culture media and have found that a 1:1 mixture of medium 199 and MCDB 104, containing 20% foetal calf serum gives the best results (Wang and Kumar, unpublished results). Once the cells become semi-confluent they can be subcultured using 0.05 % trypsin. It is essential that as soon as the cells have began to detach, the action of trypsin should be stopped, as prolonged trypsin treatment can reduce cell viability and plating efficiency.

Bovine aortic endothelial cells (BAEC)

Two methods have been employed to obtain BAEC

(a) Mechanical harvesting: The aorta is cleaned, and cut longitudinally. Endothelial cells are scraped off with a scalpel and transferred to a vial containing medium with or without collagenase. The cell pellet is resuspended in fresh medium and the contents of one aorta generally are sufficient to populate one 75cm² tissue culture flask.

(b) Enzymatic digestion, Wagner et al (1988) have obtained EC from bovine aorta and vena cava by splitting open the blood vessel and placing the luminal surface downwards in a dish containing 0.1% collagenase. After 20 minutes, the endothelial cells are mechanically removed using gentle scraping with a rubber policeman.

The above two procedures (a and b) can be used for obtaining EC from large and medium sized vessels, such as pulmonary artery and saphenous vein.

b.   Microvessels

Apart from brain, the other widely used sources of EC are bovine adrenal gland and, to a lesser degree, skin, lung, retina, liver, heart, kidney, adipose tissue etc (Folkman and Haudenschild, 1979; Davison et al, 1980; Karasek, 1989; Ryan, 1989; Pearson, 1990). A description of an isolation procedure for EC from bovine brain white matter (BBEC) is given which can be easily adapted for culturing EC from other sources.

Mammalian brain white matter is a unique source of microvessel endothelial cells which we have used since 1974 (Phillips et al, 1976). The blood supply to white matter is mainly by microvessels, brain meninges are limited to grey matter. In the adult animal glial and neuronal cells are very fastidious and do not grow under normal tissue culture conditions. To obtain endothelial cells, bovine brain white matter is finely minced with scissors, incubated with 0.25 percent trypsin for 2 minutes and pipetted vigorously. Trypsin is quickly neutralised by adding medium containing serum, cells are collected by centrifugation, resuspended in fresh medium and inoculated into tissue culture flasks. It is essential that the flasks should not be moved for one week, when the first change of medium is undertaken. Since we do not add any growth factors to the medium it is possible that white matter tissue fragments may be providing vital supplements.

(c) High endothelial cells (HEC)

The technique for the isolation of rat lymph node HEC has been published (Agar, 1990b). 1-2mm thick slices of lymph node were incubated with 50 uCi/ml[^{35}S]-sulphate for 30 minutes. Autoradiographs showed preferential localisation of sulphate in HEC; all other stromal cells remained unlabelled. Radiolabelled lymph node slices were incubated with type II collagenase for 60 minutes and passed through 100um nylon mesh. The cell suspension was resuspended in fresh medium and plated in tissue culture dishes for 60 minutes. Non-adherent cells were removed and the plates returned to an incubator.

(d) Brain tumour derived endothelial cells (BTEC)

A fresh biopsy of a brain tumour diagnosed as medulloblastoma was obtained from a child. The tissue was finely minced using a pair of scissors and vigorously sucked up and down a syringe fitted with 14 gauge needle. The cell suspension was homogenised with a Dounce homogeniser (10 times), and passed through a 150um nylon cloth. The filtrate was centrifuged, the pellet treated with collagenase, centrifuged again, and the pellet was resuspended in fresh tissue culture medium before inoculation into a T25 flask. Two days later almost the whole of the flask base was covered with flat epithelial looking cells. The cells were subcultured and grown on glass coverslips. The following day, coverslip cultures were treated with antibodies to glial fibrillary acidic protein (GFAP), cytokeratins (8 and 10), and endothelial cell specific antibodies (vWF, C3, 5.6E, E-9, fluorescently labelled ULEX europeus agglutinin I) and their uptake of Dil-Ac-LDL was also examined.

A₂ PURIFICATION OF CONTAMINATED ENDOTHELIAL CELLS

It is inevitable that primary cultures of endothelial cells will contain a varying proportion of other stromal cells like fibroblasts, pericytes and smooth muscle cells. The contaminant cells can be removed by any of the following procedures:

(a) Cloning or weeding

The cloning technique to purify EC by serial dilution as used in hybridoma production has not been successful in our hands. An alternative approach, wherein primary cultures are allowed to form microcolonies, has proved to be more practical. An islet of growing endothelial cells with

66

typical cobblestone morphology is located and using a rubber policeman or a wire loop, suspect cells are scraped off leaving the endothelial cell cluster intact. If desired, a cloning ring is placed around the endothelial cell cluster. When the cells have grown to a sufficient density, they are trypsinised and plated into fresh containers. It is a laborious technique which usually provides a limited supply of cloned cells but requires no expensive piece of equipment (for further details see Zetter, 1984).

(b)   Fluorescence activated cell sorter (FACS)

The fact that most endothelial cells are ACE positive, led Auerbach and Joseph (1984) to stain them with a fluorescently-labelled anti-ACE monoclonal antibody (Mab) and purify fluorescently labelled cells using FACS. Vyota et al (1984) realised that Dil-Ac-LDL is preferentially taken up by endothelial cells and they used this property to sort them from Dil-Ac-LDL negative cells. The yield and purity of EC cells by this method is good.

(c)   Adsorption of EC using endothelial cell specific Mab

In our laboratory Clarke et al (unpublished data) added a mixed population of cells (containing EC and non-EC) to tissue culture flasks coated with pan-endothelial cell-specific antibody [5.6E (CD31) raised by P Dias and S Kumar]. After a short incubation period, adherent EC were found to be almost 95 percent pure. When the procedure was repeated, the purity of EC cells was nearly 100 percent.

B.   2-DIMENSIONAL SDS PAGE ELECTROPHORESIS OF [^{35}S]-LABELLED EC

Semi-confluent (proliferating) and confluent (non-proliferating) endothelial cells were metabolically labelled with [^{35}S]-L-methionine. Both membrane and whole cell extracts were prepared and 2-dimensional gel-electrophoresis was carried out following the previously published procedures (Kumar et al, 1987 and Clarke et al, 1991).

Assays for angiogenic and anti-angiogenic agents

A low molecular weight angiogenesis factor from rat Walker 256 carcinoma was purified as reported (Kumar et al, 1983).

Angiogenic oligosaccharides of hyaluronate were produced as described by West et al (1985).

The chicken chorioallantoic membrane assay was used to examine the angiogenic potential of a sample. The latter was considered angiogenic if it produced a concentric growth of blood vessels radiating from the site of its application.

The effects of angiogenesis agonist and antagonist on cell proliferation and migration were examined using tissue cultured endothelial and non-endothelial cells (for details see West and Kumar 1989 and Sattar et al, 1991).

Mouse dorsal air sac was created in mice and two sealed millipore chambers containing mouse hepatoma cells, as a source of angiogenesis factor, were placed in position as shown in Figure 1. On days 1, 3 and 5, mice were given intraperitoneally either PBS or 50 mg/kg of Krestin. On day 7 the mice were sacrificed and the area under the filters exposed and examined for angiogenic response. A portion of tissue sample was fixed for light and electron microscopy and another was extracted to measure alkaline phosphatase activity.

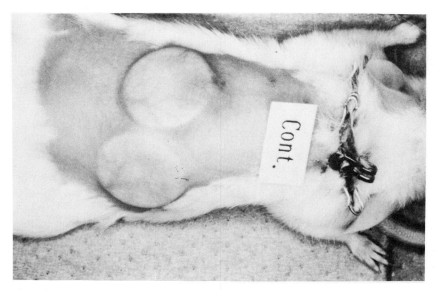

Fig 1 The mouse dorsal air sac assay: outlines of two sealed millipore chambers containing hepatoma cells as a source of angiogenesis factor can just be made out.

RESULTS

In general EC from large vessels were relatively easy to culture and recognise. In contrast, EC from microvessels showed greater variation and did not possess all the accepted hallmarks of EC (Table 2 and Fig 2). This was not due to differences in the degree of contamination with non-EC cells. Apparently the method of isolation i.e. mechanical harvesting (BAEC) vs enzymatic digestion (HUVEC) did not influence either the yield or the quality of the cells. One single factor that had the greatest effect was the batch of serum - some batches were infinitely better than others. For instance, of 20 different batches of sera from 3 different suppliers, only one was found that could sustain the growth of BBEC satisfactorily. Foetal calf serum was routinely used to culture HUVEC and BBEC, whereas new born calf serum was adequate for BAEC only.

Endothelial cells from large blood vessels (HUVEC, saphenous vein and BAEC) formed a contact inhibited cobblestone monolayer. Once confluence was reached, endothelial cells began to migrate out as sprouts. The presence of heparin (100 ug/ml) promoted such sprout formation whereas ascorbate (50 ug/ml) inhibited it.

BBEC grew as loose cell clumps (unlike BAEC they were not polygonal). After the first change of medium, growth was rapid and confluence was reached within 2-3 weeks. Irrespective of their source (rat, pig or cow) confluent cultures appeared elongated and fusiform and unlike BAEC, sprouting was uncommon.

Bovine adrenal capillary EC (BACE) were polygonal in shape and grew as pavemented-like sheets of cells. Primary cultures, especially after cloning, required the addition of tumour conditioned medium or growth factors. Even before the cells reached confluency, vessel-like tube formation was fairly frequent.

TABLE 2. Characterisation of Human Brain Tumour Derived Endothelial Cells (BTEC)

ENDOTHELIAL MARKER	CELL TYPE		
	BTEC	HUMAN UMBILICAL VEIN ENDOTHELIAL CELLS	FIBROBLASTS
vWF:Ag	− [1]	+	−
Mab 5.6E (PECAM:CD31) [2]	−	+	−
Mab C3 [3]	+	+	−
Mab E9 [4]	+	+	−
Uptake of Ac-Dil-LDL	+ [5]	+	−
Ulex Europeus lectin	+	+	−

(1) Initial culture weakly positive, but on passage became negative
(2) Produced by P Dias and S Kumar
(3) Produced by J Wang
(4) See Wang et al (1991)
(5) Faintly positive

The growth of human brain tumour derived EC cells was rapid. Within 2 days the culture dish was confluent with homogeneous epithelial-looking contact inhibited sheets of cells which covered the entire area. Initially, a few scattered adherent round cells were observed, but after a gentle wash with PBS they came off. By day 3 the sheet was confluent and the cells were subcultured. Their staining reactivity with a panel of antibodies and their ability to endocytose Dil-Ac-LDL was checked (Table 2 and Fig 3). From the results, it appears that these cells possessed many of the characteristics of endothelial cells. Since on histological examination the tumour did not contain any normal brain tissue, it is highly likely that these cells originated from the tumour vasculature.

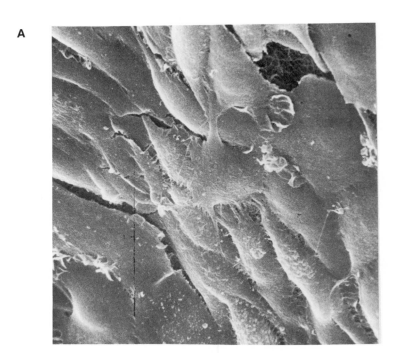

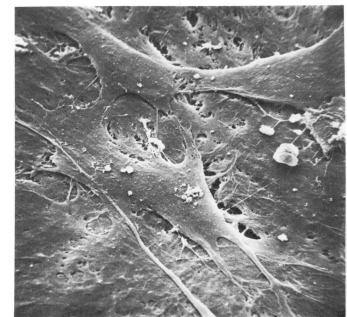

Fig 2 Scanning electron micrographs of bovine aortic endothelial cells with typical cobblestone appearance (A) and bovine brain microvascular endothelial cells (B) showing somewhat fibroblastic morphology.

The autoradiographic studies of lymph node cultures showed the presence of 3 main cell types: [$^{35}SO_4$] labelled large cells (HEC) 20-30um in diameter, unlabelled medium sized cells 10-15um diameter, (presumably macrophages) and unlabelled small lymphocytes. After 2-3 days the primary lymph node culture underwent marked changes in morphology in that the HEC flattened and adopted an epithelial morphology. At confluence HEC were nearly 99 percent pure and spindle-shaped. Binding of thoracic duct lymphocytes to these spindle-shaped HEC is approximately 50-fold greater than to flat endothelial cells. vWF was absent from HEC both *in situ* and in culture. Dil-Ac-LDL was incorporated by 2-3 day old cultures but not thereafter (Ager, 1990a).

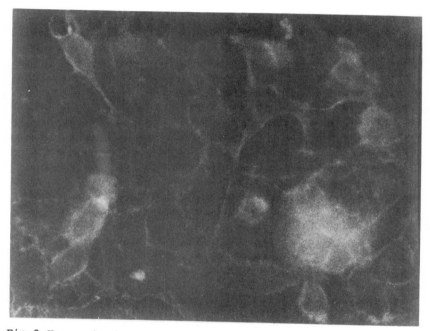

Fig 3 Human brain tumour derived endothelial cells have been stained by a monoclonal antibody E-9 (Wang et al, 1991). This Mab preferentially binds to tumour endothelial cells.

In the CAM assay both HA oligosaccharides and purified TAF induced angiogenesis. *In vitro*, HA oligosaccharides resulted in proliferation and migration of both large and microvessel derived EC. HA oligosaccharides had no effect on fibroblasts, smooth muscle cells or keratinocytes. TAF stimulated the growth of capillary but not aortic EC and only when the cells were grown on a native collagen gel. Neither TAF nor HA oligosaccharides influenced the growth of human brain tumour derived endothelial cells (Fig 4).

In the mouse dorsal air sac assay, intraperitoneal administration of Krestin significantly inhibited hepatoma induced angiogenesis: there being a reduction in the number of blood vessels (Fig 5) and a concommitant decrease in the alkaline phosphatase activity. *In vitro* Krestin had no effect on either proliferation or migration in bovine aortic endothelial cell cultures.

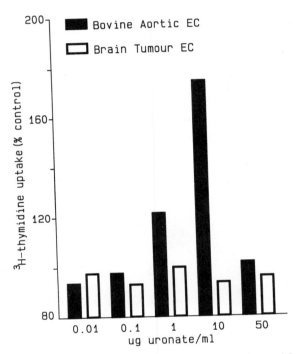

Fig 4 Angiogenic oligosaccharides of hyaluronic acid stimulated
proliferation of large and microvessel derived
endothelial cells from human and animal sources. In this
figure stimulation of bovine aortic endothelial cells is
shown. BTEC were not stimulated by HA oligosaccharides.

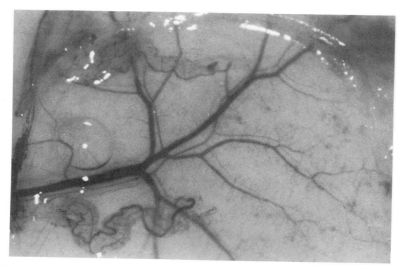

Fig 5 The mouse dorsal air sac assay: The administration of Krestin significantly inhibited the angiogenic response of hepatoma cells.

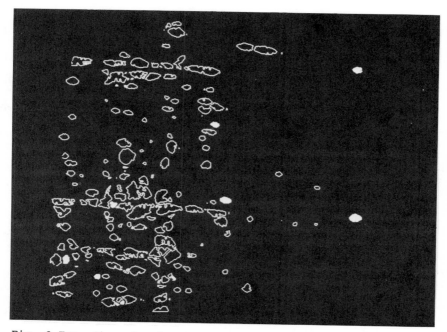

Fig 6 Two dimensional gel electrophoresis: Computer-captured picture of a fluorogram showing the presence of $[^{35}S]$-methionine labelled proteins. The solid spots show proteins which are present in proliferating, but not quiescent bovine aortic endothelial cells.

An analysis of membrane and whole cell protein profiles of [^{35}S]-L-methionine labelled proliferating and non-proliferating EC revealed dramatic differences (Fig 6). For instance, HA oligosaccharide-induced stimulation of BAEC resulted in 73 unmatched proteins compared with control untreated confluent cultures. The precise nature of these proteins has not been ascertained. Proliferation associated proteins were also observed in HUVEC cells (Clarke et al, 1991).

DISCUSSION

The aim of these studies is to understand the biology of endothelial cells. When compared to normal tissues, tumour vasculature shows gross alterations *in vivo*. One such difference is in the proliferation rate of EC. Unlike the normal vasculature which is quiescent, the proliferation rate of tumour EC is significantly higher (Denekamp, 1991). The purpose of the present summary review is to ascertain if the observed *in vivo* differences between tumour and normal vasculature induce demonstrable changes in protein profiles of tumour endothelial cells *in vitro*. This is important, as the recognition of such an activation/proliferation associated antigen may allow the development of strategies aimed at eliminating the tumour vasculature. Solid tumours are dependent on the continuous supply of new blood vessels for their growth (Folkman, 1990). The formation of new blood vessels is the ultimate result of breakdown and later synthesis of basement membrane, endothelial cell migration and proliferation. It is highly likely that these events are accompanied by the appearance of unique or significantly raised amounts of certain antigens. Indeed, the use of 2-dimensional gel electrophoresis technique has allowed recognition of 'novel' proteins unique to the proliferating endothelial cells. A somewhat surprising outcome of these studies is that many of these proteins are "common" to proliferating EC, but a few others were only induced by one or the other growth stimulant. It ought to be possible to isolate, purify, microsequence and raise monoclonal antibodies against these proteins, which can be linked to cytotoxins. This is no longer a theoretical possibility but a practical reality (Anonymous, 1991).

Two-dimensional gel electrophoresis

Two-dimensional gel electrophoresis is one of the most powerful techniques used to analyse fine changes in protein profiles of cells. The technique relies on the separation of proteins on the basis of their charge (isoelectric point) and molecular weight. It is capable of resolving two proteins that may differ only in a single charged amino acid. Lincoln et al (1984) used 2-D gel electrophoresis to demonstrate that the 'finger-prints' of EC, smooth muscle cells and fibroblasts were unique and thus could aid their identification. Wagner (1988) compared more than 1000 different proteins from bovine arterial and venous endothelial cells and identified approximately 250 proteins that differed in these two cell types. These results led the authors to conclude that differences in the EC of venous and arterial origins represented a fundamental difference between two sources. The effect of cell density and addition of exogenous growth factors on protein synthesis in EC has been investigated (Gospodarowicz, 1981; Borsum et al, 1982; Kumar et al, 1987). Gospodarowicz et al (1981) have compared the effects of EGF and FGF on protein patterns of EC and found that the two stimulants induced different changes in newly synthesised proteins. We have compared the synthesis of membrane proteins by confluent and stimulated BAEC. All 3 stimulants (serum, ECGF and angiogenic fragments of hyaluronic acid) induced common proliferation associated proteins and others which were specific for one stimulant. Hyaluronic acid fragments specifically induced 7 proteins of which two 42kDa proteins of pI 5.0 and 4.0 were particularly marked, whereas ECGF stimulated the production of a different protein (41kDa, pI 5.4). Clarke et al (1991) have examined the protein profiles of proliferating and contact inhibited (quiescent) endothelial cells of human and animal origin. Both qualitative and quantitative changes in the 'appearance' and 'loss' of certain proteins were observed. In general there was a greater similarity in protein patterns given by two endothelial cell types of same species. A few proliferation associated proteins with similar pI and molecular weight were observed in endothelial cells from various sources. Their relative positions in the fluorograms were almost identical so it is highly likely that they were the same proteins. Wagner et al (1988) found good but not absolute correlation among protein profiles of various runs using the same cell type. This is one of the limitations of the 2-D technique, wherein inter- and intra-run variability cannot be completely overcome. However, careful computer assisted analysis of individual proteins with special reference to their relative position in the gel can remove much of the uncertainty.

Some practical implications of EC heterogeneity

As mentioned earlier, the occurrence of heterogeneity among EC implies that protein(s) of tumour endothelium may be novel or their level of expression may be different from normal EC. Recently a Mab E-9 raised against activated HUVEC cells has been shown to react preferentially with tumour vasculature EC but of only a few normal tissues (Wang et al, 1991). The nature of the antigen recognised by E-9 is not known, but it is not a proliferation associated protein. Another Mab, 7.44, thought to react with proliferating and migrating EC has been produced by Sorg's group (Hagemeier et al 1986). An immunoglobulin fraction of 7.44 was labelled with ^{131}I, and injected intravenously into 2 patients with malignant gliomas. Unfortunately, there was no preferential localisation of the labelled antibody within or around the tumour mass (Kumar and Sorg, unpublished data). One reason for this failure may be that 7.44 is an IgM molecule and therefore is not an ideal antibody for immunoscintography. It is feasible that in the future, when better antibodies are available, it may be possible to undertake targeting of the tumour vasculature.

Krestin is a protein bound polysaccharide isolated from mycelia of Coriolus versicolor. Krestin retards tumour growth in animals although the mechanism of its anti-tumour effect is not known. When Millipore chambers containing tumour cells are implanted under the skin on the back of mice, they induce angiogenesis. Intraperitoneal treatment of the animals with Krestin causes a significant decrease in alkaline phosphatase activity in tissue extracts and a marked inhibition of angiogenesis. In vitro Krestin has no effect on proliferation of bovine aortic or capillary endothelial cells. Recently we found that proliferation of BTEC was inhibited by Krestin (P<0.01) (Kumar et al, unpublished data). Angiogenin induces angiogenesis in a CAM assay, but causes neither EC proliferation or migration in vitro (Anonymous, 1990). Thus in some instances an in vitro model may be of limited value for an in vivo situation. Nevertheless in the vast majority of cases, endothelial cell cultures have been successfully used both to purify and study the actions of angiogenesis agonists and antagonists. One recent example is the purification of fumagillin and the synthesis of its more potent but less toxic analogues, which inhibit angiogenesis and suppress tumour growth in animal models, and endothelial cell proliferation in vitro (Ingber et al, 1990).

REFERENCES

Ager, A. 1990a. Dynamic interactions between lymphocytes and vascular endothelial cell, in: The Endothelium: An Introduction to Current Research, p 229, J. B. Warren, Ed., Wiley-Liss, Inc, New York.

Ager, A. 1990b. Isolation and culture of high endothelial venule endothelium from rat lymph nodes in: The Endothelium: An Introduction to Current Research, p 273, J. B. Warren, Ed., Wiley-Liss, Inc, New York.

Anonymous, 1990. Exploiting angiogenesis. Lancet 1; 208.

Auerbach, R. 1991. Vascular endothelial cell differentiation: organ-specificity and selective affinities as the basis for developing anticancer strategies. Int. J. Radiol. Biol, 60:1-10.

Auerbach, R. and Joseph, J. Cell surface markers on endothelial cells: a developmental perspective. In: Biology of endothelial cells, E. A. Jaffe, ed., Martinus Nijhoff, Boston, 1984.

Blood, C.H. and Zetter, B.R. 1990. Tumour interactions with the vasculature: angiogenesis and tumour metastasis. Biochimica et Biophysica Acta, 1032:89.

Bouck, N. 1990. Tumor Angiogenesis: The role of oncogenes and tumor suppressor genes. Cancer Cells, 2:179.

Clarke, M. S. F., Kiff, R. S., Kumar, S., Kumar, P. and West, D. C. 1991. The identification of proliferation-related proteins in human endothelial cells as a possible target in tumour therapy. Int. J. Radiat. Biol 60:17.

Davison, P. M. and Karasek, M. A. 1981. Human dermal microvascular endothelial cells in vitro: effect of cyclic AMP on cellular morphology and proliferation rate. J Cell Physiol, 106:253.

Denekamp, J. 1991. The current status of targeting tumour vasculature as a means of cancer therapy: an overview. Int. J. Radiat. Biol., 60:401.

Fajardo, L. F. 1989. The complexity of endothelial cells. Am J Clin Pathol, 92:241.

Folkman, J. 1990. What is the evidence that tumours are angiogenesis dependent? J.N.C.I., 82:4-6.

Folkman, J. and Haudenschild, C. C. 1979. Long-term culture of capillary endothelial cells. Proc Natl Acad Sci USA, 76:5217.

Freshney, R. I., 1987. Culture of animal cells. Alan R Liss, New York.

Hagemeir, H., Vollmer, E., Goerdt, Schulze-Osthoff, K., Sorg, C. 1986. A Mab reacting with endothelial cells of budding vessels in tumours and inflammatory tissues and non-reactive with normal adult tissues. Int. J. Cancer, 38:481.

Hertig, A. T. 1935. Angiogenesis in the early human chorion and in the primary placenta of the macaque monkey. Contrib Emrbyol Carnegie Inst, Washington, 25: 137.

Ingber, D., Fujita, T., Kishimoto, S., Folkman, J. 1990. Synthetic analogues of fumigillin that inhibit angiogenesis and suppress tumour growth. Nature, 348:555.

Jaffe, E. A., Nachman, R. L. and Becker, C. G. 1973. Culture of human endothelial cells derived from human umbilical veins: identification by morphological and immunological criteria. J Clin Invest, 52:2745.

Karasek, M.A. 1989. Microvascular endothelial cell culture. J. Invest. Dermatol, 93:335.

Kumar, S., West, D. C., Shahabuddin, S., Arnold, F., Haboubi, N. and Reid, H. 1983. Angiogenesis factor from human myocardial infarcts. Lancet II:364.

Kumar, S., West, D. C. and Ager, A. 1987. Heterogeneity in endothelial cells from large vessels and microvessels. Differentiation, 36:57.

Le Lievre, C. S. and Le Douarin, N. M. 1975. Mesenchymal derivatives of the neural crest: analysis of chimaeric quail and chick embryos. J Embryol exp Morph, 34: 125.

Lincoln, D.W., Braunschweiger, K.I., Braunschweiger, W.R. and Smith, J.R. 1984. Cell type specific electrophoretic profiles of cultured bovine cells. J. Cell Sci. 72:135.

Paweletz, N. and Knierim, M. 1989. Tumour-related angiogenesis. Critical Revs Oncol/Hematol, 9:197.

Pearson, J. D. 1990. Isolation and Culture of Microvascular Endothelium, in: The Endothelium: An Introduction to Current Research, pp 295, J. B. Warren, Ed., Wiley-Liss, Inc, New York.

Phillips, P.J., Steward, J.K. and Kumar, S. 1976. Tumour angiogenesis factor (TAF) in human and animal tumours. Int. J. Cancer, 17:549.

Ryan, U.S. 1984. Culture of pulmonary endothelial cells on microvascular beads in Biology of Endothelial Cells (ed: E.A. Jaffe) Martinus Nijhoff, Boston pp34.

Sattar, A., Kumar, S. and West, D. C. 1991. Does hyaluronan have a role in endothelial cell proliferation of the synovium? Seminars in Arthritis and Rheumatology (in press).

Simionescu, M. 1988. Receptor-mediated transcytosis of plasma molecules by vascular endothelium, in: Endothelial Cell Biology in Health and Disease, pp 69-109. N. Simionescu and M. Simionescu, eds, Plenum Press, London.

Vyota, J.C., Via, D.P., Butterfield, C.E. and Zetter, B.R. 1984. Identification and isolation of endothelial cells based on their increased uptake of acetylated low density lipoproteins. J. Cell Biol. 99:2034.

Wagner, W.H., Henderson, R.M., Hicks, H.E., Banes, A. J. and Johnson, G. 1988. Differences in morphology, growth rate and protein synthesis between cultured arterial and venous endothelial cells. Vasc. Surg. 8:509.

Wang, J., Hunter, R. D., and Kumar, S. 1991. A monoclonal antibody (E-9) binds preferentially to the vasculature of human tumours, embryonic and regenerating tissues, in: Angiogenesis, R. Steiner et al, ed., Birkhauser Verlag, Basel, (in press).

West, D. C., Hampson, I., Arnold, F. and Kumar, S. 1985. Angiogenesis induced by degradation products of hyaluronic acid. Science, 228:1324.

West, D. C. and Kumar, S. 1989. The effect of hyaluronate and its oligosaccharides on endothelial cells: proliferation and monolayer integrity. Exp Cell Res, 183:171.

Wolinksy, H. 1980. A proposal linking clearing of circulating lipoproteins to tissue metabolic activity as a basis for understanding atherogenesis. Circ Res, 47:301.

Zetter, B.W. 1984. Culture of capillary endothelial cells. In: Jaffe EA (ed) Biology of Endothelial Cells. Martinus-Nijhoff, Boston pp14-26.

BASIC FIBROBLAST GROWTH FACTOR AND ENDOTHELIAL CELLS: RECEPTOR INTERACTION, SIGNAL TRANSDUCTION, CELLULAR RESPONSE-Dissociation of the mitogenic activity of bFGF from its plasminogen activator-inducing capacity

Marco Presta, Marco Rusnati, Jeanette A.M. Maier, Chiara Urbinati, Massimo Statuto, Anna Gualandris, Ambra Pozzi, and Giovanni Ragnotti

Unit of General Pathology and Immunology, Department of Biomedical Sciences and Biotechnology, School of Medicine, University of Brescia, 25123 Brescia, Italy

Antonella Isacchi, Laura Bergonzoni, Paolo Caccia, and Paolo Sarmientos

Department of Biotechnology, Farmitalia Carlo Erba, 20146 Milan, Italy

INTRODUCTION

Basic fibroblast growth factor (bFGF) belongs to a class of heparin-binding growth factors which includes acidic FGF (aFGF) and five other related oncogene products (Burgess and Maciag, 1989). bFGF induces cell proliferation in a variety of cell types of mesodermal and neuro-ectodermal origin and appears to be a trophic factor for cultured neuronal cells. In vivo, bFGF may play a role in embryonic development (Kimelman and Kirschner, 1987) and induces neovascularization in different angiogenesis assays (Folkman and Klagsbrun, 1987). Also, anti-bFGF antibodies have been shown to inhibit angiogenesis in normal wound repair, indicating that bFGF is intrinsically involved in normal neovascularization (Broadley et al., 1989).

Angiogenesis is a dynamic process in which changes in the composition of the extracellular matrix (ECM) and in its interaction with vascular cells are highly integrated with modifications in the behavior and differentiation of the endothelial cell. The angiogenic process is characterized by an invasive behavior of the endothelial cell, with cell proliferation, degradation of ECM, and cell movement (Folkman, 1985). bFGF has been shown to induce an "angiogenic phenotype" in cultured endothelial cells, represented by stimulation of cell proliferation, motility, and protease production (collagenase and plasminogen activator) (Moscatelli et al., 1986; Presta et al., 1986).

Angiogenesis in Health and Disease, Edited by M.E. Maragoudakis *et al.*
Plenum Press, New York, 1992

Here, we demonstrate the possibility to dissociate at the pharmacologic and structural level two of the activities exerted by bFGF in cultured endothelial cells, namely the capacity to induce cell proliferation and the capacity to induce an increase of plasminogen activator (PA) production. In theory, both activities are involved in mediating the angiogenic activity exerted by bFGF in vivo. Endothelial cell proliferation is required for the growth of the new blood vessels. On the other hand, the endothelial PA-plasmin system is involved in the invasive process occurring during angiogenesis, characterized by the degradation of the perivascular ECM and of the stroma of the tissue to be vascularized (Mignatti et al., 1989). The possibility to dissociate these bFGF activities may represent an important tool to dissect different steps of the angiogenic process and hence to study them separately in vivo.

The experiments reviewed in the present paper have been performed on cultured transformed fetal bovine aortic endothelial GM 7373 cells (Presta et al., 1989a), even though most of the findings have been confirmed on other normal fetal and/or adult bovine aortic endothelial cells.

MITOGENIC AND PA-INDUCING ACTIVITY OF bFGF

Human recombinant bFGF stimulates the proliferation of endothelial GM 7373 cells in a dose-dependent manner (Fig. 1A). The cell population doubling time is reduced from 48 h to 24 h when GM 7373 cell cultures, maintained in 0.4% fetal calf serum, are added with 10-30 ng/ml bFGF. Following bFGF treatment, GM 7373 cells divide almost syncronously and enter the S phase of the cell cycle after 12 h, cell replication beginning 4 h thereafter (Fig. 1B,C).

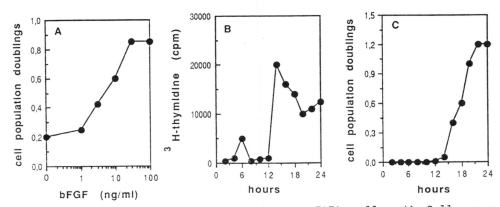

Fig. 1. Mitogenic activity of bFGF in GM 7373 cells. A) Cells were treated with bFGF in 0.4% serum. After 24 h, cells were trypsinized and counted. B and C) Cells were treated with 30 ng/ml bFGF. At different times, cultures were labelled with ^{3}H-thymidine for 2 h (B) or trypsinized and counted (C). The values are the difference between bFGF-treated and untreated cultures.

Under the same experimental conditions, bFGF induces a dose-dependent increase of GM 7373 cell-associated PA activity, as evaluated by a chromogenic plasmin assay (Fig. 2A). SDS-PAGE zymography of cell extracts identify the PA activity induced by bFGF as Mr 50,000 urokinase-type PA (uPA) (Fig. 2B). Accordingly, Northern blot analysis demonstrate an increase of uPA mRNA in bFGF-treated GM 7373 cells in respect to untreated cells (Fig. 2C).

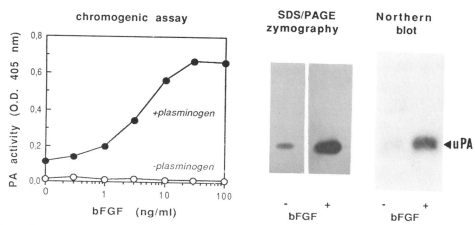

Fig. 2. PA-inducing activity of bFGF in GM 7373 cells. Cultures were treated with bFGF in 0.4% serum. After 24 h, cell lysates were assayed for PA activity with a plasmin chromogenic substrate in the absence (o) or in the presence (●) of plasminogen. Also, bFGF-treated and untreated cultures were assayed for cell-associated PA activity by casein SDS/PAGE zymography or for the levels of uPA mRNA by Northern blot analysis.

DISSOCIATION OF THE MITOGENIC AND OF THE PA-INDUCING ACTIVITY OF bFGF

Intracellular Signal Transduction: Pharmacologic Dissociation

bFGF is known to interact with different members of a family of FGF receptors endowed with tyrosine kinase activity whose prototypes are the products of the genes cek1, flg and bek (Lee et al., 1989; Ruta et al., 1989). Thus, interaction with the ligand induces the autophosphorylation of the receptor and the phosphorylation of several protein substrates, including phospholipase C-gamma (PLC-II) (Burgess et al., 1990a). This leads to an increase of the activity of PLC-II (Nishibe et al., 1990) with production of diacylglycerol and consequent activation of protein kinase C (PKC) (Nishizuka, 1986).

Different inhibitors have been used to assess whether this pathway is involved in mediating the biological activity of bFGF in GM 7373 cells: suramin, which is known to prevent the binding of bFGF to its receptors (Coffey et al., 1987); tyrosine kinase inhibitors genistein and herbimycin A (Akiyama et al., 1987; Yamashita et al., 1989); cAMP-dependent protein kinase and PKC inhibitors H-8 and H-7 (Hidaka et al., 1984). Thus, these inhibitors affect different steps along the proposed intracellular signalling pathway (see Fig. 4).

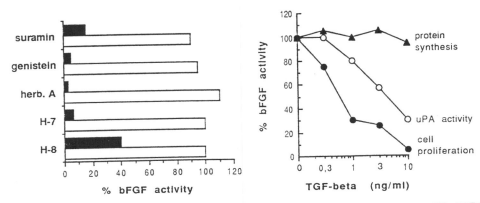

Fig. 3. Effect of different inhibitors on bFGF activity in GM 7373 cells. Cultures were treated with 10 ng/ml bFGF in the absence or in the presence of 100 ug/ml suramin, 10 ug/ml genistein, 100 ng/ml herbimycin A (herb. A), 10 uM H-7, 10 uM H-8, or increasing concentrations of TGF-beta. After 24 h, the mitogenic activity (closed symbols) and the PA-inducing capacity (open symbols) were evaluated.

As predicted by the model, all the inhibitors fully abolished the mitogenic activity of bFGF in GM 7373 cells. However, under the same experimental conditions, they did not affect the PA-inducing activity of bFGF (Fig. 3). These data indicate that different transduction mechanisms modulate the mitogenic and the PA-inducing activity of bFGF. This is in keeping with previous observations that demonstrated a role of PKC in mediating cell proliferation, but not PA production, in GM 7373 cells (Presta et al., 1989b). Interestingly, transforming growth factor beta, which is known to inhibit several biological activities exerted by bFGF in cultured endothelial cells (Saksela et al., 1987; Pepper et al., 1990), inhibits the mitogenic activity and the PA-inducing activity of bFGF with different dose-responses (Fig. 3).

To get insights into the intracellular signalling pathway involved in mediating the PA-inducing activity of bFGF in GM 7373 cells, several molecules have been used to agonize or to antagonize the PA-inducing activity of bFGF. Even though the tyrosine kinase inhibitors genistein and herbimycin A did not affect PA production induced by bFGF, the phosphatase inhibitor orthovanadate (Montesano et al., 1988) caused an increase of PA activity in GM 7373 cell cultures similar to that observed with bFGF, suggesting that a kinase is indeed involved in mediating the PA-inducing activity of bFGF. As stated above this kinase did not appear to be PKC or the cAMP-dependent kinase since the kinase inhibitors H-7 and H-8 do not affect the PA-inducing activity of bFGF. Interestingly, also the PA-inducing activity exerted by the phorbol ester TPA in GM 7373 cells does not seem to depend upon the activation of PKC (Maier et al., 1989).

Several molecules which are supposed to modulate intracellular levels of cAMP in endothelial cells [e.g. insulin, dibutyryl-cAMP, forskolin, isoproterenol, propranolol] did not affect PA production in GM 7373 cells. Accordingly, arachidonic acid metabolism does not appear to be involved in mediating the PA-inducing activity of bFGF. Indomethacin, an inhibitor of cyclooxygenase, and hydrocortisone, an inhibitor of phospholipase A, did not abolish the increase of PA production in bFGF-treated cells; arachidonic acid itself and several prostaglandins (PGE-1, PGE-2, PGA-1, PGD-2, PGF-2alfa, PGI-2) did not affect the basal levels of PA activity in bFGF-untreated cells.

Calcium represents an important intracellular signal (Rasmussen, 1986). We have shown that bFGF induces calcium influx from the extracellular environment (Presta et al., 1989b). Blockade of calcium influx by addition of EGTA to the culture medium completely abolishes the PA-inducing activity exerted both by bFGF and by the phorbol ester TPA (Presta et al., 1989b; Maier et al., 1989). Moreover, the calcium ionophore A23187 increases PA levels in GM 7373 cells (Presta et al., 1989b). Thus, intracellular levels of calcium are involved in the modulation of PA production in endothelial cells. On the other hand, calmodulin inhibitors W-13, calmidazolium, and trifluopirazine did not affect the PA inducing activity of bFGF unless they are added to GM 7373 cells at high cytotoxic doses.

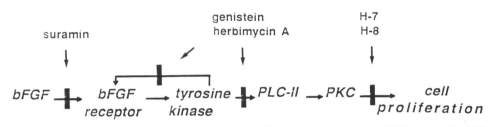

Fig. 4. Mitogenic intracellular signalling pathway of bFGF in GM 7373 cells.

In conclusion, the mitogenic activity and the PA-inducing activity of bFGF are mediated by distinct intracellular signaling pathways. This allows the pharmacologic dissociation of the two bFGF responses in cultured endothelial cells. The mitogenic activity of bFGF is triggered by the activation of the tyrosine kinase activity of the bFGF receptor. This leads to a cascade of events that ultimately will induce the endothelial cell to proliferate (Fig. 4). On the contrary, the intracellular pathway(s) involved in mediating the PA-inducing activity of bFGF is not well understood, even though some experimental data may suggest the role of a calcium-dependent kinase. The study of the trans-activating factor(s) involved in the modulation of the activity of the uPA gene promoter may help to elucidate this intracellular signalling pathway.

bFGF Mutants and Structural Dissociation

The possibility to dissociate at the structural level some of the biological activities exerted by heparin-binding growth factors has been already reported. In particular, deletion of the N-terminus of aFGF-alpha form dissociates the receptor-binding activity of the growth factor from its mitogenic activity (Imamura et al., 1990). Also, site-directed mutagenesis of Lys-132 of aFGF can dissociate the heparin-binding and the mitogenic activity of the molecule from its receptor-binding capability (Burgess et al., 1990b).

Recently, we have described the biological properties of a bFGF deletion mutant (M1-bFGF) lacking the basic amino acid cluster bFGF(27-32) Lys-Asp-Pro-Lys-Arg-Leu (Isacchi et al., 1991). This mutant exerts a mitogenic and receptor-binding activity indistinguishable from wild-type bFGF. However, M1-bFGF is at least 100 times less potent than bFGF in stimulating PA production in endothelial cells. This reduced activity does not seem to depend upon conformational changes of M1-bFGF since soluble heparin, which is known to stabilize the tertiary structure of bFGF, does not restore the PA-inducing capacity of the molecule (Isacchi et al., 1991). Thus, the deletion of the basic amino acid cluster bFGF(27-32) strongly reduces the PA-inducing activity of bFGF, without affecting its capacity to bind to and activate its plasma membrane receptor and its capacity to elicit a mitogenic response. Interestingly, the deleted amino acid sequence is similar to the basic sequence aFGF(23-28) proposed to be responsible for the nuclear targeting capacity and biological activity of aFGF (Imamura et al., 1990).

Also, we have characterized the biological properties of a bFGF mutant (M6B-bFGF) in which basic residues Lys-118, Arg-119, Arg-128, and Lys-129 have been substituted with neutral glutamine residues by site-directed mutagenesis. These residues belong to a basic region of the molecule which has been hypothesized to be involved in the heparin-binding and receptor-binding capacity of bFGF (Baird et al., 1988). In the absence of soluble heparin, this bFGF mutant was endowed with a receptor-binding capacity and a mitogenic activity for GM 7373 cells indistinguishable from wild-type bFGF. On the other hand, M6B-bFGF was 30-100 times less potent than bFGF in stimulating PA production in these cells (Fig. 5). However, at variance with M1-bFGF, addition of soluble heparin fully restored the PA-inducing activity of M6-bFGF (Fig. 5C). In conclusion, the experiments performed with both M1-bFGF and M6B-bFGF mutants indicate that the receptor-binding capacity and the mitogenic activity of bFGF can be dissociated from its PA-inducing activity at the structural level, possibly by different mechanisms.

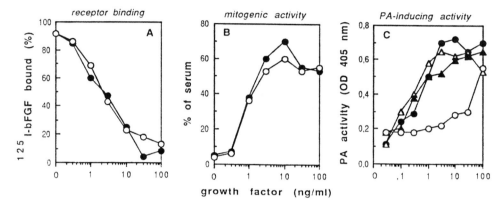

Fig. 5. Biological activity of M6B-bFGF in GM 7373 cells. Cells were incubated at 4°C with radiolabeled bFGF in the presence of unlabeled bFGF (●) or M6B-bFGF (o). After 2 h, the radioactivity associated with bFGF receptors was evaluated (A). Also, bFGF (●, ▲) and M6B-bFGF (o, △) were evaluated for mitogenic (B) and PA-inducing (C) activities in the absence (●, o) or in the presence (▲, △) of 10 ug/ml heparin.

The study of the biological properties of different aFGF and bFGF mutants has demonstrated the possibility to dissociate some of the biological activities exerted by these heparin-binding growth factors. How this dissociation occurs is still unclear. Differences in the stability and conformation of the mutagenized growth factor, in its intracellular sorting, and in its heparin-binding capacity might be all involved. In theory, dissociation of early responses (e.g., receptor-binding, tyrosine phosphorylation, protooncogene activation) from late responses (e.g., cell proliferation, PA production) may simply depend upon a reduced stability of the mutant in the cell culture medium. On the other hand, dissociation of two late responses (that is, cell proliferation from PA production) could depend upon more complex mechanism(s). One possibility is that the mutagenized growth factor is able to activate only part of the intracellular signalling pathways activated by the wild type molecule by interacting with a subset of FGF receptors. Indeed, a complex multigene family of FGF receptors exists and very little is known about the significance of this redundancy. Thus, it is possible to hypothesize that different FGF receptors may mediate different biological responses by activating different intracellular signalling pathways. A second explanation may be based on the fact that the nuclear targeting of bFGF and aFGF has been hypothesized to be involved in mediating at least some of the biological properties of these growth factors (Bouche et al., 1987; Imamura et al., 1990). A reduced stability of the tertiary structure of the mutant, an higher susceptibility to proteolytic cleavage after the internalization that follows receptor-binding, the deletion of putative nuclear targeting sequences might all affect the intracellular fate of the molecule and, consequently, its biological activity.

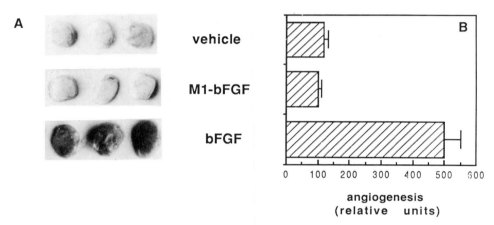

Fig. 6. Angiogenic activity of bFGF and M1-bFGF. Gelfoam sponges were absorbed with 20 ug/ml bFGF, 20 ug/ml M1-bFGF, or vehicle in the presence of 100 ug/ml heparin. Sponges were then implanted subcutaneously in female mice. After 14 days, animals were sacrificed, sponges were photographed (A) and their redness was evaluated by computerized immage analysis as an index of the degree of blood vessel infiltration (B).

CONCLUSIONS

Here we have shown the possibility to dissociate the mitogenic activity of bFGF from its PA-inducing activity both at the pharmacologic and at the structural level. These studies may help to elucidate the role exerted by endothelial cell proliferation and by the endothelial PA-plasmin system during the angiogenic process. As an example of this possibility, preliminary experiments have been performed to assess the angiogenic activity exerted by wild-type bFGF and by the bFGF mutant M1-bFGF. As stated above, this mutant shows a mitogenic activity similar to that exerted by the wild-type molecule but a reduced PA-inducing capacity (Isacchi et al., 1991). Gelatin sponges (Gelfoam, Upjohn) were adsorbed with bFGF or with M1-bFGF in the presence of a

3-fold molar excess of soluble heparin and were implanted subcutaneously in female mice. Animals were sacrificed after 14 days and the gross appearance of the sponges at the time of sacrifice is shown in Fig. 6. bFGF pellets were well vascularized while M1-bFGF pellets did not show any sign of blood vessel infiltration. We can not exclude the possibility that differences exist between the in vivo stability of the bFGF mutant and that of wild-type bFGF, nor we can exclude the possibility that M1-bFGF has different altered biological properties other than a reduced PA-inducing capacity. Nevertheless, it is tempting to hypothesize that the lack of angiogenic activity shown by M1-bFGF is related to its reduced PA-inducing capacity. If this is the case, the activation of the endothelial PA-plasmin system would represent a fundamental step of the angiogenic process and it could represent a possible target for therapeutical approaches aimed to inhibit neovascularization in different pathologic conditions.

ACKNOWLEDGEMENTS

This work was supported in part by grants from Associazione Italiana per la Ricerca sul Cancro, from C.N.R. (Progetto Finalizzato Biotecnologie, Sottoprogetto Biofarmaci), and from MURST (40%, Invecchiamento) to Marco Presta.

REFERENCES

Akiyama, T., Ishida, J., Nakagawa, S., Ogawara, H., Watanabe, S., Itoh, N., Shibuya, M., and Fukami, Y., 1987, Genistein, a specific inhibitor of tyrosine-specific protein kinases, **J. Biol. Chem.**, 262:5592.

Baird, A., Schubert, D., Ling, N., and Guillemin, R., 1988, Receptor- and heparin-binding domains of basic fibroblast growth factor, **Proc. Natl. Acad. Sci. U.S.A.**, 85:2324.

Bouche, G., Gas, N., Prats, H., Baldin, V., Tauber, J. P., Teissie, J., and Amalric, F., 1987, Basic fibroblast growth factor enters the nucleolus and stimulates the transcription of ribosomal genes in ABAE cells undergoing G_0-G_1 transition, **Proc. Natl. Acad. Sci. U.S.A.**, 84:6770.

Broadley, K. N., Aquino, A. M., Woodward, S. C., Buckley-Sturrock, A., Sato, Y., Rifkin, D. B., and Davidson, J. M., 1989, Monospecific antibodies implicate basic fibroblast growth factor in normal wound repair, **Lab. Invest.**, 61:571.

Burgess, W. H., and Maciag, T., 1989, The heparin-binding (fibroblast) growth factor family of proteins, **Annu. Rev. Biochem.**, 58:575.

Burgess, W. H., Dionne, C. A., Kaplow, J., Mudd, R., Friesel, R., Zilberstein, A., Schlessinger, J., and Jaye, M., 1990a, Characterization and cDNA cloning of phospholipase C-gamma, a major substrate for heparin-binding growth factor 1 (acidic fibroblast growth factor)-activated tyrosine kinase, **Mol. Cell. Biol.**, 10:4770.

Burgess, W. H., Shaheen, A. M., Ravera, M., Jaye, M., Donohue, P. J., and Winkles, J. A., 1990b, Possible dissociation of the heparin-binding and mitogenic activities of heparin-binding (acidic fibroblast) growth factor-1 from its receptor-binding activities by site-directed mutagenesis of a single lysine residue, **J. Cell Biol.**, 111:2129.

Coffey, R. J., Leof, E. B., Shipley, G. D., and Moses, H. L., 1987, Suramin inhibition of growth factor receptor binding and mitogenicity in AKR-2B cells, **J. Cell. Physiol.**, 132:143.

Folkman, J., 1985, Tumor angiogenesis, **Adv. Cancer Res.**, 43:175.

Folkman, J., and Klagsbrun, M., 1987, Angiogenic factors, **Science**, 235:442.

Hidaka, H., Inagaki, M., Kawamoto, S., and Sasaki, Y., 1984, Isoquinolinesulfonamides, novel and potent inhibitors of cyclic nucleotide dependent protein kinase and kinase C, **Biochemistry**, 23:5036.

Imamura, T., Englera, K., Zhan, X., Tokita, Y., Forough, R., Roeder, D., Jackson, A., Maier, J. A. M., Hla, T., and Maciag, T., 1990, Recovery of mitogenic activity of a growth factor mutant with a nuclear translocation sequence, **Science**, 249:1567.

Isacchi, A., Statuto, M., Chiesa, R., Bergonzoni, L., Rusnati, M., Sarmientos, P., Ragnotti, G., and Presta, M., 1991, A 6-amino acid deletion in basic fibroblast growth factor dissociates its mitogenic activity from its plasminogen activator-inducing capacity, **Proc. Natl. Acad. Sci. U.S.A.**, 88:2628.

Kimelman, D., and Kirschner, M., 1987, Synergistic induction of mesoderm by FGF and TGF-beta and the identification of an mRNA coding for FGF in the early Xenopus embryo, **Cell**, 51:869.

Lee, P. L., Johnson, D. E., Cousens, L. S., Fried, V. A., and Williams, L. T., 1989, Purification and complementary DNA cloning of a receptor for basic fibroblast growth factor, **Science**, 245:57.

Maier, J. A. M., Presta, M., and Ragnotti, G., 1989, Induction of plasminogen activator activity by phorbol ester in transformed fetal bovine aortic endothelial cells. Possible independence from protein kinase C, **Biochem. Biophys. Res. Commun.**, 160:682.

Mignatti, P., Tsuboi, R., Robbins, E., and Rifkin, D. B., 1989, In vitro angiogenesis on the human amniotic membrane: requirement for basic fibroblast growth factor-induced proteinases, **J. Cell Biol.**, 108:671.

Montesano, R., Pepper, M. S., Belin, D., Vassalli, J. D., and Orci, L., Induction of angiogenesis in vitro by vanadate, an inhibitor of phosphotyrosine phosphatases, 1988, **J. Cell. Physiol.**, 134:460.

Moscatelli, D., Presta, M., and Rifkin, D.B., 1986, Purification of a factor from human placenta that stimulates capillary endothelial cell protease production, DNA synthesis, and migration, **Proc. Natl. Acad. Sci. U.S.A.**, 83:2091.

Nishibe, S., Wahl, M. I., Hernandez-Sotomayor, S. M. T., Tonks, N. K., Rhee, S. G., and Carpenter, G., 1990, Increase of the catalytic activity of phospholipase C-gamma 1 by tyrosine phosphorylation, **Science**, 250:1253.

Nishizuka, Y., 1986, Studies and perspectives of protein kinase C, **Science**, 233:305.

Pepper, M. S., Belin, D., Montesano, R., Orci, L., and Vassalli, J. D., 1990, Transforming growth factor-beta 1 modulates basic fibroblast growth factor-induced proteolytic and angiogenic properties of endothelial cells in vitro, **J. Cell Biol.**, 111:743.

Presta, M., Moscatelli, D., Joseph-Silverstein, J., and Rifkin, D. B., 1986, Purification from a human hepatoma cell line of a basic fibroblast growth factor-like molecule that stimulates capillary endothelial cell plasminogen activator production, DNA synthesis, and migration, **Mol. Cell. Biol.**, 6:4060.

Presta, M., Maier, J. A. M., Rusnati, M., and Ragnotti, G., 1989a, Basic fibroblast growth factor: production, mitogenic response, and post-receptor signal trasduction in cultured normal and transformed fetal bovine aortic endothelial cells, **J. Cell. Physiol.**, 141:517.

Presta, M., Maier, J. A. M., and Ragnotti, G., 1989b, The mitogenic signalling pathway but not the plasminogen activator-inducing pathway of basic fibroblast growth factor is mediated through protein kinase C in fetal bovine aortic endothelial cells, **J. Cell Biol.**, 109:1877.

Rasmussen, H., 1986, The calcium messenger system, **N. Engl. J. Med.**, 314:1094.

Ruta, M., Burgess, W., Givol, D., Epstein, J., Neiger, N., Kaplow, J., Crumley, G., Dionne, C., Jaye, M., and Schlessinger, J., 1989, Receptor for acidic fibroblast growth factor is related to the tyrosine kinase encoded by the fms-like gene (FLG), **Proc. Natl. Acad. Sci. U.S.A.**, 86:8722.

Saksela, O., Moscatelli, D., and Rifkin, D. B., 1987, The opposing effects of basic fibroblast growth factor and transforming growth factor beta on the regulation of plasminogen activator activity in capillary endothelial cells, **J. Cell Biol.**, 105:957.

Yamashita, T., Sakai, M., Kawai, Y., Aono, M., and Takahashi, K., 1989, A new activity of herbimycin A: inhibition of angiogenesis, **J. Antibiotics**, 42:1015.

ROLE OF INTEGRINS IN ENDOTHELIAL CELL FUNCTION

Elisabetta Dejana, Adriana Zanetti, Carmen
Dominguez-Jimenez and Grazia Conforti

Istituto di Ricerche Farmacologiche Mario Negri
Via Eritrea 62,
20157 Milano, Italy

INTRODUCTION

Endothelial cell interaction with extracellular matrix proteins plays an important role in the maintenance of vascular wall integrity and in the formation of new vessels (Form et al.,1986; Ingberg and Folkman,1989).

Endothelial cell interaction with matrix proteins is also of importance in neovascularization. The components of the extracellular matrix can promote cell proliferation and motility and provide an anatomical guide for the formation of new vessels (Ingberg and Folkman,1989).

In addition, the development of endothelial cell differentiated characteristics during embryogenesis and the maintenance of polarity are regulated, as for the other cell types, by the interaction with matrix components.

The majority of the endothelial receptors for matrix proteins, characterized so far, belong to the family of "integrins". This chapter will deal with the type, structure and function of endothelial cell integrin receptors.

CLASSIFICATION OF INTEGRINS

The integrin family consists of a series of heterodimers involved in a variety of cell adhesion functions (for review see: Hynes,1987; Ruoslahti and Pierschbacher,1987; van Mourik et al.1990; Ruoslahti and Giancotti,1989; Hemler, 1990; Albelda and Buck,1990; Dejana and Lauri,1990; Marchisio,1991). All integrins are formed by two non covalently linked subunits : the larger termed α chain (m.w. ranges 150,000-200,000 in non reduced conditions) and the smaller ß chain (m.w. 90,000-110,000 non reduced). Both subunits are integral membrane proteins that present a small C-terminal cytoplasmic domain and a large N-terminal extracellular domain and a transmembrane segment.

At least six different ß chains have been characterized so far (Table 1). They present a relatively high amino acid sequence homology (40-48%) and well preserved structural features such as the presence of 56 cysteines most of which organized in four repeating units.

Eleven different α chains have been described so far (table 1). In general, the α subunits exhibit a lower sequence homology than the ß

Angiogenesis in Health and Disease, Edited by M.E. Maragoudakis *et al.*
Plenum Press, New York, 1992

Table 1 - The integrin receptor family

Subunit	I-domain/ cleaved α	Ligand/ function	RGD role	Other names
$\alpha^1\beta_1$	I	lm,coll	no	VLA-1
$\alpha^2\beta_1$	I	coll,lm,fn	no	Ia-IIa,VLA-2
$\alpha^3\beta_1$	I	fn,lm,coll,inv	yes	VLA-3
$\alpha^4\beta_1$	NC	VCAM-1,fn,inv	no	VLA-4
$\alpha^5\beta_1$	C	fn,inv	yes	Ic-IIa,VLA-5
$\alpha^6\beta_1$	C	lm,inv	no	Ic-IIa,VLA-6
$\alpha^v\beta_1$	C	fn/vn	yes	–
$\alpha^L\beta_2$	I	ICAM-1, ICAM-2	no	LFA-1, CD11a/18
$\alpha^M\beta_2$	I	iC3b,fg, ICAM-1	yes	MAC-1, CD11b/18, CR-3
$\alpha^x\beta_2$	I	iC3b	yes	p150,95, CD11c/18
$\alpha^v\beta_3$	C	vn,fg,vWf,tsp fn,lm,osp,thr	yes	vn-receptor
$\alpha^{II\ b}\beta_3$	C	fg,fn,vW,vn	yes	IIb-IIIa
$\alpha^6\beta_4$	C	basal membr,lm	no	–
$\alpha^v\beta_5$	C	vn	yes	–
$\alpha^4\beta_p$	NC	addressin (?)	no	–
$\alpha^v(?)\beta_6$	?	?	?	–

Footnote to table 1
vn, vitronectin; fn, fibronectin; vWf, vonWillebrand factor;
fg, fibrinogen; tsp, thrombospondin; thr, thrombin; osp,
osteospondin; lm, laminin; inv, invasin (Isberg and
Leong,1990); coll, collagen; I, I-domain; C, cleaved α chain;
NC, non cleaved α chain; basal membr, basal membrane. For more
information see the text and Hynes,1987; Ruoslahti and
Pierschbacher,1987; Ruoslahti and Giancotti,1989; Hemler,
1990; Albelda,1990; Marchisio,1991.

subunits but they have some common structural similarities
(Hemler,1990). The extracellular domain contains four ion binding
regions. Some α chains undergo postranslational cleavage in two subunits
linked by a disulphide bond while others present an additional 180-200
aminoacid sequence inserted between the last Ca^{2+} binding site and the
amino terminus called I-domain (Hemler,1990).

A subclassification of the integrin family has been attempted based
on the observation that some members have the same ß chain but different
α chains (Hynes,1987).

However, it was found that some α subunits have the capability to
link to more than one ß subunit. A list of the integrin molecules is
given in table 1. However, any attempt of subgrouping these receptors
has to be considered provisional since additional novel molecules or α/ß
subunits combinations might be added in the future. Some recent reviews
(Ruoslahti and Giancotti,1989; Hemler, 1990; Albelda and Buck,1990;
Dejana and Lauri,1990; Marchisio,1991) provide additionals information
on the molecular properties and function of these molecules on
different cell types.

Table 1 also reports a scheme of the principal ligands for the
listed integrin molecules. Some integrins, but not all of them,
recognize in the ligand a sequence of only three aminoacids
(arginine-glycine and aspartic acid, RGD; Ruoslahti and
Pierschbacher,1987). Many proteins containing this sequence, but not all
of them, are recognized by an integrin receptor. Despite the
similarities in the cell binding sequence in ligand proteins, the cell
can recognize them individually through specific and separate receptors.
This indicates that other amino acids surrounding the RGD sequence and
the RGD steric conformation confer specificity to the interaction
(Ruoslahti and Pierschbacher,1987). It is therefore possible to build
synthetic RGD containing peptides that express recognition specificity
for different integrin receptors on different cell types (Tranqui et
al.,1989).

Integrins expressed by endothelial cells

Endothelial cells possess five members of the $ß_1$ subfamily.
The $α^1ß_1$ was found to be present in endothelial cells for most
type of vessels (Albelda and Buck,1990) and probably acts as laminin
receptor.
The $α^2ß_1$ integrin is apparently identical to the platelet Ia-IIa
complex (table 1; Elices and Hemler,1989) which is the receptor for
collagen in platelets and in other cells of hematopoietic and
non-hematopoietic origin. In endothelial cells it behaves differently
acting as a receptor for laminin and binding less efficiently to
fibronectin and collagen (Languino et al.,1989a).
The $α^3ß_1$ receptor is in general a multifunctional integrin: it
binds to fibronectin, collagen and laminin (Wayner and Carter,1987).
The $α^5ß_1$ complex is abundant in cultured cells but is poorly
expressed in vivo (table1). $α^5ß_1$ in endothelial cells (as well as in
other cell types, Pytela et al.1985) preferentially binds fibronectin
(Conforti et al.1989).
The $α^6ß_1$ integrin is poorly expressed in cultured endothelial
cells. This molecule is the laminin receptor in platelets (Sonnenberg et
al.,1988) and probably plays an identical role in endothelial cells.
Endothelial cells exhibit only one member of the $ß_3$ integrin
subfamily this molecule has the same ß chain as the platelet complex
IIb-IIIa but a distinct α chain (Table 1). These two members of the $ß_3$
subgroup share the characteristic of having a low substratum
specificity. The $α^vß_3$ integrin in endothelial cells besides vitronectin
also recognizes fibrinogen (Dejana et al.,1987; Cheresh,1987; Languino
et al.,1989b; Dejana et al.,1990), vonWillebrand factor (Cheresh,1987;
Dejana et al.,1989), fibronectin (Smith et al.,1990; Conforti et

al.1990), thrombospondin (Lawler et al.,1988), laminin (Kramer et al.,1990) and thrombin (Bar-Shavit et al.,1991).

Different integrins can bind to the same substratum. For example endothelial cells have at least four fibronectin receptors.The reason for this redundancy is still obscure. Possibly multiple interactions with the same molecule are required for the stability of the adhesion process. In addition, different integrins may transmit distinct information from the extracellular environment. It is also possible that the cell does not use all the receptors for a given substratum at the same time. The relative biological role of each of them remains to be defined.

The diversity of the integrin system can be further augmented by a cell specific type of regulation.

Ionic concentration and the phospholipid composition of the membrane can dramatically modify integrin receptor affinity and specificity for different substrata (Gailit and Ruoslahti,1988; Conforti et al.,1990). Membrane phospholipid microenvironment might change not only in different cell types or in culture conditions but also during cell movement. This might be a very efficient and fast way the cell use for regulating receptor affinity for matrix ligands.

In general, in circulating cells, adhesion to substrata or to plasma proteins seems to be more effectively regulated by modulation of integrin activity than by changes in their number. This has been studied for platelet IIb-IIIa (Ginsberg et al.1988) and leukocyte integrins (Springer,1990; Vedder and Harlan,1988). It is, however, still unknown whether the same type of "fast" modulation of these receptors acts in other non circulating cells such as the endothelium.

Interaction with cytoskeletal proteins

Integrins may convey regulatory information to the cell through interaction with cytoskeletal proteins. This phenomenon occurs through their cytoplasmic domains and requires both subunits (Buck et al.1986).

During the first hours of endothelial cell attachment to substrata the basal surface of the cells forms several types of contacts (called focal contacts or adhesion plaques, Dejana et al.,1987; Dejana et al.,1988) which are the area of closest interaction between the substratum, the cell membrane and the membrane insertion sites of actin microfilamnet bundles (Burridge,1986). During cell adhesion integrins have been found to be clustered in focal contacts (reviwed in Burridge, 1986; Burridge et al.,1988).

It is still debated whether are the cytoskeletal proteins that regulate integrin clustering or whether are integrins that after adhesion to their specific substratum trigger cytoskeletal organization. Experimental evidence supports the last possibility (Dejana et al.,1988). This idea is also indirectly supported by the observation that integrin organization is dependent, at least during first hours of spreading, on the binding to their specific ligand (Dejana et al.,1988; Singer et al.1988).

The biological role of cytoskeleton organization is still debated. Apparently this is not required for endothelial cell adhesion. Agents that increase intracellular cAMP inhibit integrin clustering, focal contact and actin microfilament formation (Lampugnani et al.,1990). This however is not accompanied by a decrease in the ability of the cells to adhere to substrata. Similarly, deletion of the cytoplasmic domain of ß1 integrin subunit (Solowska et al.,1989; Hayashi et al.,1990) which blocks cell spreading and cytoskeletal organization does not essentially modify cell adhesion.

In contrast , cell migration is markedly inhibited in endothelial cells by agents increasing cAMP (Lampugnani et al.,1990). This suggests that the capacity of the cells of assembling cytoskeletal proteins is required for their motility.

Cytoskeletal organization might also be of importance for cell proliferation. Endothelial cell spreading is needed for induction of growth (Ingberg,1990; Ingberg et al.,1990). A possible explanation for this phenomenon is that the actin filament network could transfer mechanical signals to the nucleus that then promotes cell division .

Integrins and endothelial cell monolayer integrity

Besides their role in promoting cell attachment to matrix protein, integrins have been found to be located at cell-cell contacts in endothelial cells (Lampugnani et al,1991). The two integrin receptors $\alpha^2\beta_1$ and $\alpha^5\beta_1$, in the confluent endothelium, line the boundaries between the cells. Interestingly, α^v but not β_3 follows the same pattern of distribution, suggesting that another ß chain is involved. All the other integrin considered $\alpha^6\beta_1, \alpha^3\beta_1$ and $\alpha^v\beta_5$ remain diffuse on the cell membrane.

In epithelial cells, two β_1 integrins are located at cell-cell contacts ($\alpha^2\beta_1$ and $\alpha^3\beta_1$) (Carter et al.1990; DeLuca et al.1990; Larjava et al.1990; Marchisio et al,1991) while no such distribution was found in smooth muscle cells or skin fibroblasts (Lampugnani et al.,1991). This suggests that integrin distribution at cell-cell contacts is a specific feature of polarized cells.

In endothelial cells other molecules have been described to be localized at intercellular contacts: PECAM (or endoCAM or CD 31), a recently sequenced integral protein belonging to the immunoglobulin family (Newman et al,1990; Albelda et al.1990; Simmons et al.1990), endoglin (Gougos and Letarte,1990) and the Ca^{2+} dependent cell adhesion molecule, endocadherin (Heimark et al.,1990).

The interrelationship between these molecules and integrins at cell-cell contacts is unclear. Integrins can directly bind these proteins in a heterotypic type of interaction or just integrate and support their role in maintaining endothelial cell boundaries. This might be achieved by integrin binding of matrix or plasma proteins at intercellular level. Immunocytochemical localization of matrix proteins in endothelial cells shows that laminin, fibronectin and collagen type IV are indeed concentrated between endothelial cells in correspondence with integrin localization (Lampugnani et al,1991).

The observation that addition of antibodies to integrin receptors causes discontinuities in the endothelial cell monolayer at times of incubation too short to cause cell detachment suggests that these molecules play a role in maintaining the integrity of endothelial cell junctions. This is further documented by the fact the integrin antibodies alter endothelial cell permeability properties and induce a significant change in their capacity to restrain macromolecules at the luminal compartment (Lampugnani et al,1991).

ACKNOWLEDGEMENTS

This work was supported by the National Research Council (Progetto Finalizzato Biotecnologie e Biostrumentazione e Progetto Finalizzato Tecnologie Biomediche e Sanitarie) and by the Associazione Italiana per la Ricerca sul Cancro.

REFERENCES

Albelda, S. M., and Buck, C. A.,1990, Integrins and other cell adhesion molecules, FASEB J., 4: 2868.
Albelda, S. M., Oliver, P. D., Romer, L. H. and Buck, C.A.,1990, EndoCAM: a novel endothelial cell-cell adhesion molecule. J. Cell Biol., 110: 1227.

Bar-Shavit, R., Sabbah, V., Lampugnani, M.G., Marchisio, P.C., Fenton
II, J.W., Vlodavsky, I. and Dejana, E.,1991, An arg-gly-asp
sequence within thrombin promotes endothelial cell adhesion,
J.Cell Biol., 112:335.

Buck, C.A., Shea, E., Duggan, K. and Horwitz, A.F., 1986, Integrin (the
CSAT antigen): functionality requires oligomeric integrity,
J.Cell Biol., 103:2421.

Burridge, K.,1986, Substrate adhesions in normal and transformed
fibroblasts: organization and regulation of cytoskeletal,
membrane and extracellular matrix components at focal
contacts, Cancer Rev., 4:18.

Burridge, K., Fath, K., Kelly, T., Nuckolis, G., and Turner, C., 1988,
Focal adhesions : transmembrane junctions between the
extracellular matrix and the cytoskeleton, Ann.Rev.Cell Biol.,
4:487.

Carter, W. G., Wayner, E. A., Bouchard, T. S., and Kaur, P.,1990, The
role of integrins $\alpha_2 b_1$ and $\alpha_3 b_1$ in cell-cell and
cell-substrate adhesion of human epidermal cells, J.Cell
Biol.,110:1387.

Cheresh, D., 1987, Human endothelial cells synthesize and express an
arg-gly-asp-directed adhesion receptor involved in attachment
to fibrinogen and von Willebrand factor, Proc.Natl.Acad.Sci.
USA, 84:6471.

Conforti, G., Zanetti, A., Colella, S., Abbadini, M., Marchisio, P.C.,
Pytela, R., Giancotti, F., Tarone, G., Languino, L.R., and
Dejana, E., 1989, Interaction of fibronectin with cultured
human endothelial cells: characterization of the specific
receptor., Blood , 73:1576.

Conforti, G., Zanetti, A., Pasquali-Ronchetti, I., Quaglino Jr., D.,
Neyroz, P., and Dejana, E., 1990, Modulation of vitronectin
receptor binding by membrane lipid composition, J.Biol.Chem.,
265: 4011.

Dejana, E., Colella, S., Languino, L.R., Balconi, G., Corbascio,
G.C., and Marchisio, P.C., 1987, Fibrinogen induces adhesion,
spreading and microfilament organization of human endothelial
cells in vitro, J.Cell Biol., 104:1403.

Dejana, E., Colella, S., Conforti, G., Abbadini, M., Gaboli M., and
Marchisio, P.C., 1988, Fibronectin and vitronectin regulate
the organization of their respective arg-gly-asp adhesion
receptors in cultured human endothelial cells, J.Cell Biol.,
107:1215.

Dejana, E., Lampugnani, M.G., Giorgi, M., Gaboli, M., Federici, A.B.,
Ruggeri, Z.M., and Marchisio, P.C., 1989, von Willebrand
factor promotes endothelial cell adhesion via an arg-gly-asp
dependent mechanism, J.Cell Biol., 109:367.

Dejana, E., Lampugnani, M.G., Giorgi, M., Gaboli, M., and Marchisio,
P.C., 1990, Fibrinogen induces endothelial cell adhesion and
spreading via the release of endogenous matrix proteins and
the recruitment of
more than one integrin receptor, Blood, 75: 1509.

Dejana, E., and Lauri, D., 1990, Biochemical and functional
characteristics of integrins: a family of adhesive receptors
present in hematopoietic cells, Haematologica ,75:1.

De Luca, M., Tamura, R.N., Kajiji, S., Bondanza, S., Rossino, P.,
Cancedda, R., Marchisio, P.C., and Quaranta, V., 1990,
Polarized integrin mediates human keratinocyte adhesion to
basal lamina, Proc.Natl.Acad.Sci. USA, 87:6888.

Elices, M. J., and Hemler, M.E., 1989, The human integrin VLA-2 is a
collagen receptor on some cells and a collagen/laminin
receptor on others, Proc.Natl.Acad.Sci. USA 86:9906.

Form, D.M., Pratt, B.M., and Madri, J. A.,1986, Endothelial cell proliferation during angiogenesis. In vitro modulation by basement membrane components, Lab. Invest. 55:521.

Gailit, J., and Ruoslahti, E., 1988, Regulation of the fibronectin receptor affinity by divalent cations, J. Biol. Chem. 263:12927.

Ginsberg, M. H., Loftus, J. C., and Plow, E. F., 1988, Cytoadhesins, integrins, and platelet, Thromb. Haemostasis, 59:1.

Gougos, A., and Letarte, M., 1990, Primary structure of endoglin, an RGD-containing glycoproteins of human endothelial cells, J.Biol.Chem., 265:8361.

Hayashi, Y., Haimovich, B., Reszka, A., Boettiger D., and Horwitz, A.,1990, Expression and function of chicken integrin β_1 subunit and its cytoplasmic domain mutants in mouse NIH 3T3 cells, J.Cell Biol., 110:175.

Heimark, R. L., Degner, M., and Schwartz, S.M., 1990, Identification of a Ca^{2+}-dependent cell-cell adhesion molecule in endothelial cells J.Cell Biol., 110:1745.

Hemler, M. E., 1990, VLA proteins in the integrin family: structures, functions and their role on leukocytes, Annu. Rev. Immunol., 8: 365.

Hynes, R. O., 1987, Integrins: a family of cell surface receptors, Cell 48: 549.

Ingber, D. E., and Folkman, J., 1989, How does extracellular matrix control capillary morphogenesis, Cell, 58:803.

Ingber, D. E., 1990, Fibronectin controls capillary endothelial cell growth by modulating cell shape, Proc.Natl.Acad.Sci. USA, 87:3579.

Ingber, D. E., Prusty, D., Frangioni, J. V., Cragoe, Jr E. J., Lechene, C., and Schwartz, M.A., 1990, Control of intracellular pH and growth by fibronectin in capillary endothelial cells, J.Cell Biol., 110: 1803.

Isberg, R. R., and Leong, J.M., 1990, Multiple β_1 chain integrins are receptors for invasin, a protein that promotes bacterial penetration into mammalian cells, Cell, 60:861.

Kramer, R. H., Cheng, Y.-F., and Clyman, R., 1990, Human microvascular endothelial cells use β_1 and β_3 integrin receptor complexes to attach to laminin, J.Cell Biol., 111:1233.

Lampugnani, M. G., Giorgi, M., Gaboli, M., Dejana, E., and Marchisio, P.C., 1990, Endothelial cell motility, integrin receptor clustering, and microfilament organization are inhibited by agents that increase intracellular cAMP, Lab. Invest., 63:521.

Lampugnani, M. G., Resnati, M., Dejana, E., and Marchisio, P.C., 1991, The role of integrins in the maintenance of endothelial monolayer integrity, J.Cell Biol., 112:479.

Languino, L. R., Gehlsen, K.R., Wayner, E., Carter, W.G., Engvall, E., and Ruoslahti, E., 1989a, Endothelial cells use $\alpha_2\beta_1$ integrin as a laminin receptor, J.Cell Biol, 109:2455.

Languino, L. R., Colella, S., Zanetti, A., Andrieux, A., Ryckewaert, J. J., Charon, M.H., Marchisio, P.C., Plow, E.F., Ginsberg, M.H., Marguerie, G., and Dejana, E., 1989b, Fibrinogen-endothelial cell interaction in vitro: a pathway mediated by an arg-gly-asp recognition specificity, Blood, 73: 734.

Larjava, H., Peltonen, J., Akiyama, S. K., Yamada, S. S., Gralnick, H.R., Uitto, J., and Yamada K.M., 1990, Novel function for β_1 integrins in keratinocyte cell-cell interactions, J.Cell Biol., 110:803.

Lawler, J., Weinstein, R., and Hynes, R. O., 1988, Cell attachment to thrombospondin: the role of arg-gly-asp, calcium, and integrin receptors, J.Cell Biol., 107: 2351.

Marchisio, P. C., 1991, Integrins and tissue organization, Progr. in Neuroimmunol. in press.

Newman, P. J., Berndt, M. C., Gorski, J., White II, G. C., Lyman, S., Paddock, C., and Muller, W. A., 1990, PECAM-1 (CD31) cloning and relation to adhesion molecules of the immunoglobulin gene superfamily Science, 247:1219.

Pytela, R., Pierschbacher M.D., and Ruoslahti, E.,1985, Identification and isolation of a 140 kd cell surface glycoprotein with properties expected of a fibronectin receptor, Cell, 40:191.

Ruoslahti, E., and Pierschbacher, M.D., 1987, New perspectives in cell adhesion: RGD and integrins, Science, 238:491.

Ruoslahti, E., and Giancotti, F.G., 1989, Integrins and tumor cell dissemination, Cancer Cells, 1:119.

Simmons, D., Walker, C., Power C., and Piggott, R., 1990, Molecular cloning of CD31, a putative intercellular adhesion molecule closely related to carcinoembryonic antigen, J. Exp. Med., 171: 2147.

Singer, I. I., Scott, S., Kawka, D.W., Kazazis, D.M., Gailit, J., and Ruoslahti, E., 1988, Cell surface distribution of fibronectin and vitronectin receptors depends on substrate composition and extracellular matrix accumulation, J.Cell Biol.,106:2171.

Smith, J. W., Vestal, D. J., Irwin, S.V., Burke, T.A., and Cheresh, D.A.,1990, Purification and functional characterization of integrin $\alpha_v\beta_5$. An adhesion receptor for vitronectin, J.Biol.Chem., 265:11008.

Solowska, J., Guan, J.-L., Marcantonio, E. E., Trevithick, J. E., Buck, C.A., and Hynes, R. O.,1989, Expression of normal and mutant avian integrin subunits in rodent cells, J.Cell Biol.,109:853.

Sonnenberg, A., Modderman, P.W., and Hogervost, F.,1988, Laminin receptor on platelets is the integrin VLA-6, Nature, 336:487.

Springer, T. A., 1990, Adhesion receptors of the immune system, Nature, 346:425.

Tranqui, L., Andrieux, A., Hudry-Clergeon, G., Ryckewaert, J. J., Soyez, S., Chapel, A., Ginsberg, M. H., Plow, E. F., and Marguerie, G.,1989, Differential structural requirements for fibrinogen binding to platelets and to endothelial cells, J.Cell Biol., 108:2519.

van Mourik, J. A., von dem Borne A. E. G. Kr., and Giltay, J.G., 1990, Pathophysiological significance of integrin expression by vascular endothelial cells, Biochem. Pharm., 39:233.

Vedder, N. B., and Harlan, J. M.,1988, Increased surface expression of CD11b/CD18 (MAC-1) is not required for stimulated neutrophil adherence to cultured endothelium, J. Clin. Invest., 81:676.

Wayner, E. A., and Carter, W.G.,1987, Identification of multiple cell adhesion receptors for collagen and fibronectin in human fibrosarcoma cells possessing unique α and common β subunit, J.Cell Biol., 105:1873.

SPECIFIC LAMININ DOMAINS MEDIATE ENDOTHELIAL CELL ADHESION,

ALIGNMENT AND ANGIOGENESIS

D.S. Grant*, J. Kinsella @, M.C. Cid#, H. K. Kleinman*

*Lab. of Developmental Biology, National Institute of Dental Research, NIH Bldg. 30, NIH, Bethesda, MD, 20892- USA

@Lab of Cardiovascular Science National Institute on Aging, NIH, Baltimore, MD. 21225 USA

#Lab. of Immunoregulation National Institute of Allergy and Infectious Disease, NIH Bldg.10-11B/13,Bethesda, MD, 20892 USA. Supported by a FISS (Spanish) grant # BAE 91/5310

Abstract

The stability and integrity of blood vessels is maintained by many factors in the blood and by an important extracellular layer, the basement membrane, which underlies the endothelium of the vessels. Basement membranes are composed of an organized network of collagen (type IV), heparan sulfate proteoglycan, and glycoproteins such as entactin, fibronectin, and laminin. Laminin is one of the most important and abundant substances in basement membranes. It has a direct role in cell attachment, migration, and induction of the differentiated phenotype of many cells. We have examined and defined the role(s) of laminin and its specific cell-binding sites at the biochemical level using an in vitro angiogenic model. This model involves the differentiation of cultured endothelial cells on a laminin rich reconstituted basement membrane matrix, Matrigel, into capillary-like structures. Synthetic peptides derived from sequences in the laminin A and B1 chains (CTFAL**RGD**NP and CDP**GYIGSR**) were able to block cell attachment to Matrigel and cell-cell alignment (early events in vessel formation), respectively, and thus inhibit subsequent tube formation. The third biologically active site CSRARKQA-ASIK**V**A**V**SADR) induced the endothelial cells to become migratory and invade into the Matrigel, forming sprouts from the primary capillary-like network. This site also induces angiogenesis in the chick chorioallantoic membrane (CAM) assay. Thus, endothelial cells interact with at least three different sites in laminin and these interactions are important in vessel maintenance and repair.

Angiogenesis In Vivo

Endothelial cells form a continuous non-thrombogenic layer in both large and small vessels throughout the body. This layer is usually stable and changes only in response to injury or to extravascular stimulus (the presence of cytokines such as TNF, FGF or TGFß; Schweigerer, et al., 1987; Tsuboi, et al., 1990). In pathological conditions, this may lead to a denudation of endothelial cells and an overgrowth of smooth muscle cells in large vessels or excessive branching and capillary formation by small vessels. Upon injury or stimulation, the microvascular endothelial cells enter a migratory and proliferative phase resulting in a reorganization of the endothelium or the formation of new blood vessels. This involves a cascade of events leading to endothelial cell invasion into the underlying stroma and the formation of new vessels leading away from the parent vessel towards the stimulus (Fig 1).

This de novo angiogenic process involves the release of enzymes for degradation of the basement membrane, invasion of the surrounding matrix, and proliferation of the endothelial cells. The cells migrate into a new area, then realign, and reorganize to form a new vessel. Finally the endothelial cells secrete a new basement membrane, polarize themselves and exhibit a continuous patent vessel connected to the parent vessel. These processes are largely regulated by both soluble factors (e.g. FGF, TGF,TNF, etc..) and the insoluble factors (basement membrane matrix) associated with the endothelium (Folkman, 1985a; Maciag, 1990; Raju, et al., 1984; Sato, et al., 1988). Little is known about the specific component(s) which trigger this angiogenic process.

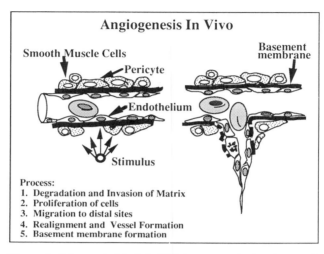

Figure 1. Schematic Model of Major events in Angiogenesis

Role of basement membrane

Basement membranes are sheets of specialized extracellular material which separate parenchymal cell types from connective tissue stroma (Martin et al., 1988). They are found on the basal surfaces of epithelial lining: in the digestive, respiratory, urinary, and reproductive tract surrounding blood vessel endothelium, skeletal and smooth muscle cells, and the nervous system. The ubiquitous nature of basement membranes indicates that they play an important role in the maintenance of normal healthy tissues.

Although most basement membranes are thought to contain over 50 different substances as well as provide an anchor for cytokines and blood proteins, we will cover only the contribution of the major and ubiquitous components which include collagen IV, laminin, and heparan sulfate proteoglycan. These components are all biologically active and bind to each other. Although heparan sulfate proteoglycan comprises less than 1% of the composition of basement membrane, by weight, it provides an important role in maintaining the ionic charge of the matrix, as observed in the glomerular basement membrane (Farquhar, et al., 1982; Kanwar, et al., 1979; Kanwar, et al., 1980) and also in binding growth factors (Rosengart, et al., 1988). One of the most abundant components of basement membranes is type IV collagen. This collagen is thought to provide the structural three-dimensional framework of basement membranes (Risteli, et al., 1984; Timpl, et al., 1981; Yurchenco, et al., 1986). Laminin (the second most abundant substance) is a large noncollagenous cross-shaped glycoprotein with a native molecular weight of approximately 1 000 KD. It was first purified from the EHS sarcoma (Timpl et al., 1979). The largest subunit, which is referred to as the A chain, has a Mr of about 400 Kd, whereas the two smaller subunits, B1 and B2 chains, have a Mr of approximately 200-220 KD (Fig 2.). The two B chains are linked to the A chain by disulfides bonds.

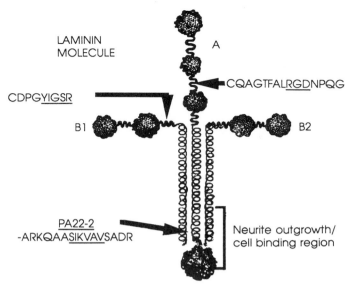

LAMININ
MOLECULE

A

CQAGTFAL<u>RGD</u>NPQG

CDPG<u>YIGSR</u>

B1

B2

PA22-2
-ARKQAA<u>SIKVAV</u>SADR

Neurite outgrowth/
cell binding region

Figure. 2   Schematic Model of Laminin showing the location and sequence of biologically
active domains.

This is the classical form of laminin but other form have been identified and described as reviewed in Kleinman, et al., 1989. Laminin's carbohydrate content accounts for 10-15% of the total weight of the molecule i.e., the B1, B2 and A chain subunits contain about 11, 13, and 14 oligosaccharide side chains respectively. The polysaccharides may account for some of the binding activity of laminin to many cells.
Laminin mediates the attachment of epithelial/endothelial cells to several different substrates. This involves integral plasma membrane proteins one of which has a molecular weight of 32/67 KD (Terranova, 1983). Other cell surface receptors have been implicated in mediating cell adhesion to laminin, such as integrins. Laminin has many biological functions and has been shown to promote growth, differentiation and maintenance of morphology of various cell types (Table 1) (reviewed by Kleinman et al., 1985). To date, specific cell binding regions have been identified on laminin and an RGD containing sequence located in the A chain designated -CTFALRGDNPQ-, an active pentapeptide on the B1 chain designated -YIGSR- , and a sequence near the carboxyl region of the A chain designated -SIKVAV- (Fig. 2).   These sites play important roles in endothelial cell attachment and differentiation.

Table 1   The Bilogical Activities of Laminin

Cell Adhesion	Cell-Cell aggregation, Cell attachment to collagen IV
Cell Morphology	Induces cell polarity in epithelial cells in culture
	Promotes neurite outgrowth
	Stabilizes endothelium
Cell Growth	Increase cell proliferation in epithelial, Schwann neurites and PC12 cells
Cell Migration	Chematactic and haptotactic
	Promotes tumor migration
Differentiation	Teratocarcinoma cell differentiation
	Endothelial cell assembly and organization
	Melanoma Cell pigmentation
	Secretion of ß casin in mammary gland cells
Metastasis	Promotes metastasis in vivo
	Promotes metastic phenotype in vitro

Relationship of Basement membrane to Endothelial Cell-Surface Receptors

Laminin and synthetic peptides derived form the laminin sequence influence endothelial cell behavior. When endothelial cells are cultured on plastic, they form a flattened cobblestone monolayer. When tissue culture plates are coated with laminin or with the synthetic laminin RGD peptide, there is an increase in the attachment rate of human umbilical vein endothelial cells (HUVEC) which flatten and form the normal cobblestone morphology (Fig. 3). If the plates are coated with the laminin B1 chain peptide YIGSR, the cells attach at a normal rate but form doughnut-like structures (Fig. 3). We believe these signet rings may participate in lumen formation during endothelial cell differentiation into capillaries. HUVEC attached to SIKVAV-coated surfaces rapidly but the cells did not form the flat cobble-stone morphology. Instead, these cells appeared more fibroblastic exhibiting elongated cell extensions (Fig. 3). It has been previously shown that endothelial cells change their appearance, becoming elongated prior to mobilizing with TNF or scatter factor (Rosen, et al., 1990) or after physical denuding of a confluent layer (Dethlefsen, et al., 1986; Groot, et al., 1987; Schröder, et al., 1987). By sending out numerous processes (spikes) and having an increased migration rate (as seen by a time-lapse video recorder, Linda Thomas, NIDR,NIH, Bethesda MD), the cells can form primitive cord-like networks. All these features mimic the initial step in angiogenesis where cells elongate and migrate into the extracellular stroma.

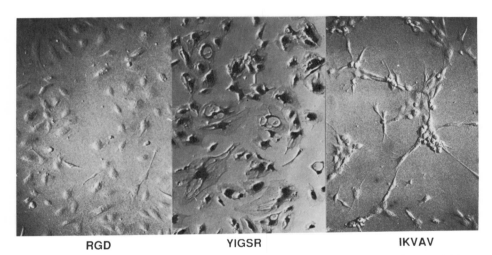

RGD   YIGSR   IKVAV

Figure 3.   The appearance of HUVEC incubated for 18 hrs on the RGD, YIGSR or SIKVAV peptides.

Endothelial cells are able to attach and spread on numerous substrates in vitro including fibronectin, gelatin, collagens, fibrinogen, laminin, vitronectin, and basement membrane coated substrata. Little is known about the receptors involved in these differentiation processes. It has been demonstrated that endothelial cells require receptors to mediated attachment to matrix and that the cells have a preference, with respect to, the substratum to which they attach. A previous investigation has demonstrated that endothelial cell attachment can be blocked by the use of antibodies to cell surface receptors or to the various matrix proteins (Basson, et al., 1990). Adhesion to fibronectin, but not to laminin can also be blocked in the presence of an Arg-Gly-Asp (RGD) peptide (Cheng, et al., 1989). We have blocked endothelial cell attachment to laminin using the laminin CTFALRGDNPQ and YIGSR peptides (Grant et al., 1989). The fibronectin peptide, GRGDS, does not block endothelial cell attachment to laminin. Therefore, endothelial cells may express cell surface molecules specific to sites in laminin and in other basement membrane proteins which are distinct from those binding fibronectin and stromal collagens.

A variety of cell surface receptors for matrix proteins have been identified on endothelial cells. In immunofluorescence studies, bovine aortic endothelial cells were shown to express at least two different integrin heterodimers of the β1 (125 Kd) and β3 (95 Kd) class of integrins (Madri, et al., 1991)(Fig. 4). When the cells are plated on laminin or fibronectin, the staining for β1 integrin aligned into stress fiber-type patterns, which co-localize to intracellular vinculin adhesion plaques (Cheng, et al., 1989). In addition, endothelial cells have also been shown to express matrix receptors to collagen (α2 (150Kd) β1), and laminin (α3 β1 (Albelda, et al., 1982) and α6 (150Kd) β4) (Fig.4). Thus, the endothelium interacts with the matrix through a variety of receptors, many of which are integrins or integrin related proteins. There are also another of set of non-integrin adhesion molecules which have been reported, which mediate matrix-endothelium interactions; many of which have not been characterized (Belloni, et al., 1988) One well characterized adhesion molecule binds to laminin and has a Mr of 67-69 Kd (Fig. 4) (Weaver et al., 1987; Runyan, et al., 1988). This laminin binding protein was identified in aortic and microvascular endothelial cells, and modulates cell attachment spreading and migration. SDS PAGE analysis of proteins from a laminin-Sepharose affinity column also indicated other minor bands at 110, 92, 85, 75, 55, and 30 Kd (Yannariello, et al., 1988). These findings suggest that the 67 Kd laminin binding protein may exist as a complex (Fig. 4) on the endothelial cell surface. We have also found that SDS-PAGE analysis of HUVEC proteins from laminin-Sepharose affinity column co-elutes proteins at 32 and 67 Kd. In addition, antibodies raised to the 32 Kd form also recognizes the 67 Kd protein suggesting that some of the receptors are related.

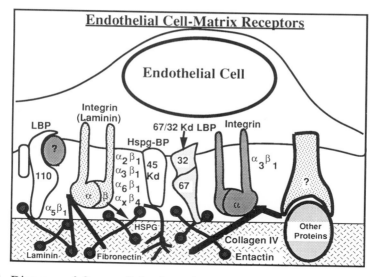

Figure 4. Diagram of Some of the Integrin and Non-Integrin Receptors used by Endothelial Cells to Attach to the Extracellular Matrix Molecules.

Models of angiogenesis

The mechanisms involved in endothelial cell attachment or vessel formation are difficult to examine in vitro due to the complex relationship of endothelial cells to both soluble and insoluble factors in situ. Endothelial cells have been isolated and maintained in culture(Jaffe, et al., 1973; Maciag, et al., 1982). If the cells are permitted to become superconfluent without change of the culture medium, structures form above the monolayer which resemble capillary-like vessels (Folkman, 1985; Madri and Williams,

1983; Maciag, 1990). The induction of capillary formation can also be initiated by the addition of phorbol esters, by incubating the cells inside a collagen I gel (Montesano, et al., 1983). These are good models of angiogenesis; but, they usually require three or more days for vessels to form, and only half the cells participate in the differentiation process (Table 2). Others have shown that if cultured endothelial cells are incubated in a fibrin clot (Diglio, et al., 1989; Nicosia, et al., 1984) with collagen IV, vessel formation will occur within a week (Madri, et al., 1983). With these systems, capillary-like structures are observed but the response is generally slow, requiring several days to weeks, and in some cases, the vessels are inside out, secreting basement membrane material into the lumen.

Table 2 In vitro models of angiogenesis

Substratum	Time	Reference
Plastic	3-6 weeks	Folkman and Haudenschild, 1980
Gelatin	3-6 weeks	Folkman and Haudenschild, 1980
Fibronectin	4-8 weeks	Maciag et al., 1982
Collagen I gel	2-8 weeks	Montesano et al., 1983
		Mori et al., 1988
Fibrin clot	4-7 days	Montesano et al., 1987
Collagen IV	4 days	Madri and Williams, 1983
Laminin/Matrigel	16-24 hours	Kubota et al., 1988
		Grant et al., 1989

Previous work in our laboratory has shown that endothelial cells attach, migrate, and assemble on a laminin-rich reconstituted matrix, Matrigel, to form tube-like structures resembling capillaries within 18 hours (Grant, et al., 1989a; Grant, et al., 1989b). Matrigel is a mixture of basement membrane components extracted from the Englebreth-Holm-Swarm (EHS) tumor and found to induce and/or maintain the differentiation of a wide variety of cells (Kleinman, et al., 1987). The components are extracted with 2.0M urea and then dialyzed into a physiological buffer. At 4°C, the components remain in solution but gel when warmed to 24-37°C. The gelled extract, Matrigel, has the appearance of authentic basement membrane. The tube formation assay using human umbilical vein endothelial cells (HUVEC) on Matrigel is carried out as previously described (Grant, et al., 1989b; Kubota, et al., 1988; Lawley, et al., 1989), by using Matrigel (15-20 mg/ml prepared from the EHS tumor), (Kleinman, et al., 1985; Kleinman, et al., 1987) to coat 16 mm Costar wells (Fig 5a). Macroscopically, differences can be seen from well to well when compounds are added which block or enhance tube formation (Fig. 5b). At a higher magnification under Hoffman modulation contrasting (Fig. 5c) the network is seen as an anastomosing series of cord-like structures. When these structures are cut in cross-section and viewed in the electron microscope, they are cylindrical form with a lumen in many regions along the cylinder; therefore we have called this network capillary-like tubes (Grant, et al., 1989a; Grant, et al., 1989b). In addition, the cells within the tubes maintain the ability to bind acetylated-LDL and continue to produce Factor VIII (unpublished observations). The degree of differentiation observed to date with most cells is beyond that which has been achieved on other matrix or artificial substrates (Table 2).

This in vitro angiogenesis model system has been employed to define the cellular interactions with the basement membrane and to identify the intracellular events occurring during capillary vessel formation. The cells use multiple receptors to interact with basement membrane components through multiple receptors. Protein synthesis, and intact cytoskeleton, and the activities of protein kinase C and collagenase IV are all required for tube formation (Grant, et al., 1991). Using this model system, both inhibitors and stimulators of angiogenesis can also be screened prior to in vivo testing.

The role of laminin peptides in tube formation on matrigel

Since laminin makes up a large proportion of Matrigel, we decided to examine the role laminin may play in tube formation. We found that this morphological change requires numerous intracellular events, many of which are receptor mediated events involving precise domains on laminin. Although HUVECs attach to peptide coated tissue culture plastic, soluble synthetic peptides prepared from the laminin A (CTFAL**RGD**NP) and B1(CDPG**YIGSR**C) chains blocked cell attachment to Matrigel and cell-cell alignment, respectively, and thus inhibit subsequent tube formation (Fig. 5d). This indicated that the sites directed towards cell adhesion and organization, may play an important role in the stabilization of blood vessel networks. A third biologically active site (CSRARKQAA-**SIKVAV**SADR) in the laminin A chain has a direct effect on endothelial cells initiating destablization and promoting angiogenesis. The presence of this peptide in the Matrigel assay caused increased endothelial migration, and invasion into the Matrigel (Grant et al.m manuscript in preparation). This resulted in short irregular tubes, and sprouting was more apparent (Fig 5e). In many cases the cells were seen inside the Matrigel, and quite often pieces of the matrix was observed floating in the medium above the cells. Based on these observations, the presence of increased enzyme activity in Zymograms preformed on the conditioned medium, and the degradation of the Matrigel by the HUVECs in the presence of this peptide (Grant et al., unpublished), we postulate that presentation of the SIKVAV site in vivo to stable cells may induce the cells to penetrate the basement membrane and to invade the underlying substratum.

A second role of the binding site may be to stimulate cells in large vessels to migrate and cover areas previously denuded by local trauma. It is not known what stimulates endothelial cells to branch from a parental vessel. Hypervascularization has been seen to occur in many diseases such as diabetic retinopathy but the underlying mechanisms are not known. The data, however, provide a potential cell-matrix inductive mechanism(s) by which the -SIKVAV- site on laminin induces increased angiogenic behavior in endothelial cells, a phenomenon similar to that observed in the vascularization of growing tumors in vivo. Furthermore, since increased angiogenesis was observed in the chorioallantoic membrane (CAM) assay, the endothelial cellular response in vitro appears to mimic steps of neovascularization in vivo.

The Role of integrins and laminin binding protein 32/67 in tube formation

The attachment, migration and organization of HUVEC to form the tube network on Matrigel involves cell-cell and cell-matrix contact adhesion. This is observed by the apparent degree of tension exerted on the matrix by the cells during and after tube formation (Fig. 6 upper left), as evidenced by the accumulation of matrix around the tubes. This behavior also indicates a possible transmition of the intracellular cytoskeletal stress to the extracellular matrix. It has been previously demonstrated that this mechanism is quite often mediated by transmembrane proteins (eg. vinculin, talin, integrin complex) and these cell membrane proteins may play an important role in our system during tube fomation.

We have studied the effect of using antibodies to cell surface receptors or to sites on the matrix molecules to see if tube formation can be blocked. The addition of an antiserum to the β1 integrin (at a 1:100 dilution) caused clumping of the endothelial cells and a partial blocking of the formation of the tube network (Fig. 5f). In previous experiments, the addition of antiserum to the laminin binding protein (32/67 Kd) also blocked tube formation (Grant, et al., 1989b). Control non-immunized rabbit serum did not show any effect. The use of antibodies to the laminin binding sites did not have a strong inhibitory effect, where as antiserum to laminin could block tube formation (Grant, et al., 1989a; Grant, et al., 1989b).

Since we observed that antibodies to either the laminin binding protein or to the β1 integrin can block tube formation, we immunostained the formed tube network with a β1 integrin antibody (obtained from Dr. Steve Akiyama). Immunostaining was observed along the length of the tubes (Fig 6, lower left). The only region where there was no staining was in the flat areas of the network in which tube formation was not observed, indicating that a correlation exists between the receptors and tube formation. In addition if we stained the tubes with an antibody to the α5 integrin, staining was also seen only around the tubes. At times this staining was patchy and incomplete, which indicated that the staining may be due to receptors which have been shed (Fig. 6, arrow). Again the flattened areas where no tube formation was observed were also unstained (Fig. 6, double arrow).

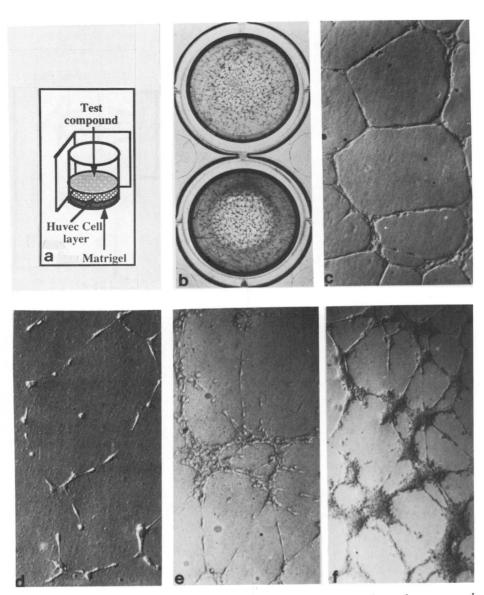

Figure 5. In Vitro (Matrigel) Angiogeneis Model. A 24-well tissue culture plate was used and each well was coated with 250 µl of Matrigel (a). Following polymerization of the Matrigel at 37°C, human umbilical vein endothelial cells (40,000 cells/0.5 ml) were added to each well, then substances to be tested in the assay (e.g. antibodies, peptides) were added to the medium . After 18hrs incubation, a network forms which can be stained using a Giemsa-like blue stain, DiffQuick (b). Panel c shows normal tube formation using Hoffman modulation contrasting Mag 175X. The endothelial cells formed a polygonal network on the surface of the Matrigel. The same contrasting method is used for panel d,e and f, but 300 µg of -RGD- (d) and -SIKVAV- (e), and anti ß1 integrin antiserum (f) was added to the wells.

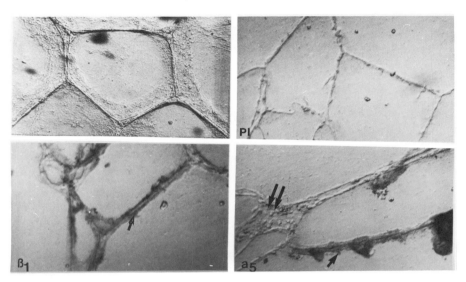

Figure 6. Tube formation on Matrigel, with Antibody staining to ß1 or α 5 Integrin. The Matrigel appears to be drawn around the tube area (upper left panel). The other panels are immunostained with control (PI), the ß1 or α5 Whereas the staining for ß1 integrin labels the length of the tube, the alpha 5 integrin staining is patchy and does not run along the tube.

Immunostaining with antibody to the laminin binding protein 32/67 also demonstrated labeling over the tube areas (data not shown).

Summary

　　　　Using human umbilical vein endothelial cells on Matrigel, we have defined the cellular interactions which the basement membrane components and the intracellular processes involved in the formation of these capillary-like structures. A number of matrix and receptor antibodies have been employed to define the cellular interactions with Matrigel. Antibodies to both laminin and collagen IV have been previously shown to block tube formation but not cell attachment (Grant, et al., 1989a). Rather, the cells remain as a flattened monolayer similar to that observed on plastic. Since the cells are not detached, it is likely that they recognize multiple compents in the Matrigel. When tested individually, endothelial cells bind to both collagen IV and laminin (Grant, et al., 1989b; Madri, et al., 1983). Antibodies against known matrix receptors 32/67 (laminin receptor) and ß1 integrin (laminin, collagen and fibronectin) also blocked tube formation. The 32/67 Kd laminin receptor antibodies blocked cell-cell alignment and a monolayer was observed, whereas antibodies to ß1 integrin blocked cell-Matrigel adhesion. These data demonstrate that endothelial cells utilize multiple interactions with Matrigel components.

　　　　We have concentrated on the role of specific laminin binding sites on endothelial cell attachment and differentiation by preparing synthetic peptides to these regions. Three sites in laminin has been identified, one on the B1 chain and two on the A chain. On the A chain, the RGD peptide promoted the strongest HUVEC cell adhesion and spreading but had little effect on cell morphology on plastic (Table 1). The YIGSR peptide on the B1 chain promoted moderate adhesion and induced the formation of lumen-like structures in the cells. This peptide may play and important role in the lumen formation in vessels. The third peptide, SIKVAV, on the A chain of laminin, promoted a fibroblastoid morphology in HUVECs on plastic. On Matrigel the peptide inhibited the normal tube formation and stimulated sprouting of the endothelial cells into the matrix. The presence of this peptide in the Matrigel assay caused increased migration, cell surface expression of integrins, invasion and collagenase IV activity. These events mimic the initial events leading to de novo angiogenesis. The expression of this site in laminin may be a signal to initiate new blood vessel formation.

Table 3   Summary of the activities of the laminin binding sites

	YIGSR	RGD	SIKVAV
Promoting adhesion	++	++++	++
Inducing morphological change	+++	+	+++
Stimulating migration	+	+	++++
Growth induction	0	0	0
Promoting angiogenesis	+++	0	++
Collagenase activity	++	0	++++
Integrin expression	++	++	++++

Conclusions

The ability of endothelial cels to form capillary-like structures in vitro can serve as a rapid method for identifying new angiogeneic or anti-angiogenic components. This method may be useful, in that, it reduces the lengthy and expensive animal testing such as the rabbit eye model assay for angiogenesis. Recently, two cytokines, inteferon alpha and gamma, have been found to promote and inhibit tube formation respectively in the in vitro assay described above(Maheshwari, et al., 1991). Both are active in a dose dependent manner and do not affect cell adhesion. The in vivo activities of these cytokine on angiogenesis are similar; since interferon can inhibit vascular endothelial cell proliferation and subsequent tumor growth(Maheshwari, et al., 1991). We have also used synthetic peptides to discern the role of the sites in laminin in tube formation on Matrigel. These results indicate that the laminin molecule has multiple activites, all of which may either stablize or stimulate new vessel formation. It is unclear whether all three chains of the laminin molecule are present in all basement membranes (Tokida et al., 1990). Differential expression of individual chains has been documented. Therefore, this may suggest that basement membranes may exert a different effect on endothelial cell behavior. Finally, the rapid response observed with this angiogenesis model may serve as an important in vitro model to further define endothelial cell differentiation and angiogenesis both at the cell and molecular level, as well as serve as a screen for factors which affect vessel formation.

References

Albelda, S. M., Daise, M., Levine, E. M., and Buck, C. A. ,1982, Identification and characterization of cell-substratum adhesion receptors on cultured human endothelial cells, J. Clin. Invest., 83: 1992-2002.

Basson, C. T., Knowles, W. J., Bell, L., Albelda, S. M., Castronovo, V., Liotta, L. A., and Madri, J. A. ,1990, Spatiotemporal segregation of endothelial cell integrin and nonintegrin extracellular matrix-binding proteins during adhesion events, J Cell Biol., 110: 789-801.

Belloni, P. N., and Nicolson, G. L. ,1988, Differential expression of cell surface glycoproteins on various organ-derived microvascular endothelia and endothelial Cell Cultures, J. Cell. Physiol., 136: 398-410.

Cheng, Y.-F., and Kramer, R. ,1989, Human microvascular endothelial cells express integrin-related complexes that mediate adhesion to the extracellular matrix., J. Cellul. Phys., 139: 275-286.

Diglio, C. A., Grammas, P., Giacomelli, F., and Weiner, J. ,1989, angiogenesis in rat aorta ring explant cultures, Lab. Invest. 60: 523-531.

Farquhar, M. G., Courtoy, P. J., Lemkin, M. C., and Kanwar, J. S. ,1982, Current knowledge of the functional architecture of the glomerular basement membrane, New Trends in BM Research.

Folkman, J. ,1985, Toward an understanding of angiogenesis: search and discovery, Perspect. Biol. Med.29: 10-36.

Grant, D. S., Kleinman, H. K., and Martin, G. R., 1989a, The role of basement membranes in vascular development., in:"Embryonic origins of defective heart development" D.E. Bockman and M.L. Kirby eds.,. New York Acad. Sci., New York.

Grant, D. S., Leldes, P. I., Fukuda, K., and Kleinman, H. K. ,1991, Intracellular mechamisms involved in basement membrane induced blood vessel differentiation(in vitro, In Vitro., June(in Press):

Grant, D. S., Tashiro, K.-I., Segui-Real, B., Yamada, Y., Martin, G. R., and Kleinman, H. K. ,1989b, Two different laminin domains mediate the differentiation of human endothelial cells into capillary-like structures in vitro, Cell., 58: 933-943.

Jaffe, E. A., Nachman, R. L., Becker, C. G., and Minick, C. R. ,1973, Culture of human endothelial cells derived from umbilical veins-Identification by morphological and immunological criteria, J. Clin. Invest., 52: 2745-2756.

Kanwar, Y. S., and Farquhar, M. A. ,1979, Presence of heparan sulfate in the glomerular basement membrane, Proc. Natl. Acad. Sci. USA., 76: 1303-1307.

Kanwar, Y. S., Linker, A., and Farquhar, M. G. ,1980, Increased permeability of the glumerular basement membrane to ferritin after removal of glycosaminoglycans (heparan sulfate) by enzyme digestion, J. Cell Biol., 86: 688.

Kleinman, H. K., Cannon, F. B., Laurie, G. W., Hassell, J. R., Aumailley, M., Terranova, V. P., Martin, G. R., and Dalaq, M. D. B. ,1985, Biological activities of laminin, J. Cell Biol., 27: 317-325.

Kleinman, H. K., Graf, J., Iwamoto, Y., Kitten, G. T., Ogle, R. C., Sasaki, M., Yamada, Y., Martin, G. R., and Luckenbill-Edds, L. ,1987, Role of Basement Membranes in Cell Differentiation, Ann. N. Y. Acad. of Sci., 513: 134-145.

Kleinman, H. K., Ogle, R. C., Cannon, F. B., Little, C. D., Sweeney, T. M., and Luckenbill-Edds, L. ,1988, Laminin receptors for neurite formation, Proc. Natl. Acad.Sci. USA., 85: 1282-6.

Kleinman, H. K., and Weeks, B. S. ,1989, Laminin: structure, functions and receptors, Current Opinion in Cell Biology., 1: 964-967.

Kubota, Y., Kleinman, H. K., Martin, G. R., and Lawley, T. J. ,1988, Role of laminin and basement membrane in the morphological differentiation of human endothelial cells into capillary-like structures, J. Cell Biol., 107: 1589-1598.

Lawley, T. J., and Kubota, Y. ,1989, Induction of Morphologic Differentiation of endothelial cells in culture, J. Invest. Dermatol.,93: 59s-61s.

Maciag, T. ,1990, Molecular and cellular mechanisms of angiogenesis, Important Adv Oncol., 1990: 85-98.

Maciag, T., Kadish, J., Wilkins, L., Stemerman, M. B., and Weinstein, R. ,1982, Organization behavior of human umbilical vein endothelial cells, J. Cell Biol., 94: 511-520.

Madri, J. A., Bell, L., Marx, M., Merwin, J. R., Basson, C., and Prinz, C. ,1991, Effects of soluble factors and extracellular matrix components on vascular cell behavior in vitro and in vivo: Models of de-endothelialization and repair., J. Cellul. Biochem., 45: 123-130.

Madri, J. A., and Williams, S. K. ,1983, Capillary endothelial cell cultures: Phenotypic modulation by matrix components, J. Cell Biol., 97: 153-165.

Maheshwari, R. K., Srikantan, V., Bhartiya, D., Kleinman, H. K., and Grant, D. S.,

1991, Differential effects of inteferon gamma and alpha on in vitro model of angiogeneisis, J. Cellul. Phys., 146: 164-169.

Martin, G. R., Timpl, R., and Kuhn, K. ,1988, Basement membrane proteins : Molecular structure and function, 39: 1-50.

Montesano, R., Orci, L., and Vassalli, P. ,1983, In Vitro Rapid Organization of endothelial cells into capillary-like networks is promoted by collagen matrices, J. Cell Biol., 97: 1648-1652.

Nicosia, R. F., McCormick, J. F., and Bielunas, J. ,1984, The formation of endothelial webs and channels in plasma clot culture, Scan. Electron Micros. 793-799.

Raju, K. S., Alessandri, G., and Gullino, P. M. ,1984, Characterization of a chemoattractant for endothelium induced by angiogenesis, Cancer Res., 44: 1579-1584.

Risteli, J., Foidart, J. M., Ruteli, L., Boniver, J., and Goffinet, G. ,1984, The basement membrane proteins laminin and type IV collagen is isolated villi in pre-eclampsia, Placenta., 5: 541-550.

Rosengart, T. K., Johnson, W. V., Friesel, R., Clark, R., and Maciag, T. ,1988, Heparin protects heparin-binding growth factor-I from proteolytic inactivation in vitro, Biochem Biophys Res Commun., 152: 432-40.

Runyan, R. B., Versalovic, J., and Shur, B. D. ,1988, Functionally distinct laminin receptors mediate cell adhesion and spreading: the requirement for surface galactosyltransferase in cell spreading, J Cell Biol., 107: 1863-71.

Sato, Y., and Rifkin, D. B. ,1988, Autocrine activities of basic fibroblast growth factor: regulation of endothelial cell movement, plasminogen activator synthesis, DNA synthesis, J. Cell Biol., 107: 1199-1205.

Schweigerer, L., Neufeld, G., and Gospodarowicz, D. ,1987, Basic fibroblast growth factor as a growth inhibitor for cultured human tumor cells, J. Clin. Invest., 80: 1516-1520.

Tokida, Y., Aratan, Y., Morita, A., and Kitagawa, Y., 1990, Production of two variant laminin forms by endothelial cells and shift of their relative levels by angiostatic steroids, J. Biol. Chem., 265: 18123-18129.

Timpl, R., Wiedemann, H., Delden, V. V., Furthmayr, H., and Kuhn, K. ,1981, A network model for the orginization of type IV collagen molecules in basement membrane, Euro. J. of Biochem., 120: 203-211.

Tsuboi, R., Sato, Y., and Rifkin, D. B. ,1990, Correlation of cell migration, cell invasion, receptor number, proteinase production, and basic fibroblast growth factor levels in endothelial cells, J. Cell Biol., 110: 511-517.

Yannariello, B. J., Wewer, U., Liotta, L., and Madri, J. A. ,1988, Distribution of a 69-kD laminin-binding protein in aortic and microvascular endothelial cells: modulation during cell attachment, spreading, and migration, J Cell Biol., 106: 1773-86.

Yurchenco, P. D., Tsilibary, E. C., Charonis, A. S., and Furthmayer, H. ,1986, Models of the self-assembly of basement membrane, J. Histochem. and Cytochem., 34: 93-102.

110

PRODUCTION OF PLASMINOGEN ACTIVATORS AND MATRIX METALLOPROTEINASES BY

ENDOTHELIAL CELLS: THEIR ROLE IN FIBRINOLYSIS AND LOCAL PROTEOLYSIS

Victor W.M. van Hinsbergh and Pieter Koolwijk

Gaubius Laboratory
IVVO-TNO
Leiden, The Netherlands

INTRODUCTION

Matrices composed of fibrous proteins play an important role in the growth and repair of tissues. The complex extracellular matrix not only anchors the cell in the body, but it also provides the cell - via interaction with receptors and the cytoskeleton - with information about its position in the tissue. Various proteases, in particular plasmin, plasminogen activators (PAs) and matrix degrading metalloproteinases (MMPs) are involved in the regulation of the turnover and destruction of such matrices [Danø et al., 1985; Saksela, 1985; Mignatti et al., 1986; Liotta et al., 1991].

The best studied matrix is the "temporary" fibrin matrix, which is formed after injury to prevent blood loss. However, when deposited within the blood vessel, fibrin can obstruct the vessel and impair the supply of oxygen and nutrients to the distal tissue. Proper timing of the dissolution of the fibrin matrix is essential to prevent blood loss, as well as to maintain the patency of blood vessels and to prevent the formation of scar tissue. In fibrinolysis, an enzymatic process that results in the removal of fibrin, the action of plasmin and plasminogen activators has clearly been demonstrated. Furthermore, strong evidence has been presented that PAs, together with MMPs, are also involved in the remodelling of the extracellular matrix in a variety of biological processes, such as cell invasion in developmental processes, in pannus formation in rheumatoid arthritis [Hart and Fritzler, 1989], in tumour metastasis [Danø et al., 1985; Goldfarb et al., 1986] and in tumour neovascularisation, which is important for the nourishment and growth of many tumours. In all these processes, the action of the various proteinases is finely tuned by a series of inhibitors, the PA inhibitors (PAIs), α2-antiplasmin and tissue inhibitors of metalloproteinases (TIMPs), as well as by specific cellular receptors [Andreasen, 1990; Dochtery and Murphy, 1989; Blasi, 1988; Miles and Plow, 1988].

This chapter focusses on the production of PAs and MMPs by endothelial cells and their role in fibrinolysis and angiogenesis. Angiogenesis implies that capillary endothelial cells alter the interaction with their environment, find their way through or replace their original basal membrane and bud into sprouts. PAs and MMPs (collagenases and stromelysins) have been observed to play a role in angiogenesis [Gross et al., 1983; Mignatti et al., 1989; Pepper et al., 1990]. In fibrinolysis, the endothelium provides

Angiogenesis in Health and Disease, Edited by M.E. Maragoudakis *et al.*
Plenum Press, New York, 1992

the key regulator proteins for the generation of plasmin, the major fibrin degrading enzyme. These two processes partly reflect the polarity of the endothelium, which faces the blood on its luminal side and the tissue at its basolateral side.

THE COMPONENTS OF THE FIBRINOLYTIC AND COLLAGENOLYTIC SYSTEMS

Plasminogen Activation

Fibrin degradation and the activation of several matrix degrading metalloproteinases can be accomplished by the serine protease plasmin, which is formed from its zymogen plasminogen by plasminogen activators (PAs) (Figure 1). Two types of PAs are presently known: tissue-type PA (t-PA) and urokinase-type PA (u-PA) [Wallén, 1987; Bachman, 1987]. The actual activity of these PAs is regulated not only by their concentration, but also by their interaction with PA inhibitors (PAIs), cellular receptors and matrix proteins. Some of the properties of the proteins involved in plasminogen activation are given in Table 1.

Proteases of the fibrinolytic system. The cDNAs and genes of plasminogen, t-PA and u-PA have been cloned and characterized [Petersen et al., 1990; Pennica et al., 1983; Friezner Degen et al., 1986; Riccio et al., 1985]. All three proteins are synthesized as a single polypeptide chain. The polypeptide is converted by proteolytic cleavage to a molecule with two polypeptide chains connected by a disulphide bond. The carboxy-terminal part of the molecule, the so-called B-chain, contains the proteolytic active site. It is made up by three amino acids: serine, aspartic acid and histine (the "catalytic triad"), which are located separately on the B-chain polypeptide, but interact by a specific folding of the protein. This structure is a general feature of proteins belonging to the serine protease family. The amino-terminal parts of the molecules, the A-chains, are build up of domains that determine the interaction of the proteases with matrix

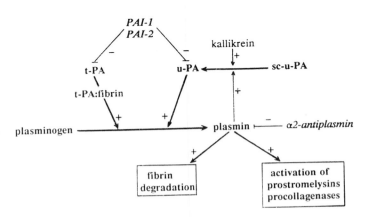

Fig. 1. Schematic representation of the plasminogen activation system. +: activation; -: inhibition. Inhibitors are indicated in italics. PA: plasminogen activator; t-PA: tissue-type PA; u-PA: urokinase-type PA; scu-PA: single-chain u-PA; PAI: PA inhibitor.

Table 1.   Properties of protein involved in plasminogen activation

Protein	MW (kD)	Number of amino acids	Plasma concentration (mg/l)	Produced by
Plasminogen	92	791[a]	200	hepatocytes
t-PA	68	530[a]	0.005	endothelium, mesothelium
scu-PA	54	411	0.008[b]	monocytes/mø renal tubuli, activated endothelial cells
α_2-Antiplasmin	70	452	70	hepatocytes
PAI-1	50	379	0.05	smooth muscle cells, activator endothelium, ?liver
PAI-2	46/56[c]	393	<0.005	Macrophages, placenta, mesothelial cells
u-PA Receptor	55	313/282-284[d]		present on monocytes-mø endothelial cells and other cells

[a] The numbering of amino acid residues is usually based on a total of 790 residues for plasminogen (Lys-plasminogen) and of 527 residues for t-PA.
[b] Sum of scu-PA and u-PA: inhibitor complexes.
[c] 46 kD: intracellular non-glycosylated form, 60 kD excreted glycosyated form.
[d] 313: postulated transmembrane form; 282-284: PI-linked form, which has been processed on its COOH-terminal side.

Abbreviations: mø: macrophages; MW: molecular weight.

proteins and cellular receptors. Plasminogen contains five so-called kringle structures in its A-chain; one or more are involved in the interaction with fibrin, presumably via lysine-binding sites [Müllertz, 1987]. The A-chain of t-PA consists of a finger and growth factor domain and two kringle structures. The finger and kringle 2 domain of t-PA are involved in the interaction of this PA with fibrin [van Zonneveld et al., 1986; Verheijen et al., 1986] and the finger and growth factor domain of t-PA have been indicated to be involved in the binding of t-PA to cells [Beebe et al., 1989]. The growth factor domain of u-PA A-chain binds to a cellular receptor of u-PA [Blasi, 1988].

   Activation. The cleavage of plasminogen and single-chain u-PA is necessary to disclose the proteolytic active site and to activate the molecule. On the contrary, the activity of t-PA depends not on its conversion in a two-chain form, but on its interaction with fibrin [Rijken et al., 1982; Rånby et al., 1982]. Once bound to this substrate both the single-chain and the two-chain forms of t-PA are active. Interestingly, several other matrix proteins, in particular thrombospondin, fibronectin and laminin [Silverstein et al., 1986; Salonen et al., 1984, 1985], can also activate PAs, albeit to a lesser degree. Whether such proteins may play a role in the activation of PAs in the extracellular space remains to be established.

Inhibitors. The various proteases of the fibrinolytic system are counter-regulated by potent inhibitors, which are members of the serine protease inhibitor superfamily (serpins) [Holmes te al., 1987; Ny et al., 1986; Ginsburg et al., 1986; Pannekoek et al., 1986]. Plasmin is almost instantaneously inhibited by α2-antiplasmin, but this reaction is attenuated when plasmin is bound to fibrin. Although several inhibitors can interact with t-PA or u-PA, it is generally accepted that two PA inhibitors, PAI-1 and PAI-2, are the predominant regulators of PA activity. PAI-1 is a 50 kD glycoprotein present in blood platelets and is synthesized by endothelial cells, smooth muscle cells and many cell types in culture [Sprengers and Kluft, 1987; Schleef and Loskutoff, 1988; Andreasen et al., 1990]. The PAI activity in human plasma is usually only PAI-1. It is bound to and stabilized by vitronectin [DeClerck et al., 1988; Preissner et al., 1990]. PAI-2 is secreted as a 56 kD glycoprotein by monocytes/macrophages and probably represents an important regulator of fibrinolytic activity of invading monocytes and tissue macrophages [Wohlwend et al., 1987]. A non-glycosylated PAI-2 molecule is found intracellularly. PAI-2 is encountered in blood only under special conditions, such as pregnancy and the presence of certain tumours [Kruithof et al., 1987, 1988].

The serine protease inhibitors are characterized by a specific reactive site peptide bond, which consists of Arg-Met for α2-antiplasmin and PAI-1 and Arg-Thr for PAI-2. This reactive site and the neighbouring amino acids determine the specificity for serine proteases. From studies with mutants of PAI-1 it has been established that other parts of the molecule are involved in the interaction of PAI-1 with t-PA.

Cell interaction via receptors. High-affinity binding sites for plasminogen, t-PA and u-PA are found on various types of cells including endothelial cells [Miles and Plow, 1988; Beebe et al., 1989; Hajjar and Hamel, 1990; Barnathan et al., 1990a]. As mentioned above, a specific u-PA receptor has been identified, which binds both scu-PA and tcu-PA. This u-PA receptor has recently been cloned [Roldan et al., 1990]. It is heavily glycosylated [Behrendt et al., 1990] and is encountered on the surface of cells as a phosphatidyl-inositol-anchored glycoprotein [Plough et al., 1991]. Bound to its receptor, scu-PA can still be activated. Hence, the receptor may be involved in the localized expression of u-PA activity. In favour of this suggestion, Estreicher et al. [1990] recently demonstrated that the receptor of u-PA polarizes u-PA activity to the leading edge of migrating monocytes. In other cell types u-PA, probably bound to its receptor [Hebert and Baker, 1988], was found specifically in the focal attachment beneath the cells [Pollänen et al., 1987]. In addition to a directional control on u-PA activity, the u-PA receptor also plays a role in the internalization of complexes of u-PA with PAI-1 [Cubellis et al., 1990] or PAI-2 [Estreicher et al., 1990]. Other molecules have also been indicated to play a role in the interaction of u-PA with cells. They include thrombospondin [Harpel et al., 1989] and gangliosides [Miles et al., 1989]. Gangliosides have also been indicated to bind glu-plasminogen to the cell surface [Miles et al., 1989]. An putative interplay between cell-bound plasminogen and receptor-bound PA would enable the cell to control tightly the plasmin generation on its surface [Miles and Plow, 1988; Hajjar and Nachman, 1988]. Because the lipoprotein Lp(a) can compete for plasminogen binding sites on endothelial cells, this lipoprotein may interfere with the plasminogen activation route on the surface of these cells [Miles et al., 1989; Hajjar et al., 1989].

Matrix Metalloproteinases (MMPs) and Tissue Inhibitors of Matrix Metalloproteinases (TIMPs)

The fibrous proteins of the basal membrane include various types of collagens, proteoglycans, laminin and fibronectin. Remodelling of the matrix

Table 2. Mammalian matrix-degrading metalloproteinases

Group	Name	Size (kDa)	Other names	Substrate specificity	Produced by e.g.
I	MMP-1	52-57	interstitial collagenase type I collagenase fibroblast collagenase	collagen type I, II, III VII, and X (gelatins?)	endothelial cells chondrocytes PMN, tumour cells keratinocytes monocytes, mø fibroblasts
	MMP-8	75	PMN collagenase	collagen type I, II, III	PMN
II	MMP-2	72	72-kDa type IV collagenase 72-kDa gelatinase	collagen type IV, V, VII and XI, fibronectin, gelatins (type I, II, III) elastin	endothelial cells chondrocytes PMN, tumour cells monocytes, mø fibroblasts
	MMP-9	92	92-kDa type IV collagenase 92-kDa gelatinase	collagen type IV, V, gelatins	endothelial cells SV40 transformed fibroblasts, PMN chondrocytes keratinocytes tumour cells monocytes, mø
III	MMP-3	57	stromelysin transin proteoglycanase	collagen type I, II, III and IV, IX, X laminin, fibronectin elastin, gelatins proteoglycans	endothelial cells chondrocytes keratinocytes fibroblasts mø
	MMP-10	53	stromelysin-2	collagen type III, IV, V	tumour cells
	MMP-7	28	PUMP-1	gelatins, fibronectin proteoglycans	rat uterus cells

involves the proteolytic degradation of these proteins. Collagenases and glycanases (stromelysins) belong to the family of matrix metalloproteinases [Docherty and Murphy, 1990] and are responsible for the breakdown of many fibrous matrix proteins. The mammalian matrix metalloproteinases (MMPs) and their substrate specificities are summarized in Table 2. They are all secreted in a zymogen form and depend on zinc and calcium ions for activity [Liotta et al., 1990; Woessner et al., 1991]. Activation of these proteins can proceed by the action of poteases, of which plasmin and stromelysin have received particular attention. This results in the loss of a pro-peptide and the reduction of the molecular weight by about 10 kD. The MMPs are inhibited by tissue inhibitors of metalloproteinases, the so called TIMPs. Two TIMPs have been characterized: TIMP-1 [Docherty et al., 1985] and TIMP-2 [Stetler-Stevenson et al., 1989]. Hence, MMP activities are regulated by (a) the activation of the zymogen, (b) the presence and interaction with TIMPs and (c) probably by the presence of binding and competing substrates.

REGULATION OF PLASMINOGEN ACTIVATION BY ENDOTHELIAL CELLS

Two important physiological processes are generally assumed to be regulated by the production of plasminogen activators by endothelial cells (Figure 2). The production and release of t-PA regulates fibrinolysis, a process necessary to prevent fibrin deposition in blood vessels. In

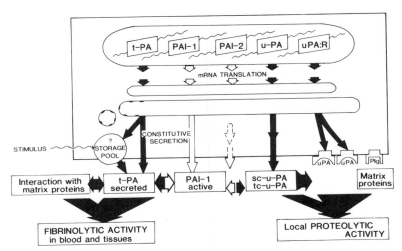

Fig. 2. Schematic picture of the production of plasminogen activators and their inhibitors and receptors by human endothelial cells in vitro (from Van Hinsbergh, 1990, with permission).

addition, the production of PAs, in particular u-PA, is assumed to play a role in the control of local proteolysis. Proteolysis of the endothelial extracellular matrix is necessary to allow the generation of capillary sprouts and the development of new capillaries. Direction of PA activity to specific sites of the cell or its matrix can be accomplished by a cellular receptor for u-PA [Miles et al., 1988] and high affinity binding sites for glu-plasminogen and t-PA, which are present on endothelial cells.

In addition to the production of PAs, endothelial cells synthesize and secrete PAI-1 [Sprengers and Kluft, 1987; Schleef and Loskutoff, 1988]. This PAI-1 can modulate the fibrinolysis process, as platelet PAI-1 does, in a thrombus, but it also can regulate PA activation in the extracellular matrix at the basolateral side of the endothelium.

The production of protease nexin and the mRNA of another PA inhibitor, PAI-2, have been occasionally demonstrated in endothelial cells, but it remains uncertain whether they have a physiological meaning for the endothelium.

THE SYNTHESIS AND RELEASE OF t-PA BY ENDOTHELIAL CELLS: ITS ROLE IN FIBRINOLYSIS

The fibrinolytic activity in the blood is largely determined by the concentration of t-PA [Wun and Capuano, 1985], which is synthesized by endothelial cells. Although the plasma t-PA concentration is rather low under resting conditions (5 ng/ml), the daily amount of t-PA synthesized comes to several milligrams, because the turnover of t-PA in blood is very short (several minutes). This is due to the rapid clearance of t-PA by specific receptors in the liver. The endothelial cells not only synthesizes t-PA, but they are also able to release rapidly and locally a large amountof t-PA after exposure to a stimulus. Many agents act as a stimulus of acute release of t-PA. They include thrombin, platelet activating factor, bradykinin, and endothelin [Emeis, 1988; Tranquille and Emeis, 1990]. By the acute release reaction a large amount of t-PA (often more than 100-fold the basal concentration) becomes available at a site of initial fibrin formation.

Besides the acute release of t-PA, which proceeds within minutes, the rate of t-PA synthesis is an important determinant in the actual levels of t-PA under basal and stimulated conditions. Most of the studies on the regulation of t-PA synthesis and gene regulation have been done on endothelial cells in vitro, mainly from human and bovine origin.

A variety of vasoactive agents including thrombin and histamine have been shown to increase the synthesis of t-PA by endothelial cells, but only recently some initial insight into the intracellular regulatory mechanisms responsible for altered t-PA expression has been obtained. Activation of protein kinase C has been implicated in the induction of t-PA expression in human endothelial cells [Levin and Santell, 1988; Levin et al., 1989; Grülich-Henn and Müller-Berghaus, 1990; Kooistra et al.,1991a]. This induction of t-PA expression was potentiated by simultaneously increasing the intracellular cyclic AMP concentration, although cyclic AMP alone did not affect t-PA synthesis [Santell and Levin, 1988; Levin et al., 1990; Kooistra et al., 1991a]. The stimulatory effect of histamine and thrombin on t-PA synthesis is probably related to the activation of protein kinase C by these agents. The involvement of transcription factors c-fos and c-jun in the protein kinase C dependent induction of t-PA has recently been suggested [Kooistra et al., 1991a].

Several other substances have been reported to increase the production of t-PA by cultured human endothelial cells. Retinoids induce the transcription and synthesis of t-PA molecules. The size of the response depends on the structural properties of these retinoids [Kooistra et al., 1991b]. Retinoid acid has been demonstrated to affect the synthesis of t-PA in the rat in vivo [Kooistra et al., 1991b]. Another differentiating agent, butyrate, stimulates the synthesis and secretion of t-PA antigen and t-PA mRNA level in human endothelial cells 10- to 20-fold [Kooistra et al., 1987]. Shear forces (15 and 25 dynes.cm^2) also increase t-PA production by endothelial cells in vitro [Diamond et al., 1989].

Acidic and basic fibroblast growth factors (a-FGF, b-FGF) induce no or only a small increase in t-PA production in human endothelial cells [Konkle et al., 1990; Verbeek and van Hinsbergh, unpublished]. On the other hand, b-FGF is a strong inducer of u-PA and t-PA in bovine aorta and adrenal microvascular endothelial cells [Moscatelli, 1987; Saksela et al., 1988]. After discussing the effects of cytokines on the production of fibrinolysis proteins, we shall focus on the effect of b-FGF on PA production and its putative role in the regulation of local proteolytic events and angiogenesis.

CYTOKINE REGULATION OF THE FIBRINOLYTIC SYSTEM

Further insight into the regulation of the fibrinolytic system has been obtained from studies with bacterial lipopolysaccharide (LPS) and inflammatory mediators. When LPS or the cytokines interleukin-1 (IL-1), tumour necrosis factor-α (TNFα) or lymphotoxin are added to endothelial cells in culture, an increase in the synthesis and secretion of PAI-1 is found, whereas t-PA synthesis does not change or even decreases [Collucci et al., 1985; Emeis and Kooistra, 1986; Schleef et al., 1988; van Hinsbergh et al., 1988]. The induction of PAI-1 may contribute to the severe thrombotic complications in endotoxinemia. Studies in rats also demonstrated that plasma PAI activity increased upon administration of LPS, IL-1 or TNFα [Emeis and Kooistra, 1986; van Hinsbergh et al., 1988] and that PAI-1 mRNA was elevated upon LPS infusion [Quax et al., 1990]. The latter study, as well as studies by Keeton et al [1991] in mice, suggests that PAI-1 mRNA was predominantly increased in the endothelium. However, in addition to increasing PAI-1 mRNA, LPS infusion also induced t-PA mRNA [Quax et al., 1990]. A similar increase in t-PA was found in man. After administration of LPS or TNFα to healthy volunteers or patients with active malignancies the plasma concentration of PAI-1 markedly increases [Suffredini et al., 1989; van Deventer et al., 1990; van Hinsbergh et al., 1990b]. This increase in PAI activity is preceded by a rapid increase in t-PA antigen and an initial rise in fibrinolytic activity. The rapid increase in t-PA is probably induced by another mediator that is generated by the administration of LPS or TNFα. A good candidate may be thrombin, which is generated by the extrinsic coagulation pathway initiated by LPS- or TNFα-induced tissue factor. Indeed, thrombin was generated in the plasma of volunteers and patients that received LPS and TNFα, respectively [van Deventer et al, 1990; Bauer et al., 1990]. Therefore, it is possible that the evaluation of the effect of TNFα in vivo by systemic addition of TNFα may be complicated by an early generation of thrombin. Despite this complication, the concomitant increase of t-PA and PAI-1, resulting in an excess of PAI activity, gave us the opportunity to evaluate whether fibrinolysis still proceeds under these conditions. From the generation of fibrin degradation products and the generation of plasmin:α2-antiplasmin complexes, we have concluded that fibrinolysis still proceeds [van Hinsbergh et al., 1990b]. However, it is likely that the presence of PAI-1 may have attenuated the fibrinolysis process.

In addition to a role in the regulation of fibrinolysis at the luminal side of the endothelium, the increase in PAI-1 may also be an effective modulator of proteolysis of the abluminally located extracellular matrix. It is generally believed that urokinase plays a role in pericellular proteolysis. Normally, human endothelial cells do not produce significant amounts of u-PA. However, TNFα, as well as LPS, lymphotoxin and IL-1α, induced u-PA mRNA and u-PA synthesis in human endothelial cells in vitro [van Hinsbergh et al., 1990a]. Although the in vivo induction of endothelial u-PA has still to be demonstrated in man, immunolocalisation studies have shown the presence of u-PA in/on endothelial cells of microvascular endo-thelium of inflamed human appendices, whereas it is absent in normal appen-dices [Grøndahl-Hansen et al., 1989]. The enhanced secretion of u-PA induced by TNFα was vectorial. With endothelial cells cultured on porous filters it has been demonstrated that nearly all u-PA was secreted to the basolateral side of the cell [van Hinsbergh et al., 1990a]. This favours the idea that the function of this u-PA is not in the circulation but in proteolytic events in the basal membrane of the cells. The concomitant secretion of PAI-1 may therefore have an additional function: to protect the extracellular matrix from excessive u-PA action. This mechanism may be a human counterpart of a similar regulatory mechanism induced by b-FGF and other mediators in bovine endothelial cells, and which has been related to the regulation of endothelial cell migration and capillary formation (see below).

THE PRODUCTION OF MATRIX METALLOPROTEINASES AND THEIR INHIBITORS BY ENDOTHELIAL CELLS

Several MMPs and their inhibitors have been demonstrated in the conditioned media of cultured endothelial cells (see Table 2). MMP-1, in a proenzyme form, has been demonstrated in the conditioned media of endothelial cells from human umbilical vein [Moscatelli et al., 1985], bovine adrenal capillaries [Gross et al., 1982; 1983] and rabbit brain microvessels [Herron et al., 1986a,b]. The production of procollagenase was markedly enhanced after addition of phorbol ester, an activator of protein kinase C, and was acccompanied by an increase in the production of u-PA [Gross et al., 1982]. MMP-3 (prostromelysin) is also synthesized [Herron et al., 1986a,b] and its synthesis and mRNA are elevated after treatment with phorbol ester [Frisch et al., 1987]. In bovine aorta endothelial cells a profound inhibition of collagenase was found [Gross et al., 1982], due to the production of two inhibitors, TIMP-1 and a protein that is identical to TIMP-2 [DeClerck et al., 1989]. MMP-2 (72 kD type IV collagenase) and MMP-9 (92 kD type IV collagenase) can be demonstrated in the conditioned medium of bovine and rabbit endothelial cells [Unemori et al., 1990; Tsuboi et al., 1990; Murphy et al., 1989] after SDS-polyacrylamide gel electrophoresis and using gelatin zymography, an assay of gelatinase activity. These MMPs are also present in the conditioned media of human umbilical vein endothelial cells [Koolwijk, unpublished]. Unemori et al. [1990] reported that the gelatinase activity (MMP-2) was secreted predominantly to the basolateral side of the endothelial monolayer, while the secretion of TIMP displayed no polarity. This suggests an involvement of this enzyme in the turnover of the matrix proteins. The presence of MMP-3 in/on endothelial cells from microvessels of inflamed synovium in rheumatoid arthritis [Case et al., 1990] further supports the suggestion that a scala of matrix proteases and protease inhibitors are involved in the control of endothelial matrix turnover.

REGULATION OF LOCAL PROTEOLYSIS: ITS PUTATIVE ROLE IN ANGIOGENESIS

Proteolysis of the basal matrix of endothelial cells is a prerequisite for capillary endothelial cells to migrate and to form capillary sprouts. It is generally believed that plasminogen activators and matrix metalloproteinases interact in this process, which may proceed in a complex way similar to the proteolytic cascade of coagulation.

A direct correlation between the expression of PA activity and the migration and sprouting of endothelial cells has recently been obtained from in vitro studies with bovine adrenal microvascular endothelial cells. Upon wounding of a monolayer of these endothelial cells, Pepper et al [1988] observed that, during the repair of the wound, the migrating endothelial cells produced u-PA activity, whereas before wounding and after closure of the gap no PA activity was present. These studies were confirmed by Sato and Rifkin [1988], who subsequently demonstrated that both the migration of the cells and the production of u-PA could be specifically inhibited by anti-b-FGF immunoglobulins. Saksela et al. [1988] had shown earlier that b-FGF increases PA activity in these bovine endothelial cells and that transforming growth factor-β (TGF-β) decreased the basal and b-FGF-stimulated PA activity. Cultured endothelial cells produce TGF-β but usually in a latent form. The finding that plasmin activates latent TGF-β [Sato et al., 1990] suggests that TGF-β acts as a counter-regulatory signal. Upon expression to b-FGF, which may have become available from broken endothelial cells, the cells start to produce u-PA and to activate plasminogen. Thus, plasmin not only degrades fibrin and activates pro-collagenases and pro-stromelysin, but it can also activate TGF-β, which induces PAI-1 production and shuts off the PA activity. Subsequently, Pepper et al [1990] have demonstrated that the formation of sprouts from an endothelial cell monolayer into a fibrin gel is stimulated by b-FGF and inhibited by TGF-β. Furthermore their studies have shown that both b-FGF and TGF-β induce u-PA and PAI-1 in bovine endothelial cells. However, the kinetics and the degree of induction of u-PA and PAI-1 are different for the two mediators, as a result of which u-PA activity was generated by b-FGF and PA inhibition by TGF-β.

Phorbol ester and b-FGF, factors that can induce angiogenesis, increase the expression of u-PA receptors on endothelial cells [Barnathan et al., 1990b; Mignatti et al., 1991]. This suggests that the regulation of proteolytic activation and activity is more complex than depicted above. Interaction of u-PA and PAI-1 with receptors or matrix proteins may largely affect their effectivity. Because of that, an agent that acts as an angiogenesis inhibitor in a certain cirumstance may act as an activator in another environment. This may explain the contradictory results obtained with TGF-β in various angiogenesis assays. Furthermore, it is likely that under various conditions, in which the formation of capillary sprouts occurs, different mediators are involved. From our data with human endothelial cells one may consider that the concomitant induction of u-PA and PAI-1 by TNFα may result in a similar mechanism as described above for b-FGF. Although several experiments have indicated the potency of TNFα to induce endothelial cell migration in vitro and angiogenesis in vivo [Leibovich et al., 1987; Fràter-Schröder et al., 1987], further experiments have to verify this hypothesis.

REFERENCES

Andreasen, P. A., Georg, B., Lund, L. R., Riccio, A., and Stacey, S.N., 1990, Plasminogen activator inhibitors: hormonally regulated serpins, Molec. Cell. Endocrinol., 68:1.
Bachmann, F., 1987, Fibrinolysis, in: "Thrombosis and Haemostasis 1987," M. Verstraete, J. Vermylen, R. Lijnen, and J. Arnout, eds., Leuven University Press, Leuven, p. 227.
Barnathan, E. S., Kuo, A., Rosenfeld, L., Karikó, K., Leski, M., Robbiati, F., Nolli, M. L., Henkin, J., and Cines, D. B., 1990a, Interaction of single-chain urokinase-type plasminogen activator with human endothelial cells, J. Biol. Chem., 265:2865.
Barnathan, E. S., Kuo, A., Karikó, K., Rosenfeld, L., Murray, S. C., Behrendt, N., Rønne, E., Weiner, D., Henkin, J., and Cines, D. B., 1990b, Characterization of human endothelial cell urokinase-type plasminogen activator receptor protein and messenger RNA, Blood, 76:1795.
Bauer, K. A., Ten Cate H, Barzegar, S., Spriggs, D. R., Sherman, M. L., and Rosenberg, R. D., 1989, Tumor necrosis factor infusions have a procoagulant effect on the hemostatic mechanism of humans, Blood, 74: 165.
Beebe, D. P., Miles, L. A., and Plow, E. F., 1989, A linear amino acid sequence involved in the interaction of t-PA with its endothelial cell receptor, Blood, 74:2034.

Behrendt, N., Rønne, E., Ploug, M., Petri, T., Løber, D., Nielsen, L. S., Schleuning, W.-D., Blasi, F., Appella, E., and Danø, K., 1990, The human receptor for urokinase plasminogen activator. NH_2-terminal amino acid sequence and glycosylation variants, J. Biol. Chem., 265: 6453.

Blasi, F., 1988, Surface receptors for urokinase plasminogen activator, Fibrinolysis, 2:73.

Case, J. P., Lafyatis, R., Remmers, E. F., Kumkumian, G. K., and Wilder, R. L., 1989, Transin/stromelysin expression in rheumatoid synovium. A transformation-associated metalloproteinase secreted by phenotypically invasive synoviocytes, Am. J. Pathol., 135:1055.

Colucci, M., Paramo, J. A., and Collen, D., 1985, Generation in plasma of a fast-acting inhibitor of plasminogen activator in response to endotoxin stimulation, J. Clin. Invest., 75:818.

Cubellis, M. V., Wun, T.-C., and Blasi, F., 1990, Receptor-mediated internalization and degradation of urokinase is caused by its specific inhibitor PAI-1, EMBO J., 9:1079.

Danø, K., Andreasen, P. A., Grøndahl-Hansen, J., Kristensen, P., Nielsen, L. S., and Skriver, L., 1985, Plasminogen activators, tissue degradation and cancer, Adv. Cancer Res., 44:139.

DeClerck, Y. A., Yean, T.-D., Ratzkin, B. J., Lu, H. S., and Langley, K. E., 1989, Purification and characterization of two related but distinct metalloproteinase inhibitors secreted by bovine aortic endothelial cells, J. Biol. Chem., 264:17445.

Diamond, S. L., Eskin, S. G., and McIntire, L. V., 1989, Fluid flow stimulates tissue plasminogen activator secretion by cultured human endothelial cells, Science, 243:1483.

Docherty, A. J. P., and Murphy, G., 1990, The tissue metalloproteinase family and the inhibitor TIMP: a study using cDNAs and recombinant proteins, Ann. Rheumatic Diseases, 49:469.

Emeis, J. J., 1988, Mechanisms involved in short-term changes in blood levels of t-PA, in: "Tissue-Type Plasminogen Activator (t-PA): Physiological and Clinical Aspects," C. Kluft, ed., CRC Press Inc., Boca Raton, FL), Vol. II, p. 21.

Estreicher, A., Mühlhauser, J., Carpentier, J.-L., Orci, L., and Vassalli, J.-D., 1990, The receptor for urokinase type plasminogen activator polarizes expression of the protease to the leading edge of migrating monocytes and promotes degradation of enzyme inhibitor complexes, J. Cell Biol., 111:783.

Fràter-Schröder, M., Risau, W., Hallmann, R., Gautschi, P., and Böhlen P, 1987, Tumor necrosis factor type α, a potent inhibitor of endothelial cell growth in vitro, is angiogenic in vivo, Proc. Nat. Acad. Sci. USA., 84:5277.

Friezner Degen, S. J., Rajput, B., and Reich, E., 1986, The human tissue plasminogen activator gene, J. Biol. Chem., 261:6972.

Frisch, S. M., Clark, E. J., and Werb, Z., 1987, Coordinate regulation of stromelysin and collagenase genes determined with cDNA probes, Proc. Natl. Acad. Sci. USA, 84:2600.

Ginsburg, D., Zeheb, R., Yang, A. Y., Rafferty, U. M., Andreasen, P. A., Nielsen, L., Danø, K., Lebo, R. V., and Gelehrter, T. D., 1986, cDNA cloning of human plasminogen activator-inhibitor from endothelial cells, J. Clin. Invest., 78:1673.

Goldfarb, R. H., Ziche, M., Murano, G., and Liotta, L. A., 1986, Plasminogen activators (urokinase) mediate neovascularization: possible role in tumor angiogenesis, Semin. Thromb. Hemostas., 12:337.

Gross, J. L., Moscatelli, D., Jaffe, E. A., and Rifkin, D. B., 1982, Plasminogen activator and collagenase production by cultured capillary endothelial cells, J. Cell Biol., 95:974.

Gross, J. L., Moscatelli, D., and Rifkin, D. B., 1983, Increased capillary endothelial cell protease activity in response to angiogenic stimuli in vitro, Proc. Nat. Acad. Sci. USA, 80:2623.

Grülich-Henn, J., and Müller-Berghaus, G., 1990, Regulation of endothelial tissue plasminogen activator and plasminogen activator inhibitor type 1 synthesis by diacylglycerol, phorbol ester, and thrombin, Blut, 61:38.

Hajjar, K. A., and Nachman, R. L., 1988, Endothelial cell-mediated conversion of glu-plasminogen to lys-plasminogen. Further evidence for assembly of the fibrinolytic system on the endothelial cell surface, J. Clin. Invest., 82:1769.

Hajjar, K. A., Gavish, D., Breslow, J. L., and Nachman, R. L., 1989, Lipoprotein(a) modulation of endothelial cell surface fibrinolysis and its potential role in atherosclerosis, Nature, 339:303.

Hajjar, K. A., and Hamel, N. M., 1990, Identification and characterization of human endothelial cell membrane binding sites for tissue plasminogen activator and urokinase, J. Biol. Chem., 265:2908.

Harpel, P. C., Silverstein, R. L., Pannell, R., Gurewich, V., and Nachman, R. L., 1990, Thrombospondin forms complexes with single-chain and two-chain forms of urokinase, J. Biol. Chem., 265:11289.

Hart, D. A., and Fritzler, M. J., 1989, Regulation of plasminogen activators and their inhibitors in rheumatic diseases: new understanding and the potential for new directions, J. Rheumatol., 16:1184.

Hébert, C. A., and Baker, J. B., 1988, Linkage of extracellular plasminogen activator to the fibroblast cytoskeleton: colocalization of cell surface urokinase with vinculin, J. Cell Biol., 106:1241.

Herron, G. S., Werb, Z., Dwyer, K., Banda, M. J., 1986a, Secretion of metalloproteinases by stimulated capillary endothelial cells. I. Production of procollagenase and prostromelysin exceeds expression of proteolytic activity, J. Biol. Chem., 261:2810.

Herron, G. S., Banda, M. J., Clark, E. J., Gavrilovic, J., and Werb, Z., 1986b, Secretion of metalloproteinases by stimulated capillary endothelial cells. II. Expression of collagenase and stromelysin activities is regulated by endogenous inhibitors, J. Biol. Chem., 261:2814.

Holmes, W. E., Nelles, L., Lijnen, H. R., and Collen, D., 1987, Primary structure of human α_2-antiplasmin, a serine protease inhibitor (Serpin), J. Biol. Chem., 262:1659.

Konkle, B. A., Kollros, P. R., and Kelly, M. D., Heparin-binding growth factor-1 modulation of plasminogen activator inhibitor-1 expression. Interaction with cAMP and protein kinase C-mediated pathways, J. Biol. Chem., 265:21867.

Kooistra, T., Van den Berg, J., Töns, A., Platenburg, G., Rijken, D. C., and Van den Berg, E. A., 1987, Butyrate stimulates tissue-type plasminogen-activator synthesis in cultured human endothelial cells, Biochem. J., 247:605.

Kooistra, T., Bosma, P. J., Toet, K., Cohen, L. H., Griffioen, M., Van den Berg, E., le Clercq, L., and Van Hinsbergh, V. W. M., 1991a, Role of protein kinase C and cAMP in the regulation of tissue-type plasminogen activator, plasminogen activator inhibitor 1 and platelet-derived growth factor mRNA levels in human endothelial cells. Possible involvement of proto-oncogenes c-jun and c-fos, Arteriosclerosis and Thrombosis, 11:1042.

Kooistra, T., Opdenberg, J. P., Toet, K., Hendriks, H. F. J., Van den Hoogen, R. M., and Emeis, J. J., 1991b, Stimulation of tissue-type plasminogen activator synthesis by retinoids in cultured human endothelial cells and rat tissues in vivo, Thromb. Haemostas., 65:565.

Kruithof, E. K. O., Tran-Thang, C., Gudinchet, A., Hauert, J., Nicoloso, G.,

Genton, C., Welti, H., and Bachmann, F., 1987, Fibrinolysis in pregnancy: a study of plasminogen activator inhibitors, <u>Blood</u>, 69: 460.

Kruithof, E. K. O., Gudinchet, A., and Bachmann, F., 1988, Plasminogen activator inhibitor 1 and plasminogen activator inhibitor 2 in various disease states, <u>Thromb. Haemostas.</u>, 59:7.

Leibovich, S. J., Polverini, P. J., Shepard, H. M., Wiseman, D. M., Shively, V., and Nuseir, N., 1987, Macrophage-induced angiogenesis is mediated by tumour necrosis factor-α, <u>Nature</u>, 329:630.

Levin, E. G., and Santell, L., 1988, Stimulation and desensitization of tissue plasminogen activator release from human endothelial cells, <u>J. Biol. Chem.</u>, 263:9360.

Levin, E. G., Marotti, K. R., and Santell, L., 1989, Protein kinase C and the stimulation of tissue plasminogen activator release from human endothelial cells. Dependence on the elevation of messenger RNA, <u>J. Biol. Chem.</u>, 264:16030.

Liotta, L. A., Steeg, P. S., and Stetler-Stevenson, W. G., 1991, Cancer metastasis and angiogenesis: an imbalance of positive and negative regulation, <u>Cell</u>, 64:327.

Medcalf, R. L., Rüegg, M., and Schleuning, W.-D., 1990, A DNA motif related to the cAMP-responsive element and an exon-located activator protein-2 binding site in the human tissue-type plasminogen activator gene promoter cooperate in basal exxpression and convey activation by phorbol ester and cAMP, <u>J. Biol. Chem.</u>, 265:14618.

Mignatti, P., Robbins, E., and Rifkin, D.B., 1986, Tumor invasion through the human amniotic membrane: requirement for a proteinase cascade, <u>Cell</u>, 47:487.

Mignatti, P., Tsuboi, R., Robbins, E., and Rifkin, D.B., 1989, In vitro angiogenesis on the human amniotic membrane: requirement for basic fibroblast growth factor-induced proteinases, <u>J. Cell Biol.</u>, 108: 671.

Mignatti, P., Mazzieri, R., and Rifkin, D. B., 1991, Expression of the urokinase receptor in vascular endothelial cells is stimulated by basic fibroblast growth factor, <u>J. Cell Biol.</u>, 113:1193.

Miles, L. A., and Plow, E. F., 1988, Plasminogen receptors: ubiquitous sites for cellular regulation of fibrinolysis, <u>Fibrinolysis</u>, 2:61.

Miles, L. A., Levin, E. G., Plescia, J., Collen, D., and Plow, E. F., 1988, Plasminogen receptors, urokinase receptors, and their modulation on human endothelial cells, <u>Blood</u>, 72:628.

Miles, L. A., Dahlberg, C. M., Levin, E. G., and Plow, E. F., 1989a, Gangliosides interact directly with plasminogen and urokinase and may mediate binding of these fibrinolytic components to cells, <u>Biochemistry</u>, 28:9337.

Miles, L. A., Fless, G. M., Levin, E. G., Scanu, A. M., and Plow, E. F., 1989b, A potential basis for the thrombotic risks associated with lipoprotein(a), <u>Nature</u>, 339:301.

Montesano, R., Pepper, M. S., Vassalli, J.-D., and Orci, L., 1987, Phorbol ester induces cultured endothelial cells to invade a fibrin matrix in the presence of fibrinolytic inhibitors, <u>J. Cell. Physiol.</u>, 132:509.

Moscatelli, D., Rifkin, D. B., and Jaffe, E. A., 1985, Production of latent collagenase by human umbilical vein endothelial cells in response to angiogenic preparations, <u>Exp. Cell Res.</u>, 156:379.

Moscatelli, D., 1987, High and low affinity binding sites for basic fibroblast growth factor on cultured cells: absence of a role for low affinity binding in the stimulation of plasminogen activator production by bovine capillary endothelial cells, <u>J. Cell. Physiol.</u>, 131:123.

Müllertz, S., 1987, Fibrinolysis. General aspects, characteristic features and perspectives, <u>Fibrinolysis</u>, 1:3.

Ny, T., Sawdey, M., Lawrence, D., Millan, J. L., and Loskutoff, D. J., 1986, Cloning and sequence of a cDNA coding for the human β-migrating endothelial-cell-type plasminogen activator inhibitor, Proc. Natl. Acad. Sci. USA, 83:6776.

Pannekoek, H., Veerman, H., Lambers, H., Diergaarde, P., Verweij, C. L., Van Zonneveld, A.-J., and Van Mourik, J. A., 1986, Endothelial plasminogen activator inhibitor (PAI): a new member of the serpin gene family, EMBO J., 5:2539.

Pennica, D., Holmes, W. E., Kohr, W. J., Harkins, R. N., Vehar, G. A., Ward, C. A., Bennett, W. F., Yelverton, E., Seeburg, P. H., Heyneker, H. L., and Goeddel, D. V., 1983, Cloning and expression of human tissue-type plasminogen activator cDNA in E.coli, Nature, 301:214.

Pepper, M. S., Vassalli, J.-D., Montesano, R., and Orci, L., 1987, Urokinase-type plasminogen activator is induced in migrating capillary endothelial cells, J. Cell Biol., 105:2535.

Pepper, M. S., Belin, D., Montesano, R., Orci, L., and Vassalli, J.-D., 1990, Transforming growth factor-beta 1 modulates basic fibroblast growth factor-induced proteolytic and angiogenic properties of endothelial cells in vitro, J. Cell Biol., 111:743.

Petersen, T. E., Martzen, M. R., Ichinose, A., and Davie, E. W., 1990, Characterization of the gene for human plasminogen, a key proenzyme in the fibrinolytic system, J. Biol. Chem., 265:6104.

Ploug, M., Rønne, E., Behrendt, N., Jensen, A. L., Blasi, F., and Danø, K., 1991, Cellular receptor for urokinase plasminogen activator. Carboxyl-terminal processing and membrane anchoring by glycosyl-phosphatidylinositol, J. Biol. Chem., 266:1926.

Pöllänen, J., Hedman, K., Nielsen, L. S., Danø, K., and Vaheri, A., 1988, Ultrastructural localization of plasma membrane-associated urokinase-type plasminogen activator at focal contacts, J. Cell Biol., 106:87.

Preissner, K. T., Grulich-Henn, J., Ehrlich, H. J., Declerck, P., Justus, C., Collen, D., Pannekoek, H., and Müller-Berghaus, G., Structural requirements for the extracellular interaction of plasminogen activator inhibitor 1 with endothelial cell matrix-associated vitronectin, J. Biol. Chem., 265:18490.

Quax, P. H. A., Van den Hoogen, C. M., Verheijen, J. H., Padró, T., Zeheb, R., Gelehrter, T. D., Van Berkel, Th. J. C., Kuiper, J., and Emeis, J. J., 1990, Endotoxin induction of plasminogen activator and plasminogen activator inhibitor type 1 mRNA in rat tissues in vivo, J. Biol. Chem., 265:15560.

Rånby, M., 1982, Studies on the kinetics of plasminogen activation by tissue plasminogen activator, Biochim. Biophys. Acta, 704:461.

Riccio, A., Grimaldi, G., Verde, P., Sebastio, G., Boast, S., and Blasi, F., 1985, The human urokinase plasminogen activator gene and its promoter, Nucl. Acids Res., 13:2759.

Rijken, D. C., Hoylaerts, M., and Collen, D., 1982, Fibrinolytic properties of one-chain and two-chain human extrinsic (tissue-type) plasminogen activator, J. Biol. Chem., 257:2920.

Roldan, A. L., Cubellis, M. V., Masucci, M. T., Behrendt, N., Lund, L. R., Danø, K., Appella, E., and Blasi, F., 1990, Cloning and expression of the receptor for human urokinase plasminogen activator, a central molecule in cell surface, plasmin dependent proteolysis, EMBO J., 9:467.

Saksela, O., 1985, Plasminogen activation and regulation of pericellular proteolysis, Biochim. Biophys. Acta, 823:35.

Saksela, O., Moscatelli, D., and Rifkin, D. B., 1987, The opposing effects of basic fibroblast growth factor and transforming growth factor beta on the regulation of plasminogen activator activity in capillary endothelial cells, J. Cell Biol., 105:957.

Saksela, O., and Rifkin, D. B., 1990, Release of basic fibroblast growth factor-heparan sulfate complexes from endothelial cells by plasminogen activator-mediated proteolytic activity, J. Cell Biol., 110:767.

Salonen, E.-M., Zitting, A., and Vaheri, A., 1984, Laminin interacts with plasminogen and its tissue-type activator, FEBS Lett., 172:29.

Salonen, E.-M., Saksela, O., Vartio, T., Vaheri, A., Nielsen, L. S., and Zeuthen, J., 1985, Plasminogen and tissue-type plasminogen activator bind to immobilized fibronectin, J. Biol. Chem., 260:12302.

Santell, L., and Levin, E. G., 1988, Cyclic AMP potentiates phorbol ester stimulation of tissue plasminogen activator release and inhibits secretion of plasminogen activator inhibitor-1 from human endothelial cells, J. Biol. Chem., 263:16802.

Sato, Y., and Rifkin, D. B., 1988, Autocrine activities of basic fibroblast growth factor: regulation of endothelial cell movement, plasminogen activator synthesis, and DNA synthesis, J. Cell Biol., 107:1199.

Sato, Y., Tsuboi, R., Lyons, R., Moses, H., and Rifkin, D. B., 1990, Characterization of the activation of latent TGF-β by co-cultures of endothelial cells and pericytes or smooth muscle cells: a self-regulating system, J. Cell Biol., 111:757.

Schleef, R. R., Bevilacqua, M. P., Sawdey, M., Gimbrone, M. A., Jr., and Loskutoff, D. J., 1988, Cytokine activation of vascular endothelium. Effects on tissue-type plasminogen activator and type 1 plasminogen activator inhibitor, J. Biol. Chem., 263:5797.

Schleef, R. R., and Loskutoff, D. J., 1988, Fibrinolytic system of vascular endothelial cells. Role of plasminogen activator inhibitors, Haemostasis, 18:328.

Silverstein, R. L., Leung, L. L. K., and Nachman, R. L., 1986, Thrombospondin: a versatile multifunctional glycoprotein, Arteriosclerosis, 6:245.

Sprengers, E. D., and Kluft, C., 1987, Plasminogen activator inhibitors, Blood, 69:381.

Suffredini, A. F., Harpel, P. C., and Parrillo, J. E., 1989, Promotion and subsequent inhibition of plasminogen activation after administration of intravenous endotoxin to normal subjects, New Engl. J. Med., 320: 1165.

Tsuboi, R., Sato, Y., and Rifkin, D. B., 1990, Correlation of cell migration, cell invasion, receptor number, proteinase production, and basic fibroblast growth factor levels in endothelial cells, J. Cell Biol., 110:511.

Unemori, E. N., Bouhana, K. S., and Werb, Z., 1990, Vectorial secretion of extracellular matrix proteins, matrix-degrading proteinases, and tissue inhibitor of metalloproteinases by endothelial cells, J. Biol. Chem., 265:445.

Van Deventer, S. J. H., Büller, H. R., Ten Cate, J. W., Aarden, L. A., Hack, E., and Sturk, A., 1990, Experimental endotoxemia in humans: analysis of cytokine release and coagulation, fibrinolytic, and complement pathways, Blood, 76:2520.

Van Hinsbergh, V. W. M., 1988a, Regulation of the synthesis and secretion of plasminogen activators by endothelial cells, Haemostasis, 18:307.

Van Hinsbergh, V. W. M., 1988b, Synthesis and secretion of plasminogen activators and plasminogen activator inhibitor by endothelial cells, in: "Tissue-Type Plasminogen Activator (t-PA): Physiological and Clinical Aspects," C. Kluft, ed., CRC Press Inc., Boca Raton, FL, Vol. II, p. 3.

Van Hinsbergh, V. W. M., Kooistra, T., Van den Berg, E. A., Princen, H. M. G., Fiers, W., and Emeis, J. J., 1988, Tumor necrosis factor increases the production of plasminogen activator inhibitor in human endothelial cells in vitro and in rats in vivo, Blood, 72:1467.

Van Hinsbergh, V. W. M., 1990, Coagulation and fibrinolysis in vascular occlusion and repair, in: "Applied Cardiovascular Biology 1989," P. Zilla, R. Fasol, and A. Callow, eds., Int. Soc. Appl. Cardiovasc. Biol. Karger, Basel, Vol. 1, p. 73.

Van Hinsbergh, V. W. M., Van den Berg, E. A., Fiers, W., and Dooijewaard, G., 1990a, Tumor necrosis factor induces the production of urokinase-type plasminogen activator by human endothelial cells, Blood, 75: 1991.

Van Hinsbergh, V. W. M., Bauer, K. A., Kooistra, T., Kluft, C., Dooijewaard, G., Sherman, M. L., and Nieuwenhuizen, W., 1990b, Progress of fibrinolysis during tumor necrosis factor infusions in humans. Concomitant increase in tissue-type plasminogen activator, plasminogen activator inhibitor type-1, and fibrin(ogen) degradation products, Blood, 76:2284.

Van Zonneveld, A. J., Veerman, H., and Pannekoek, H., 1986, On the interaction of the finger and the kringle-2 domain of tissue-type plasminogen activator with fibrin, J. Biol. Chem., 261:14214.

Verheijen, J. H., Caspers, M. P. M., Chang, G. T. G., De Munk, G. A. W., Pouwels, P. H., and Enger-Valk, B. E., 1986, Involvement of finger domain and kringle 2 domain of tissue-type plasminogen activator in fibrin binding and stimulation of activity by fibrin, Eur. Mol. Biol. Organisat. J., 5:3525.

Wallén, P., 1987, Structure and function of tissue plasminogen activator and urokinase, in: "Fundamental and Clinical Fibrinolysis," Castellino, P. J. Gaffney, M. M. Samama, and Takada, eds., Elsevier, Amsterdam, p. 1.

Wohlwend, A., Belin, D., and Vassalli, J.-D., 1987, Plasminogen activator-specific inhibitors produced by human monocytes/macrophages, J. Exp. Med., 165:320.

Wun, T.-C., and Capuano, A., 1985, Spontaneous fibrinolysis in whole human plasma. Identification of tissue activator-related protein as the major plasminogen activator causing spontaneous activity in vitro, J. Biol. Chem., 260:5061.

Ye, R. D., Wun, T.-C., and Sadler, J. E., 1987, cDNA cloning and expression in Escherichia coli of a plasminogen activator inhibitor from human placenta, J. Biol. Chem., 262:3718.

HEPARIN COORDINATELY SUPPRESSES ENDOTHELIAL CELL PLASMINOGEN ACTIVATOR

INHIBITOR-1, FIBRONECTIN AND THROMBOSPONDIN mRNA LEVELS

Bernadette Lyons-Giordano+, Jane M. Brinker* and
Nicholas A. Kefalides*

Connective Tissue Research Institute and Departments of
Medicine*, and Dermatology+, University of Pennsylvania*+
and University City Science Center*
Philadelphia, PA

INTRODUCTION

The studies described herein were undertaken to define the effects
of heparin, an angiogenic factor, on endothelial cell (EC) expression of
extracellular matrix components. In vitro heparin is known to enhance EC
chemotactic and proliferative responses to endothelial cell growth factor
(ECGF), a member of the fibroblast growth factor family of mitogens (1,2).
While numerous studies have demonstrated the importance of matrix composition
in the regulation of EC attachment (3), migration (4), proliferation (5),
and multicellular organization (6), little is known about the effects of
angiogenic factors on EC biosynthesis of matrix molecules. In culture, EC
synthesize a matrix which contains proteoglycans (7), collagens (8) and
other glycoproteins including fibronectin (FN) (9), laminin (10), thrombospondin (11), and plasminogen activator inhibitor-1 (PAI-1) (12). In addition, EC synthesize fibroblast growth factor which they sequester in their
matrix (13). In this report, we present data which demonstrate that heparin,
in the presence of ECGF, coordinately suppresses EC steady-state mRNA levels
for FN, TSP and PAI-1. The suppression of these messages correlates with
the potentiation of EC proliferative response to ECGF by heparin. These
data suggest a potential role for heparin in the remodeling of extracellular
matrix concomitant with neovascularization and further highlight the potential role of matrix components in the autocrine regulation of cell growth.

MATERIALS AND METHODS

Cell Culture

Human umbilical vein EC were obtained as described previously (14).
Primary and subcultured EC cultures were grown in complete culture medium
(CCM): Modified M-199 growth medium containing 16.5% FCS, 2 mM glutamine,
15 mM HEPES, 10 µg/ml gentamycin, 0.4 µg/ml amphotericin B, and 30 µg/ml
ECGF (Meloy Laboratories, Inc., Springfield, VA) in the absence and presence
of heparin from porcine intestinal mucosa (Sigma, H3125, St. Louis, MO).
EC were maintained at 37°C in a humidified 5% CO_2 atmosphere. Analyses
were conducted using cells passaged one to three times.

Angiogenesis in Health and Disease, Edited by M.E. Maragoudakis *et al.*
Plenum Press, New York, 1992

RNA Analysis

Total cellular RNA was prepared as described previously (15). For quantitative analyses, total RNA from control cultures and cultures treated with heparin was applied to nitrocellulose using the procedure and apparatus of Schleicher and Schuell, Inc., Keene, NH. The cDNA probes were nick-translated using [^{32}P]CTP and [^{32}P]dTTP (3000 Ci/mmole, Amersham, Arlington Heights, IL) to a specific activity of $5-8 \times 10^8$ cpm/µg without excision from the plasmid. The filters were pre-hybridized overnight and then hybridized with the denatured labeled probes for 22h at 42°C in a solution containing 50% formamide, 5x standard saline citrate (1x SSC; 0.15 M NaCl, 0.015 trisodium citrate), 1 mM EDTA, 2x Denhardt's solution, 100 µg/ml salmon sperm DNA, and 0.1% SDS. For hybridizations with the cDNAs for FN, TSP and actin the filters were washed with a final stringency of 0.5x SSC containing 0.1% SDS at 65°C. For hybridizations with the cDNA for PAI-1, the filters were washed with a final stringency of 0.2x SSC containing 0.1% SDS at 60°C. In all cases, the specificity of the hybridization conditions was verified by Northern analyses of total RNA from control cultures and cultures treated with 90 µg/ml heparin. The filters were exposed to Kodak XAR-5 film at -70°C for varying lengths of time. Radioactivity in the probes hybridizing to the filters was quantified by scanning densitometry of the autoradiograms. The levels of radioactivity were linear with repect to the quantity of total RNA applied to the filters.

cDNA Probes

The probes used include the human cDNA clone pFH1 for FN obtained from Dr. Francisco Baralle, University of Oxford (16) and the TSP clone from Dr. Vishva Dixit, University of Michigan (17). Human fibroblast actin and tubulin cDNA clones were provided by Dr. Larry Kedes, Stanford University. The actin cDNA clone includes sequences from the coding region up to amino acid 145 plus the complete 3' untranslated region of the message and hybridizes to both the γ and β-actin mRNA (18). The tubulin cDNA clone pHFBT-1, with its 2.1 kb insert, is virtually a full length clone specific for β-tubulin mRNAs (19). The PAI-1 (20) clone was generously provided by T. C. Wun (Monsanto, St. Louis, MO).

RESULTS

EC cultured in CCM synthesize large quanities of FN and TSP mRNAs. The sizes of these messages, 8.0 and 5.5 kb for FN and TSP respectively, agree with those previously reported for these molecules (16,21). The effects of heparin on EC expression of TSP and FN mRNAs were investigated. Slot blot hybridization analyses of total RNA prepared from EC grown in CCM without exogenous heparin and from EC cultured in CCM supplemented with heparin (0.09 to 90 µg/ml) were conducted using [^{32}P]-labeled cDNA probes specific for FN or TSP. Heparin caused a concentration dependent suppression of both TSP and FN message levels (Fig. 1). Maximal heparin-induced suppression of these messages was apparent in EC treated with 9 µg/ml heparin. Maximal suppression of FN mRNA level was approximately 85%, whereas, that of TSP was 40%. By interpolation, the concentrations of heparin expected to yield half-maximal suppression of TSP and FN mRNA levels were the same (2.5 µg/ml). Autoradiograms of duplicate slot blots hybridized with radiolabeled cDNA probes specific for actin or for tubulin revealed no differences between control and heparin-treated EC. Thus, the effect of heparin on FN and TSP message levels is not reflective of a general decrease in mRNA levels. Additionally, hybridization to the probe pA, a genomic clone encoding most of the 28S ribosomal DNA gene (22) confirmed that equal amounts of RNA from control and heparin-treated cultures were bound to the nitrocellulose filters.

EC express high levels of both the 3.0 and 2.2 kb transcripts specific for PAI-1 (20); Northern analyses of RNA from our cultures confirmed these findings. Heparin in the presence of ECGF, a component of our CCM, has been reported to decrease EC PAI-1 mRNA levels (23). The effect of heparin on EC PAI-1 mRNA levels relative to its effect on FN and TSP mRNA levels was quantified by slot blot hybridization analyses. Equivalent amounts of total RNA from control cultures and EC grown in the presence of heparin concentrations ranging from 0.09 µg/ml to 90 µg/ml were hybridized to a [^{32}P]-labeled cDNA specific for PAI-1. The heparin-induced decrease in PAI-1 message levels was coordinate with the diminution in TSP and FN message levels with respect to concentration dependence (Fig. 1). Maximal suppression was observed in EC treated with 9.0 µg/ml heparin in which PAI-1 mRNA levels were approximately 15% of those of control cultures.

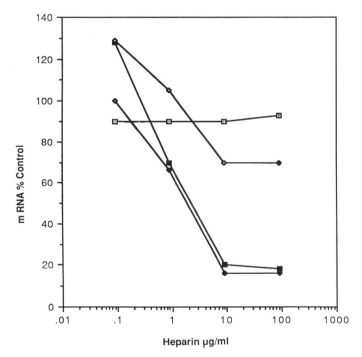

Figure 1. Quantitation of the concentration dependence of heparin suppression of FN, TSP and PAI-1 mRNA levels in EC. The mRNA levels of FN (■), TSP (◇), PAI-1 (◆) and β-tubulin (▫) in EC grown in the presence of increasing concentrations of heparin were determined by scanning densitometry of the autoradiograms following hybridization of total cellular RNA to the specific [^{32}P]-labeled cDNA probes. The percentage of each message relative to that in control cells is plotted as a function of heparin concentration present in the culture medium.

The concentration dependence of heparin-induced EC proliferation is shown in Fig. 2. Maximal stimulation of EC growth was evidenced in cultures treated with 9 µg/ml heparin. The concentration of heparin expected to yield half-maximal stimulation of proliferation was 2.5 µg/ml. Thus, there is a high correlation between heparin effects on EC growth and expression of TSP, FN and PAI-1 message levels.

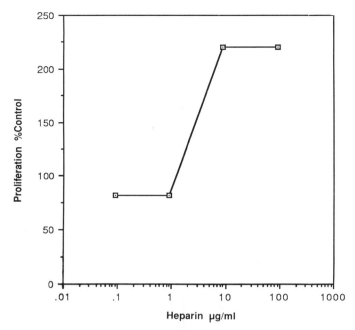

Figure 2. Concentration dependence of heparin effects on EC proliferation.
EC seeded at a density of 7×10^3 cells/cm^2 in CCM were treated with heparin
concentrations ranging from 0.09 to 90 μg/ml for 3 days and then trypsinized
and counted. Control cultures were maintained in CCM without exogenous
heparin. Proliferation, defined as the net increase in cell number, was
determined for control and heparin-treated EC. Data are presented as a percentage of control cell proliferation.

DISCUSSION

     Enhanced migration and proliferation of the endothelium are fundamental
events in the process of angiogenesis. Despite the current view that FN
and other extracellular matrix molecules play an important role in the regulation of cell behaviors including migration and growth, limited data are
available concerning the effects of angiogenic factors on EC synthesis of
extracellular matrix components. Our studies of the effects of heparin, an
angiogenic factor, demonstrate a high correlation between the suppression
of steady-state mRNA levels for FN, TSP, and PAI-1 and enhanced EC proliferation. Biosynthetic studies from our laboratory have shown that the dramatic suppression of FN mRNA levels is reflected in a coordinate decrease
in EC synthesis of FN, whereas, TSP synthesis is not modulated with heparin
treatment (15). PAI-1 levels in EC cultures treated with heparin are known
to be reduced (23). Notably, EC message levels for tissue plasminogen activator and synthesis of this protein are not changed with heparin treatment (23). Preliminary studies in our laboratory provide evidence that EC
synthesize message for urokinase and the steady-state mRNA levels for the
enzyme are not modulated with heparin treatment. Similar to its effects on
EC growth, the modulation of EC synthesis of FN and PAI-1 are contingent upon
the presence of exogenous ECGF in the culture medium (23). The potential
significance of the heparin-induced alterations in EC secretory phenotype
coordinate with its enhancement of EC proliferative response to ECGF remains
to be elucidated. It may be that FN and/or PAI-1 function in an autocrine
manner in regulating EC growth.

In vivo the combined effects of heparin and ECGF on EC growth and secretory phenotype may be important in the repair of vessel wall following injury or in the events leading to neovascularization. ECGF is synthesized by EC and is known to be present in basement membrane (24). Our studies with EC *in vitro* suggest that in addition to promoting EC growth and migration, local increases in heparin concentration at a site of vessel injury may result in profound changes in the subendothelial matrix consequent to the reduction in EC biosynthesis of matrix molecules and the increased activity of plasminogen activator enzymes.

SUMMARY

Since matrix composition and organization profoundly influence cell behavior, the effect of heparin, an angiogenic factor, on the expression of extracellular matrix components by endothelial cells (EC) was investigated. EC cultured in the presence of 90 μg/ml heparin and 30 μg/ml endothelial cell growth factor (ECGF) contained significantly lower levels of mRNA for plasminogen activator inhibitor-1 (PAI-1), fibronectin (FN) and thrombospondin (TSP) compared to EC cultured with ECGF alone. EC steady-state mRNA levels for actin and tubulin were not modulated by heparin treatment. The suppression of PAI-1, FN and TSP message levels by heparin was coordinate with respect to concentration dependence. Maximal suppression of these transcripts was observed in EC treated with 9 μg/ml heparin; half-maximal suppression was estimated to require 2.6 μg/ml heparin. Maximal suppression of PAI-1 and FN steady-state mRNA levels was the same, 85%. In contrast, maximal suppression of TSP transcript levels was only 40%. These data imply that in addition to promoting EC proliferation, local increases in heparin concentration at a site of vessel injury may result in profound changes in the subendothelial matrix through effects on EC biosynthesis of matrix molecules and the increased activity of plasminogen activator enzymes.

ACKNOWLEDGEMENTS

This work was supported by NIH Grants AR-20553 and HL-29492.

REFERENCES

1. S. C. Thornton, S. N. Mueller, and E. M. Levine, Human endothelial cells: Use of heparin in cloning and long-term serial cultivation, Science 222:623 (1983).
2. V. P. Terranova, R. M. DiFlorio, S. Hic, R. Friesel, R. M. Lyall, and T. Maciag, Human endothelial cells are chemotactic to endothelial cell growth factor and heparin, J. Cell Biol. 101:2330 (1985).
3. E. J. Macarak, and P. S. Howard, Adhesion of endothelial cells to extracellular matrix proteins, J. Cell. Physiol. 116:76 (1983).
4. W. C. Young, and I. M. Herman, Extracellular matrix modulation of endothelial cell shape and motility following injury in vitro, J. Cell Sci. 73:19 (1985).
5. D. Gospodarowicz, and C. R. Ill, Do plasma and serum have different abilities to promote cell growth?, Proc. Natl. Acad. Sci. USA 77:2726 (1981).
6. T. Maciag, S. Kadish, L. Wilkins, M. B. Stemerman, and R. Weinstein, Organizational behavior of human umbilical vein endothelial cells, J. Cell Biol. 94:511 (1982).

7. J. R. Hassell, P. Gehron Robey, H.-J. Barrach, J. Wilczek, S. I. Rennard, and G. R. Martin, Isolation of heparan sulfate-containing proteoglycan from basement membrane, Proc. Natl. Acad. Sci. USA 77:4494 (1980).

8. B. V. Howard, E. J. Macarak, D. Gunson, and N. A. Kefalides, Characterization of collagen synthesized by endothelial cells in culture, Proc. Natl. Acad. Sci. USA 73:2361 (1976).

9. E. A. Jaffe, and D. F. Mosher, Synthesis of fibronectin by cultured human endothelial cells, J. Exp. Med. 147:1779 (1978).

10. D. Gospodarowicz, G. Greenburg, J. M. Foidart, and N. Savion, The production and localization of laminin in cultured vascular and corneal endothelial cells, J. Cell. Physiol. 107:171 (1981).

11. J. McPherson, H. Sage, and P. Bornstein, Isolation and characterization of a glycoprotein secreted by aortic endothelial cells in culture: Apparent identity with platelet thrombospondin, J. Biol. Chem. 256:11330 (1981).

12. D. J. Loskytoff, J. A. Van Mourik, L. A. Erickson, and D. Lawrence, Detection of an unusually stable fibrinolytic inhibitor produced by bovine endothelial cells, Proc. Natl. Acad. Sci. USA 80:2956 (1983).

13. I. Vlodavsky, J. Folkman, R. Sullivan, R. Fridman, R. Ishai-Michaeli, J. Sasse, and M. Klagsbrun, Endothelial cell-derived basic fibroblast growth factor: Synthesis and deposition into the subendothelial extracellular matrix, Proc. Natl. Acad. Sci. USA 84:2292 (1987).

14. M. A. Gimbrone, Jr., R. S. Cotran, and J. Folkman, Human vascular endothelial cells: Growth and DNA synthesis, J. Cell Biol. 60:673 (1974).

15. B. Lyons-Giordano, J. M. Brinker, and N. A. Kefalides, The effect of heparin on fibronectin and thrombospondin synthesis and mRNA levels in cultured human endothelial cells, Exp. Cell Res. 186:39 (1990).

16. A. R. Kornblihtt, K. Vibe-Pedersen, and F. E. Baralle, Human fibronectin: Molecular cloning evidence for two mRNA species differing by an internal segment coding for a structural domain, EMBO 3:221 (1984).

17. V. M. Dixit, S. W. Hennessy, G. A. Grant, P. Rotwein, and W. A. Frazier. Characterization of a cDNA encoding the heparin and collagen binding domains of human thrombospondin, Proc. Natl. Acad. Sci. USA 83:5449 (1986).

18. P. Gunning, P. Ponte, H. Okayama, J. Engel, H. Blau, and L. Kedes, Isolation and characterization of full-length cDNA clones for human α-, β- and γ- mRNAs: Skeletal but not cytoplasmic actins have an amino terminal cysteine that is subsequently removed, Mol. Cell. Biol. 3:787 (1983).

19. J. L. Hall, L. Dudley, P. R. Dobner, S. A. Lewis, and N. J. Cowan, Identification of two human β-tubulin isotypes, Mol. Cell. Biol. 3:854 (1983).

20. T. C. Wun, and K. K. Kretzner, cDNA cloning and expression in E. Coli of a plasminogen activator inhibitor (PAI) related to PAI produced by Hep G2 hepatoma cell, FEBS Letters 210:11 (1987).

21. R. A. Majack, J. Milbrandt, and Y. M. Dixit, Induction of thrombospondin messenger RNA levels occurs as an immediate primary response to platelet-derived growth factor, J. Biol. Chem. 262:8821 (1987).

22. J. M. Erickson, C. L. Rushford, D. J. Dorney, G. N. Wilson, and R. D. Schmickel, Structure and variation of human ribosomal DNA: Molecular analysis of cloned fragments, Gene 16:1 (1981).

23. B. A. Konkle, and D. Ginsburg, The addition of endothelial cell growth factor and heparin to human umbilical vein endothelial cell cultures decreases plasminogen activator inhibitor-1 expression, J. Clin. Invest. 82:579 (1988).

24. J. Folkman, M. Klagsburn, J. Sasse, M. Wadzinski, D. Ingber, and I. Vlodavsky, A heparin-binding angiogenic protein - Basic fibroblast growth factor - is stored within basement membrane, Am. J. Pathol. 130:393 (1988).

PROPERTIES OF ENDOTHELIAL AND SMOOTH MUSCLE CELLS GROWN IN 3-DIMENSIONAL COLLAGEN LATTICES

Robert Alper[*] and Nicholas A. Kefalides[*+]

Connective Tissue Research Institute,
University City Science Center[*] and Department
of Medicine, University of Pennsylvania[+],
Philadelphia, PA

INTRODUCTION

Several years ago, Weinberg and Bell (1, 2) demonstrated the feasibility of developing models of the vascular wall in which smooth muscle cells (SM) were embedded within a 3-dimensional collagen lattice and the lattice then overlaid with a layer of endothelial cells (EC). They went on to form a model containing adventitial cells surrounding a tubular composite lattice and showed that under these conditions, the cells maintained certain of their phenotypic properties. Since that time, it has become evident that the composition of the extracellular environment may have important effects upon the properties of cells and that cell-cell and cell-matrix interactions are of vital importance in the physiology of most tissues (3, 4).

The present study was undertaken in an attempt to examine more closely the influences of the cellular environment on the properties of EC and SM grown on and within 3-dimensional collagen lattices. These studies demonstrate that SM and EC morphology, replication and migration can be influenced by the nature of their interactions with the collagen lattice and with each other.

Materials and Methods

Human umbilical vein endothelial cells (EC) were prepared according to the procedure of Gimbrone et al. (5). and were used between the 3rd and 6th passages. Human saphenous vein smooth muscle cells (SM) were a gift of Dr. Robert Kaner, Sloan-Kettering Memorial Hospital. Culture media were purchased from Gibco, Grand Island, N.Y. Cellmatrix 1A, a sterile preparation of acid extracted porcine dermal collagen in 1mM HCl (3mg/mL), was purchased from Nitta Gelatin Inc., Osaka, Japan.

The culture medium used in these studies was M199 supplemented with HEPES, crude endothelial cell growth factor (ECGF) prepared according to Maciag et al. (6), glutamine,

Angiogenesis in Health and Disease, Edited by M.E. Maragoudakis *et al.*
Plenum Press, New York, 1992

heparin (90ug/mL), 20% fetal bovine serum and gentamycin. Collagen lattices were prepared by modifications of the procedures of Sarber et al. (7). For every 10 volumes of lattice, 5 volumes of Cellmatrix 1A were mixed at 0° C with 2 volumes of 5X M199, 2 volumes of fetal calf serum and 1 volume of either sterile, 1X M199 or cells suspended in M199 at the appropriate density. The mixture was poured into sterile 12 well Costar culture dishes which were then placed in a 37°C incubator. Thermal gelation was complete within 5-10 minutes. For these studies, EC were always seeded on top of preformed lattices, whereas, SM were seeded within the lattices, either uniformly suspended as described above, or in a "sandwich" culture in which the SM were seeded onto a preformed lattice, allowed to adhere for 2 hours, after which a second lattice was formed over the cell layer. Composite cultures involving both EC and SM were formed by seeding EC onto collagen gels containing SM either in uniform or sandwich culture. Cells (EC or SM) were seeded at a density of 10^5 cells per well. The total volume of the collagen lattices in every well was approximately 2 mL. For sandwich cultures, the SM were seeded between two 1 mL collagen lattices. Importantly, the collagen lattices formed in this manner were not contracting lattices inasmuch as they were formed in standard culture dishes, to which the lattices adhere, rather than in bacteriologic dishes to which they do not. Media were replaced every 24 hours. Lattices were fixed after 3, 7 or 10 days in phosphate buffered, 4% formaldehyde at 4°C for 1 hour and stored in 70% ethanol prior to embedding in paraffin, sectioning and staining with hematoxylin-eosin.

RESULTS

The collagen lattices seeded either with EC or SM did not undergo any contraction whereas cells seeded into bacteriologic dishes contracted to a fraction of the original diameter of the well within 24 hours after seeding. As a result, the non-contracting lattices remained transparent for up to 30 days in culture and the cells could be continuously observed by phase contrast microscopy, whereas the contracting lattices became opaque within the first 48 hours in culture.

Smooth muscle cells seeded in collagen lattices as uniform cultures showed little tendency to replicate as evidenced by their appearance as mostly single cells after 3 or 7 days in culture (Figure 1 A,B). On the other hand, SM seeded as sandwich cultures grew in the plane of seeding to form confluent layers after 3 days in culture (Figure 1C) and then showed a tendency to form contiguous layers several cells thick (Figure 1D). Occasionally, cells were seen that migrated away from the cell layer (not shown). When this occurred, the cells were always single cells and showed no signs of cell-cell attachment.

EC seeded onto collagen lattices grew to confluence within 5-7 days whether in the presence or absence of SM in the lattices and maintained a "cobblestone" morphology (not shown). In composite cultures in which SM were seeded uniformly, the SM showed signs of replication after 7 days in culture, particularly those cells closest to the EC layer (Figure 2 A,B). After 10 days in culture, the SM seemed to form layers parallel to the EC layer and frequent cell-cell contacts were

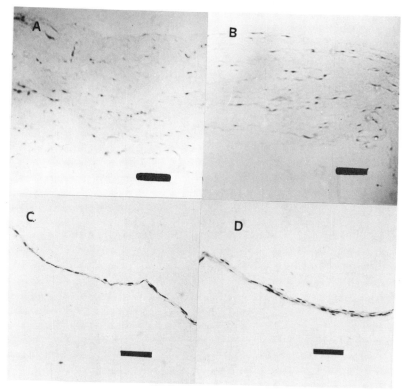

Figure 1. Appearance of smooth muscle cells seeded in 3-dimensional collagen lattices. Cells were seeded uniformly (A,B) or as a layer sandwiched between two collagen lattices (A,D). A,C, after 3 days in culture; B,D, after 7 days in culture Note the absence of colony formation in the uniform cultures and the growth to confluence with subsequent overgrowth after 7 days in the sandwich cultures. Bars = 100μ.

observed (Figure 2 C). In composite cultures in which the SM were seeded as sandwiches, the cells showed scant tendency to migrate away from the cell layer after 7 days in culture (Figure 3 A,B) but after 10 days, cells frequently were observed migrating away from the SM layer towards the EC layer (Figure 3C). No cells were found migrating in the opposite direction. Cells migrating usually were single cells and there were much fewer instances of cell-cell contact than observed in 10 day uniform cultures.

DISCUSSION

The 3-dimensional collagen lattice system described here has a number of unique features. First is the use of non-contracting lattices. This was achieved by using conventional cell culture dishes which has the apparent effect of anchoring the lattices, thereby preventing contraction from taking place. This probably created some tension within the lattices due to forces generated by the cells in their normal attempt to cause

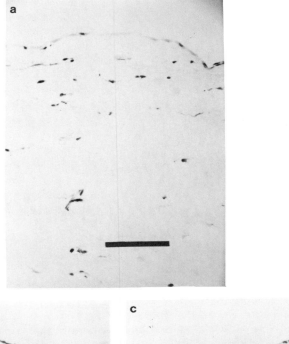

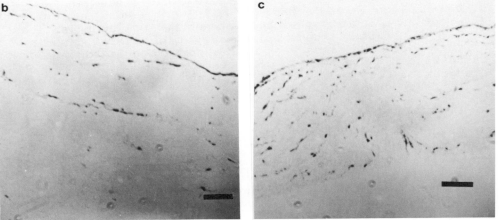

Figure 2. Appearance of composite cultures of EC and SM; uniform seeding of SM. a) 3 days after seeding; b) 7 days; c) 10 days. Note the beginning concentration of cells near the EC layer after 7 days and the formation of cell-cell contacts and a layered appearance after 10 days. Bar = 100μ.

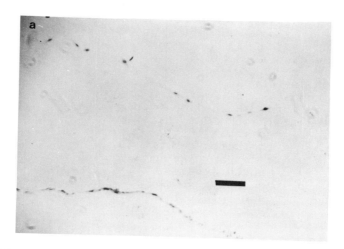

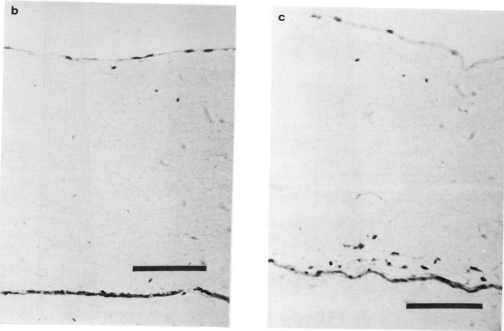

Figure 3. Appearance of composite cultures of EC and SM; seeding of SM in sandwich culture. a) 3 days after seeding; b) 7 days; c) 10 days. Note the absence of significant migration after 3 and 7 days and the migration of SM only towards the EC layer after 10 days in culture. Bar = 100μ.

shrinkage, as seen in lattices cast in bacteriologic dishes. However, the advantages of the transparency of the non-contracting system, the greater ease of constructing a composite system as well as the probability that, in vivo, vascular cells also are under a degree of tension, led us to adopt this as our working model. A second feature is the use of non-pepsinized collagen preparations. Preliminary studies (unpublished) indicated that dermal collagen preparations that had been extracted from the tissue after pepsinization caused changes in the morphology and growth patterns of both SM and EC when these cells were seeded onto preformed lattices whereas, the usual morphology, as seen with monolayers grown on gelatin coated dishes, were retained on preparations of unpepsinized, acid extracted collagen. The third feature is the apparent ability of SM to replicate in the presence of heparin, which has been reported to inhibit SM growth (8, 9). This difference is apparentlyly due to the presence of ECGF in these cultures. We have performed growth studies (unpublished) indicating that ECGF can reverse the inhibition of SM growth induced by heparin. The inclusion of ECGF was prompted by the intent to prepare composite cultures of EC and SM which necessitated the inclusion of ECGF in the medium.

A fourth feature is the lack of cell replication seen when cells were uniformly dispersed within the collagen lattice in the absence of other cell types. This is shown here for SM, but it seems to hold as well for fibroblasts and EC (1, 7). When, in the present studies, SM were seeded as sandwich cultures, the cells grew to confluence in the plane of the seeding and then proceeded to overgrow in a manner similar to the "hill and valley" patterns seen in SM monolayers on plastic surfaces. If EC were seeded within collagen gels as sandwich layers, we found that they adopted the morphology of "sprout" cells (not shown) and went on to form tubular structures as has been reported by Montesano et al. (10). It is possible that these effects are related to the symmetry of the distribution of cell attachment receptors on the cell surface. Under this hypothesis, uniformly dispersed cells can bind to the collagen lattice along their entire surfaces. On the other hand, cells seeded as sandwiches may concentrate their receptors at the basal surfaces of the cells. This may endow the cells with sufficient motility to undergo cell division and/or cause morphologic changes as the binding to the substratum triggers intracellular events. The stated observations that SM cells in sandwich culture grow in contiguous layers and are seen as isolated cells upon migration from these layers are consistent with this hypothesis.

Unlike what was found when SM only were seeded in collagen lattices, uniformly seeded SM grown in the presence of EC showed signs of directional proliferation and migration, particularly those cells situated close to the EC layer. Furthermore, cells in sandwich culture demonstrated a time dependent migration towards the EC layer. Together, these observations suggest that EC may exert a paracrine effect on SM replication and motility. It cannot be determined from the present studies, however, if these are direct effects of EC-derived factors on SM or are indirect effects in which EC factors stimulate the production of SM autocrine factors nor can it be determined if these observations are due to alterations, e.g. induced proteolysis, of the collagen lattice.

These studies indicate that the 3-dimensional collagen lattice system may provide a unique and useful model for the in vitro study of vascular cell biology. However, more work has to be done to study the effects of cell and matrix interactions as well as the interplay of exogenous factors.

ACKNOWLEDGEMENTS

The authors wish to express their appreciation to Dr. Robert Kaner for his donation of smooth muscle cells, to Dr. Peter S. Amenta for his assistance in the performance of the histological studies and to Won K. Han for his excellent technical assistance. This study was supported by a grant from the W.W. Smith Charitable Trust and by NIH grants AR-20553 and HL-29492.

REFERENCES

1. C. B. Weinberg and E. J. Bell, Regulation of proliferation of bovine aortic endothelial cells, smooth muscle cells and adventitial fibroblasts, _ J. Cell. Physiol. 122:410 (1985).

2. C. B. Weinberg and E. J. Bell, A blood vessel model constructed from collagen and cultured vascular cells, Science 231:397 (1986).

3. D. Gospodarowicz, I. Vlodavsky and N. Savion, The extracellular matrix and the control of proliferation of vascular endothelial and vascular smooth muscle cells, J. Supramolec. Struct. 13:339 (1980).

4. R. A. Majack, S. C. Cook and P. Bornstein, Control of smooth muscle cell growth by components of the extracellular matrix: autocrine role for thrombospondin, Proc. Natl. Acad. Sci. USA 83:9050 (1986).

5. M. A. Gimbrone, Jr., R.S. Cotran and J. Folkman, Human vascular endothelial cells: Growth and DNA synthesis, J. Cell Biol. 60:673 (1974).

6. T. Maciag, J. Cerundolo, S. Ilsley, P. R. Kelley and R. Foraand, An endothelial growth factor from bovine hypothalamus: Identification and partial characterization, Proc. Natl. Acad. Sci. USA 76:5674 (1979).

7. R. Sarber, B. Hull, C. Merrill, T. Soranno and E. Bell, Regulation of proliferation of fibroblasts of low and high population doubling levels grown in collagen lattices, Mech. Aging and Develop. 17:107 (1981).

8. J. J. Castellot, M. L. Addonizio, R. D. Rosenberg and M. J. Karnovsky, Cultured endothelial cells produce a heparin-like inhibitor of smooth muscle cell growth, J. Cell Biol. 90:372 (1980).

9. R. A. Majack and P. Bornstein, Heparin regulates the collagen phenotype of vascular smooth muscle cells. Induced synthesis of an Mr 60,000 collagen, J. Cell Biol. 100:613 (1985).

10. R. Montesano, L. Orci and P. Vassalli, In vitro rapid organization of endothelial cells into capillary-like networks is promoted by collagen matrices, J. Cell Biol. 97:1648 (1983).

EFFECT OF ISCHAEMIA ON ENDOTHELIAL CELLS

O. Hudlicka , S. Egginton and J.M. Dawson

Department of Physiology
University of Birmingham Medical School
Birmingham B15 2TJ, UK

INTRODUCTION

Acute ischaemia produces marked changes in capillary endothelium in skeletal muscle (Strock & Majno, 1969), in the heart (Kloner et al, 1974, Arminger & Gavin, 1975), brain (Chaing et al, 1968) and kidney (Johnston & Latta, 1977) which are characterized by cytoplasmic swelling, protrusions into the lumen, and decreased number of pinocytic vesicles. These changes are even more apparent during reperfusion injury (Gidlof et al 1988, Suval et al, 1987), and it is supposed that they are a cause of the no-reflow phenomenon after ischaemia. In addition to endothelial swelling, ischaemia increases leucocyte adherence to endothelium in postcapillary venules; leucocytes depletion attenuated vascular injury in postischaemic skeletal muscles (Korthuis et al, 1988). These ultrastructural changes are usually accompanied by an increased permeability to macromolecules such as albumin or dextran (Korthuis et al, 1985, Suval et al, 1987).

Much less is known about the chronic effect of ischaemia on capillary endothelium. Sjostrom (1982) noticed some swollen capillary endothelial cells in muscle specimens from patients with peripheral vascular disease, while Shearman et al (1988) described microalbuminuria in patients with peripheral vascular diseases which was increased after physical exertion (Hickey et al, 1990) indicating an increased permeability in kidney glomeruli. Thus muscle ischaemia can have deleterious effect not only on capillaries in the affected muscles but also elsewhere in the body. This paper summarizes our evidence for the effects of chronic ischaemia on perfusion in muscles with limited blood supply, and capillary ultrastructure in affected, as well as muscles remote from the site of ischaemia.

METHODS AND RESULTS

Ischaemia was induced by unilateral ligation of the common iliac artery in Sprague Dawley rats. In some animals,the effect of ischaemia was accentuated by chronic electrical stimulation applied at 10 Hz for 10-15 min 7 times/day (Hudlicka et al, 1991). The tibialis anterior muscle was exposed for observation of microcirculation using reflected

Angiogenesis in Health and Disease, Edited by M.E. Maragoudakis *et al.*
Plenum Press, New York, 1992

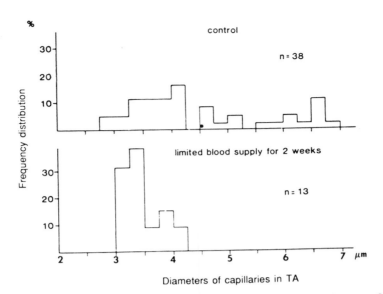

Fig.1.   Capillary diameters in the superficial layer of
         rat tibialis anterior muscle.

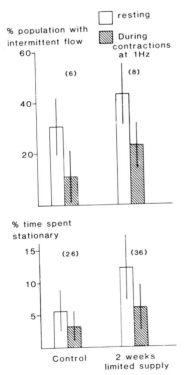

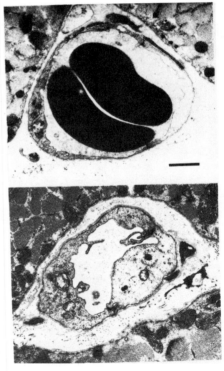

Fig.2.   Top: % of capillaries
         with intermittent flow
         bottom: % of time red
         blood cells spend
         stationary.

Fig.3.   Capillary from normal (top)
         and ischaemic stimuluted
         (bottom) rat extensor digit-
         orum longus Bar = 0.94μm

light (Dawson et al, 1987), at rest and during contractions. Adherence of leucocytes was assessed after labelling with acridine orange and observing their passage through postcapillary venules over a period of 1 minute (Dawson et al, 1990). Samples from unexposed muscles (extensor digitorum longus from ischaemic and contralateral leg, diaphragm and papillary muscles) were immersion fixed in buffered glutaraldehyde, resin embedded and viewed under electron microscope.

Most capillaries in normal muscles have red blood cells (RBC) present at all times. However, the velocity of RBC varies and flow is thus intermittent in up to 30% of capillaries. The proportion of time the red cells spend stationary is normally rather low, between 2-7 %. This proportion increased to 5-18% one week after ligation of the common iliac artery. The proportion of capillaries with intermittent flow and the proportion of time the RBC were not moving was still elevated 2 weeks after ligation,both at rest and during muscle contractions at 1 Hz (Fig.2). At this time, the capillary lumen diameters were significantly smaller (Fig.1). This may be due to swelling of capillary endothelium: 52% of capillaries had swollen endothelium over some part of circumference (Fig 3) two weeks after ligation, compared to 27 % in control muscles. Endothelial swelling was accentuated by chronic electrical stimulation and was then also observed in contralateral muscles (57%), diaphragm (62%) and papillary muscle (44%). Stimulation of ischaemic muscles also resulted in increased permeability for albumin and increased the relative adherence of leucocytes from 8% to 14 % (Hickey & Hudlicka, 1991).

DISCUSSION

As mentioned earlier, capillary swelling and increased permeability has been observed during ischaemia, and particularly during reperfusion, in many organs. The exact cause of these changes is not known. They may be due to oxygen free radicals, since perfusion with autologous blood equilibrated with nitrogen limited the increase in permeability (Korthuis et al, 1989). It is, however, unlikely that capillary damage produced in the present experiments would be due to the oxygen free radicals as it was present two weeks after limitation of the blood supply when total resting blood flow was almost normal (Hudlicka & Price, 1990). Morover, capillary swelling was observed in muscles which were never exposed to ischaemia.

It is thus more probable that ischaemia, particularly in conjunction with muscle contractions, produces some metabolic changes that can be transmitted to remote sites. One such change could be accumulation of adenosine which may facilitate calcium entry into endothelial cells (Rall 1985). This hypothesis is supported by the findings that increased capillary permeability and endothelial swelling during reperfusion could be reduced by calcium entry blockers (Paul et al, 1990); calcium entry blocker also decreased accumulation of leucocytes and improved perfusion in ischaemic and reperfused cardiac muscle(Rousseau et al, 1991). Calcium is also necessary for activation of leucocytes (Gallin & Rosenthal, 1974). There are a number of other substances that can modify the endothelial cell surface, produce endothelial-leucocyte adhesion molecules (ELAMs) and thus cause leucocyte adhesion and activation. These includeas cytokines (Benjamin et al, 1990), prostacyclins (Schade, 1989) or tumour necrosis factor (Vadas & Gamble, 1990). Activated leucocytes can then interact with endothelium elsewhere. Our preliminary results (Hickey & Hudlicka, 1991) support such a possibility, since increased leycocyte adherence and increased permeability was found in muscles contralateral to ischaemic stimulated ones.

REFERENCES

Arminger, L.C. & Gavin, J.B. Changes in the microvasculature of ischemic and infarcted myocardium. Lab. Invest. **33**, 51–56, 1975.

Benjamin, C., Dougas, I., Chi-Rosso, G., Luhowskyj, S., Goelz, S., McCarthy, K. & Cobb, R. A blocking monoclonal antibody to endothelial- leukocyte adhesion molecule 1. Biochem. Biophys. Res. Comm **171**, 348–353, 1990.

Chaing, J., Kowada, M., Ames, A III. Wright, R.L. & Majno, G. Cerebral Ischemia III. Vascular changes. Am. J. Pathol. **52**, 455–476, 1968.

Dawson, J.M., Tyler, K.R. & Hudlicka, O. A comparison of microcirculation in rat fast glycolytic and slow oxidative muscles at rest and during contractions. Microvasc. Res. **33**, 167–182, 1987.

Dawson, J.M., Okyauz-Baklouti, I. & Hudlicka, O. Skeletal muscle microcirculation: the effects of limited blood supply and treatment with torbafylline. Int. J. Microcirc. Clin. Exper **9**, 385–400, 1990.

Gallin, J.I. & Rosenthal, A.S. The regulatory role of divalent cations in human granulocyte chemotaxis: evidence for an association between calcium and microtubule assembly. J. Cell. Biol. **62**, 594–609, 1974.

Gidlof, A., Lewis, D.H. & Hammersen, F. The effect of prolonged total ischemia on the ultrastructure of human skeletal muscle capillaries. Int. J. Microcirc. Clin. Exper. **7**, 67–86, 1988.

Hickey, N.C. & Hudlicka, O. White cell mediated systemic vascular injury in intermittent claudication: evidence using a rat claudicant model. Int. J. Microcirc. Clin. Exper. **10**, 96, 1991.

Hickey, N.C., Shearman, C.P., Gosling, P. & Simms, M.H. Assessment of intermittent claudiaction by quantitation of exercise-induced microalbuminuria Eur. Vasc. Surg. **4**, 603–606, 1990.

Hudlicka, O. & Price, S. The role of blood flow and/or muscle hypoxia on capillary growth in chronically stimulated fast muscles. Pflugers Arch **417**, 67–72, 1990.

Hudlicka, O., Egginton, S., Brown, M.D. & Dawson, J.M. Blood flow, performance and capillary ultrastructure in ischaemic rat fast muscles: effect of chronic electrical stimulation. J. Physiol (London) **435**, 15P, 1991.

Johnston, W.H. & Latta, H. Glomerulal mesangial and endothelial cell swelling following temporary renal ischemia and its role in no-reflow phenomenon. Am. J. Pathol. **89**, 153–156, 1977.

Kloner, R.A., Ganote, C.E. & Jennings, R.B., The no-reflow phenomenon after temporary coronary occlusion in the dog. J. Clin. Invest **54**, 1496–1508, 1974.

Korthuis, R.J., Granger, D.N., Townsley, M.I. & Taylor, A.E. The role of oxygen-derived free radicals in ischemia induced increase in canine skeletal muscle vascular permeability Circ. Res. **57**, 599–609, 1985.

Korthuis, R.J., Grisham, M.B. & Granger, D.N., Leukocyte depletion attenuates vascular injury in postischemic skeletal muscle. Am. J. Physiol **254**, H823–H827, 1988.

Korthuis, R.J., Smith, K. & Carden, D.L., Hypoxic reperfusion attenuates postischemic vascular injury. Am. J. Physiol 256, H315–H319, 1989.

Paul, J., Bekker, A.Y. & Duran, W.N. Calcium entry blockade prevents leakage of macromolecules induced by ischemia–reperfusion in skeletal muscle. Circ. Ress. 66, 1636–1642, 1990.

Rall, T.W. Evolution of the mechanism of action of methylxanthines: from calcium mobilizers to antagonists of adenosine receptors. Pharmacologist 24, 277–287, 1985.

Rousseau, G., Stjaen, G., Latour, J.G., Merhi, Y., Nattel, S. & Waters, D. Diltiazem at reperfusion reduces neutrophil accumulation and infarct size in dogs with ischaemic myocardium. Cardiovasc. Res. 25, 319–329, 1991.

Schade, U.F. The role of prostacyclin in the protective effects of pentoxifylline and other xanthinbe derivatives in endotoxin action in mice. Eicosanoids 2, 183–188, 1989.

Shearman, C.P., Gosling, P., Gwynn, B.R. & Simms, M.H. Systemic effects associated with intermittent claudication. A model to study biochemical aspects of vascular disease. Eur. J. Vasc. Surg. 2, 401–404, 1988.

Sjostrom, M. Muscle fine structure and less advanced ischaemia. In Induced skeletal muscle ischaemia in man. Ed. D. Lewis. Karger, Basel,
p 80–100, 1982.

Strock, P.E. & Majno, G. Microvascular changes in acutely ischemic rat muscle. Surg. Gyn. Obst. 129, 1215–1224, 1969.

Suval, W.D., Duran, W.N., Boric, M.P., Hobson, R.W., Berensen, P.B. & Ritter, A.B. Microvascular transport and endothelial cell alterations preceeding skeletal muscle damage in ischemia and reperfusion injury. Am. J. Surg. 154, 211–218, 1987.

Vadas, M.A. & Gamble, J.R. Regulation of the adhesion of neutrophils to endothelium. Biochem Pharmacol. 40, 1683–1688, 1990.

ANGIOGENESIS AND NEOPLASTIC TRANSFORMATION

Pietro M. Gullino

Dept. Biomedical Sciences
Via Santena 7
University of Torino, ITALY

The major symptom of the neoplastic diseases is the unrestrained growth of cell populations with formation in most cases of solid tumors. Passage of cells into circulation and metastases formation completes the lethal outcome of the disease. When an agglomerate of cells is being formed in vivo, the increment in volume is conditioned by the supply system that sustains the metabolic needs. Thus, tumor formation requires not only the presence of a neoplastic cell population but also the participation of the host tissues supplying the vascular network. If this does not occur, the formation or growth of a solid tumor does not occur either. The validity of this conclusion was demonstrated several years ago[1].

Neoplastic cells derived from V2 carcinoma were implanted in the rabbit's eye at the center of the vitreous body. A nodule about 1 mm in diameter was formed within a few days, then growth ceased and the animal had no clinical symptoms for weeks. However, if this small nodule drifted toward the retina, newly formed vessels developed from the retinal network to colonize the neoplastic cell population. Explosive growth ensued and a solid tumor destroyed the eye within a week. Similar data were obtained with a glioma[2].

A reasonable interpretation of these observations was that neoplastic cells produced "angiogenesis factors" able to influence capillary endothelium if released at a sufficiently close distance from microvessels.

Neoplastic cell populations release molecules able to stimulate angiogenesis into the surroundings

The following experiments were performed to demonstrate the validity of the above hypothesis. A micropore chamber was built cutting a 3-4 mm-thick ring from a standard polystirene tubing. Two millipore filters (0.45 μ pore size) were sealed to the opposite surfaces of the ring to form a round chamber with a cavity containing about 100 μl of fluid. A catheter forced into a hole drilled through the polystyrene ring permitted withdrawal of the fluid from the chamber[3].

In the same animal two micropore chambers were implanted in the subcutaneous tissue, one in the interscapular, the other in the sacral region. 5-6 million neoplastic cells were implanted around one chamber. Within 8-10 days a tumor grew to envelope the chamber and a model was available to analyze and compare the difference in composition between interstitial fluid sampled from the interscapular subcutaneous tissue and that obtained from the tumor. Indeed, the micropore chamber introduced in the subcutaneous pouch is rapidly surrounded by normal or neoplastic cells. The pore size of the filter keeps the cells outside the chamber's cavity but the fluid surrounding the cells penetrates within the chamber and is sampled. The chamber does not induce formation of granulation tissue and the fluid sampled is believed to reflect the situation present in vivo since large molecules of dextran (350000 MW) present in the environment can pass through the filter and be sampled[3,4].

The tumor interstitial fluid (TIF) was lyophilized, the powder (0.1-1.0 mg) was compressed into a pellet about 1 mm in diameter and implanted within the cornea. TIF was found to be strongly angiogenic but also inflammatory. However, when TIF was treated with equal amounts of 95% ethanol, the lyophilized supernatant was angiogenic but not inflammatory[5]. This observation is interpreted to show that in vivo neoplastic cells discharge within the surrounding fluid molecules that are soluble in ethanol and have angiogenic capacity.

Several authors have utilized tumor extracts to demonstrate that neoplastic tissues contain angiogenic molecules but usually the nature of these molecules was not clearly established following extraction procedures[6,7,8,9]. In our work we had observed that TIF contained about twice as much type E prostaglandins as those present in the normal subcutaneous interstitial fluid (about 1.0 versus 0.4 ng/ml)[5]. In a study on angiogenesis occuring in ophthalmological diseases, Ben Ezra demonstrated that prostaglandin E1 (PGE1) was especially neovasculogenic[10].

We confirmed the work of Ben Ezra and established as a routine procedure the induction of a corneal angiogenic response utilizing PGE1 as an angiogenesis factor[5] delivered with a slow-release pellet according to Langer and Folkman[11]. In our experience PGE1 and PGE2 are both angiogenic in the rabbit cornea, but 250 ng of PGE1 incorporated with a 1 mm pellet of ethylene – vinyl acetate copolymer[11] induced angiogenesis in about 70% of implants versus about 10% of implants for the same dose of PGE2. The release of prostaglandins by the copolymer pellet as we prepare it, is in the order of a few ng/h. Under these conditions PGE1 appears to be more effective than PGE2 as an angiogenesis trigger. Whether the increased content of type E prostaglandins formed in TIF is sufficient to explain the angiogenic capacity of this fluid remains an open question. Two points should be stressed here: (1) Lyophilized TIF is angiogenic, and this can be taken as evidence that in vivo neoplastic cells release angiogenic molecules in the surrounding fluid, and (2) PGE1 is a reliable trigger of an angiogenic response that, under proper conditions of dosage and delivery, occurs without an inflammatory response.

The angiogenic capacity is acquired by a cell population in the course of neoplastic transformation

Since neoplastic tissues release angiogenic molecules in the interstitial fluid in vivo, the acquisition of angiogenic capacity should be one new property that reveals progression toward neoplastic transformation. Two approaches were followed to test this hypothesis.

Fibroblasts were removed from the subcutaneous tissue of 6 week old Balb/c male mice and cultured in DMEM + 10% fetal calf serum. The first 3 passages of these implants were not angiogenic in the rabbit cornea. At passage 4 some cultures induced angiogenesis and by passage 10 all cultures were angiogenic. None of the fibroblast cultures of passage 10 were able to produce a sarcoma when injected subcutaneously in syngenic mice; tumors were obtained only after about 20 passages[12].

This experimental approach was based on the well known fact that mouse fibroblasts cultured for many passages acquire the ability to produce a sarcoma. The data obtained indicate that during "spontaneous" neoplastic transformation in vitro the fibroblasts gained angiogenic capacity long before they were able to produce a tumor upon transplantation.

When the experiment was repeated utilizing human mammary epithelium cultivated in vitro, basically the same results were obtained i.e. angiogenic capacity was present in cultures unable to produce a tumor when injected into athymic nude mice[12].

The relationship between acquisition of angiogenic capacity and neoplastic behaviour was also tested with an in vivo experiment. Several years ago, Brand and co-workers demonstrated that plastic coverslips made of rigid, unplasticized vinyl chloride acetate produced sarcomas when implanted subcutaneously in adult CBA mice[13,14]. In these experiments the pertinent observation for our purpose was the relationship between time of sarcoma initiation and size of the plastic coverslip. Large coverslips (22 x 15 x 0.2 mm) left in situ for the life span of the animal produced sarcomas in almost 100% of mice by the 16th month after S.C. implant. Identical coverslips but one half the size or less, still produced sarcomas when left in situ, but the frequency of tumors was only about one third, and the first sarcomas appeared only after 15 months or later. In fact about 2/3 of the animals died of old age or intercurrent diseases before any sarcoma was evident. These observations suggested that the cell population surrounding the large coverslip was at higher risk of neoplastic transformation as compared to the population surrounding the smaller coverslip.

In our experiment two coverslips were implanted in the subcutaneous tissue of the same CBA mouse, one in the right, the other in the left lumbar region. After an interval between 1 and 16 weeks, both coverslips were removed, the plastic was cut into pieces of about 1 mm^2 each, and one piece of the large coverslip and one piece of the small coverslip were implanted respectively in the right and left cornea of the same rabbit. The objective was to compare the angiogenic response induced by the cell population attached to the plastic fragments derived from the coverslips of different sizes.

We found that the plastic material was not angiogenic, the cells attached to the fragments from large coverslips were angiogenic in 50 out of 117 implants (43%) and the cells adhering to the fragments from small coverslips were angiogenic in only 11 out of 117 implants (9%)[15]. At the time the plastic fragments were removed, no tumor was present nor expected to develop because the interval between plastic implantation and removal was too short. Therefore the results indicate that the cell population attached to the larger coverslips and able to form sarcomas more rapidly and in larger numbers than cells attached to the smaller coverslips also showed a higher frequency of angiogenic response.

If the experiments in vitro and in vivo reported above indicated that in the course of neoplastic transformation the angiogenic capacity appears before a cell population becomes capable of forming a tumor, one can hypothesize that detection of angiogenic capacity in a tissue normally deprived of it, suggests that the risk for the tissue of becoming neoplastically transformed is increased.

Angiogenic capacity and neoplastic tranformation

As it became clear that a neoplastic tissue is usually angiogenic while a normal tissue is not, the acquisition of angiogenic capacity by a tissue normally deprived of it could be regarded as a marker of an increased risk. We utilized the mammary tissue to analyze this possibility because the gland offers experimental models of "preneoplastic conditions" that permit evaluating a possible correlation in vivo between neoplastic transformation and appearance of angiogenic capacity.

C3HA^{vy} is a strain of mice characterized by high frequency of "hyperplastic alveolar nodules" in the mammary tissue[16]. These nodules are constituted by mammary epithelium that still maintains a regular structure and does not present signs of neoplastic transformation. Transplantation of these nodules in the fat pad of the mammary gland sometimes originates a tissue morphologically similar to the mammary gland but prone to neoplastic transformation. Medina[17] developed two lines of transplants (D1 and D2) characterized by an identical morphology but a different frequency of neoplastic transformation. D1 originated mammary carcinomas in less than 5% of transplants but the D2 line about 45%. We tested the angiogenic capacity of both lines before any neoplastic transformation was evident and compared the results with normal resting or lactating mammary gland. We observed that only 4 out 63 (6%) corneal implants of normal resting mammary gland gave a positive angiogenic response while 11 out of 50 (22%) transplants of tissue from pregnant mice and 13 out of 49 (27%) implants from lactating mice showed angiogenic capacity. The D1 line with low potential for neoplastic transformation was angiogenic in 19 out of 59 implants (32%) while the D2 line with high frequency of neoplastic transformation gave an angiogenic response in 83 out of 109 implants (76%). Primary hyperplastic alveolar nodules taken at random from untreated virgin C3HA^{vy} females showed an angiogenic response in 7 out of 23 implants (30%). Mammary carcinomas developed within hyperplastic nodules were angiogenic in 89 out of 98 implants (90%)[18,19].

When the same experimental approach was followed for fragments of human mammary gland, the following results were obtained: normal lobules induced an angiogenic response of the rabbit cornea in 5 out of 146 transplants (3%), hyperplastic lobules 14 out of 50 (35%), fibrocystic disease 0 out of 96, fibroadenomas 0 out of 28, carcinomas in situ 8 out of 12 (66%), infiltrating carcinomas 41 out of 63 transplants (65%)[20].

On the whole the results obtained in the mouse and human mammary tissues support the generalization that "normal tissues" are usually not angiogenic but neoplastic tissues usually are.

Specimens classified as "hyperplastic" have a higher frequency of angiogenic response than the normal counterpart in both mice and humans. In mice where transplantation permits comparing the evolution of several explants, the high risk tissues showed a higher frequency of angiogenic response than low risk tissues. However, specimens from pregnant as well as lactating mice also showed an angiogenic response that was not substantially different from the hyperplastic tissue. Therefore, at least for the mammary gland, the assessment of angiogenic capacity as a probe to evaluate risk of neoplastic transformation is at this time not promising. This is particularly true if one considers that the methods available for measuring the angiogenic response are not quantitative. Consequently, one must add the limited precision of the methodology to the uncertainty of the interpretation of the results.

Concluding remarks

There is reasonably convincing evidence that neoplastic cell populations release in vivo angiogenic molecules which appear before the neoplastic transformation has reached the point of solid tumor formation.

The obvious deduction that appearance of angiogenic capacity in a tissue normally deprived of it should indicate increased risk of neoplastic transformation has two limitations under present conditions. One is that the capacity to elicit neovascularization is not exclusive to neoplastic tissues and the reasons why some normal or non neoplastic tissues may be occasionally angiogenic are unknown. The second limitation is due to present methodology that only permits evaluation of the angiogenic response in a descriptive manner. There is, however one observation that suggests a possibility of utilizing the angiogenic response as a diagnostic tool. Fluids in contact with tumor surfaces such as urine for bladder carcinomas[21] or aqueous humor for eye retinoblastomas[22] may acquire angiogenic capacity and therefore may reveal presence or regrowth of the neoplasia.

References

1. S. Brem, The role of vascular proliferation in the growth of brain tumors, Clinical Neurosurgery, 23:440-453 (1976).

2. S. Brem, H. Brem, J. Folkman, D. Finkelstein and A. Patz, Prolonged tumor dormancy by prevention of neovascularization in the vitreous, Cancer Res., 36:2807-2812 (1976).

3. P. M. Gullino, S. H. Clark and F. H. Grantham, The interstitial fluid of solid tumors, Cancer Res., 24:780-798 (1964).

4. P. M. Gullino, F. H. Grantham and S. H. Smith, The interstitial water space of tumors, Cancer Res., 25:727-731 (1965).

5. M. Ziche, J. Jones and P. M. Gullino, Role of prostaglandin E1 and copper in angiogenesis, J. Natl. Cancer Inst., 69:475-482 (1982).

6. D. Tuan, S. Smith, J. Folkman and I. Merler, Isolation of the non histone proteins of rat Walker carcinoma 256 their association with tumor angiogenesis, Biochemistry, 12:3159-3165 (1973).

7. B. R. McAuslan and H. Hoffman, Endothelium stimulating factor from Walker carcinoma cells. Relation to tumor angiogenic factor, Exp. Cell Res. 119:181-190 (1979).

8. J. B. Weiss, R. A. Brown, S. Kumar and P. Phillips, An angiogenic factor isolated from tumors. A potent low molecular weight compound, Br. J. Cancer, 40:493-496 (1979).

9. P. M. Gullino, Angiogenesis factors, in "Handbook of Exp. Pharmacology", R. Baserga, ed., Springer Verlag Berlin, Vol. 57, pp. 427-449 (1981).

10. D. Ben-Ezra, Neovasculogenic ability of prostaglandins, growth factors and synthetic chemoattractants, Am. J. Ophthalmol., 86:455-461 (1978).

11. R. Langer and J. Folkman, Polymers for the sustained release of proteins and other macromolecules, Nature, 263:797-800 (1976).

12. M. Ziche and P. M. Gullino, Angiogenesis and neoplastic progression in vitro, J. Natl. Cancer Inst., 69:483-487 (1982).

13. G. K. Brand, Foreign body induced sarcomas, in: "Cancer: a comprehensive treatise," F. F. Becker ed., Plenium Publ. Corp., New York, Vol. 1, pp. 485-511 (1975).

14. G. K. Brand, L. C. Buoen and I. Brand, Carcinogenesis from polymer implants. New aspects from chromosomal and transplantation studies during premalignancy, J. Natl. Cancer Inst., 39:663-679 (1967).

15. M. Ziche and P. M. Gullino, Angiogenesis and prediction of sarcoma formation by plastic, Cancer Res., 41:5060-5063 (1981).

16. W. E. Heston and G. Vlahakis, Mammary tumors, plaques and hyperplastic alveolar nodules in various combinations of mouse inbred strains and the different lines of mammary tumor virus, Int. J. Cancer 36:2505-2610 (1976).

17. D. Medina, Preneoplastic lesions in mouse mammary tumorigenesis, in: "Methods of Cancer Research", H. Busch, ed., Academic Press, New York, Vol. 7, pp. 3-53 (1973).

18. M. Gimbrone and P. M. Gullino, Angiogenic capacity of preneoplastic lesions of murine mammary gland as a marker of neoplastic transformation, Cancer Res., 36:2611-2620 (1976).

19. M. Gimbrone and P. M. Gullino, Neovascularization induced by intraocular xenografts of normal, preneoplastic and neoplastic mouse mammary tissue, J. Natl. Cancer Inst. 56:305-318 (1976).

20. S. S. Brem, H. M. Jensen and P. M. Gullino, Angiogenesis as a marker of preneoplastic lesions of the human breast, Cancer, 41:239-244 (1978).

21. G. W. Chodak, C. J. Scheiner and B. R. Zetter, Urine from patients with transitional cell carcinoma stimulated migration of capillary endothelial cells, New Engl. J. Med., 305:869-874 (1981).

22. D. Tapper, R. Langer, A. R. Bellow and J. Folkman, Angiogenic capacity as a diagnostic marker for human eye tumors, Surgery, 86:36-40 (1979).

VASCULAR PROLIFERATION IN TUMOURS AND ITS INFLUENCE ON CANCER THERAPY

Juliana Denekamp

CRC Gray Laboratory
Northwood
Middlesex, U.K.

INTRODUCTION

The focus of most cancer research today relates to the alterations resulting from the malignant transformation of individual cells, and the key to new developments seems to lie in the subcellular technologies of molecular biology and gene expression. The rationale for the design of anti-cancer therapies therefore usually relates to perceived differences between normal and transformed cells. These differences may involve altered rates (e.g. of proliferation), or they may relate to expression of surface markers that are generally absent in adult normal tissues (e.g. growth factors and their receptors or surface antigens). Apparently in contrast, the concept of targeting the stromal support of the tumour requires an understanding of **systems**; of the morphology of vascular networks, the factors that control vascular tone, permeability, cardiac output, tissue distribution, blood viscosity, coagulation, tissue specific inflammation and cytokine mediated immunity. To obtain the complete picture of such a "system" we will need studies of the whole animal, as well as those at the macroscopic tissue, microscopic, cellular and molecular level. The main difference in philosophy with this approach is that we need to study the system as a whole and not focus on features of the individual transformed malignant cells (Denekamp 1982). Indeed it is the "activated" normal (vascular) cells that will be the target of this approach and the fact that activation has resulted from a stimulus provided by malignant cells may be of secondary importance. Some (though not necessarily all) features of vessels in tumours may be shared with the new vessels induced in other normal or non malignant pathophysiologies, e.g. in placenta, wound healing, arthritic or diabetic angiopathies. However, the tendency towards an increased coagulative status of the tumour-bearing host may be a specific effect of the type of tumour growing in that host (Clauss et al. 1990).

The breadth of the subjects covered in vascular targeting therefore provides an exciting challenge and the need for interaction between scientists and clinicians interested in many different disease states. If the general concept of targeting solid tumours by attacking their vessels can be developed, it is so important that it is well worth the intellectual efforts needed to cross the subject boundaries.

The strategy for design and screening of new agents, for scheduling of existing agents and particularly the sequencing of adjunctive therapies are likely to be completely

Angiogenesis in Health and Disease, Edited by M.E. Maragoudakis *et al.*
Plenum Press, New York, 1992

different for the "direct" tumour cell or "indirect" vascular-mediated approaches. It may be appropriate to <u>combine</u> vascular manipulation with direct cytotoxicity aimed at malignant cells but the two mechanisms must be recognised as distinct entities and considered separately before attempting to coordinate them. It is important therefore to identify the "hallmarks" of vascular mediated injury and the means by which this can be distinguished from direct cell kill (Table 1).

TUMOUR GROWTH AND ANGIOGENESIS

A few tumours can grow by invading normal tissues, e.g. leukaemic infiltration of liver, but solid masses only appear if the tumour cells have arisen (or settled) in a particular location and then forced the adjacent vascular network to expand in order to supply their needs. The production of factors that stimulate angiogenesis has been studied for many

Table 1. Hallmarks of vascular-mediated injury

Focal necrosis	This is inconsistent with random cell kill
Massive necrosis with individual cords of intact cells	Indicates individual vessels are functional or have failed
Haemorrhage	Evidence of vessel wall damage
Emboli	Evidence of coagulopathy
Isotope uptake ^{57}Cr RBC; ^{86}RbCl	Reductions in these values show reduced vascular volume or relative blood perfusion
Morphometry of vascular space	Reduced volumes can indicate closure of vessels of specific diameters
Greater effect *in vivo* than *in vitro*	Indicates a host-mediated effect that can represent immunity or vascular damage
Greater effect in vascularised than non-vascularised mini tumour spheroids	Indicates a host-mediated effect
Reduced cell survival after delayed excision for plating	Indicates rapid host mediated effect
More effect on solid than on ascitic or disseminated tumours	Indicates an effect that depends on structural organisation
More effect on large tumours than on small	May correlate with deteriorating (and hence more susceptible) vasculature as tumours enlarge
Toxicities involving permeability syndromes, altered coagulation patterns, blood pressure changes etc.	Indicates agent is influencing cardiovascular system

years, with lactic acid or other by-products of anaerobic glycolysis being considered initially as strong contenders for the angiogenic stimulus. In the early seventies, however, Folkman and colleagues (1971, 1974, 1983) showed that crude extracts of tumours could have mitogenic properties for endothelial cells *in vitro* and in other *in vivo* models. Their crude extract was termed TAF (Tumour Angiogenesis Factor). Many groups have since worked to purify components of the complex, or have extracted factors from other tissues or cell types. Some have been identified at the molecular level and produced as a cloned product, e.g. angiogenin, and others are still a chemical mystery. They cover a wide spectrum of molecular weights, from large molecules, e.g. the Fibroblast Growth Factors (MW ~ 20,000d) to the small Endothelial Cell Stimulating Angiogenic Factors (MW ~ 500d). Table 2 illustrates the wide spectrum of angiogenic agents, which may have independent actions or may be co-factors in a complex chain of events. A growing panel of anti-angiogenic agents is also being identified, with an equally wide spectrum of molecular structures. It is quite evident that angiogenesis is a complex response and it is perhaps not surprising that it can be blocked at many levels.

The concept of anti-angiogenic therapy, promoted by Folkman and others, relates to interfering with the stimulating substances that cause new vessel formation. This would be important in preventing establishment of solid tumours and may be most effective on small tumour masses, or in preventing metastases. By contrast, vascular targeting, i.e. attacking the existing neovasculature, is likely to be most effective on large tumours, where the vasculature is already compromised.

Table 2. Substances with effects on vascular proliferation

ANGIOGENIC ACTIVITY	ANTI-ANGIOGENIC ACTIVITY
aFGF	TGFβ
bFGF	α interferon
ESAF	TNF γ
TNF	TIMP 1
TAF	TIMP 2
Prostaglandins	Protamine
PGE1, PGE2	Cartilage extract
Angiotropin	Hydrocortisone
Angiogenin	Heparin
Angiotensin	Retinoids
Heparin	Methotrexate
Histamine	Mitoxantrone
Lactic Acid	Penicillamine
Leukotriene B4	Fumagillin

In order to initiate angiogenesis a capillary bud must be formed or the tumour must incorporate a nearby capillary by engulfing it. Budding involves three discrete events: localised dissolution of the basement membrane, migration of the endothelial cells towards the angiogenic stimulus, and cell proliferation to provide the necessary extra endothelial cells. It now seems likely that most of the identified angiogenic factors cause one or two but not all three of these responses. When the vessels have become incorporated in the tumour mass they must then elongate and branch and this requires proliferation of the constituent cells of the vessel wall.

VASCULAR PROLIFERATION

The thin-walled vessels that are present in tumours consist almost entirely of basement membrane with a single cell layer, i.e. endothelium. Studies with tritiated thymidine, which is a precursor of DNA and is incorporated prior to mitosis, have been performed on blood vessels in a wide range of tumours and normal tissues. These studies have shown that normal vessel endothelium is remarkably stable, with an uptake of the flash label into only one in a thousand or ten thousand cells. By contrast in tumour vessels, on average one in ten cells is labelled. This indicates a very different rate of cell production. Replacement of cells lost by wear and tear in mature vessels occurs very rarely (turnover time is months or years). Production is very fast in tumours, with a doubling of the cell number occurring in days. This is illustrated in Figure 1 for rodent tumours and normal tissues. Human tumours are very similar to the experimental tumours.

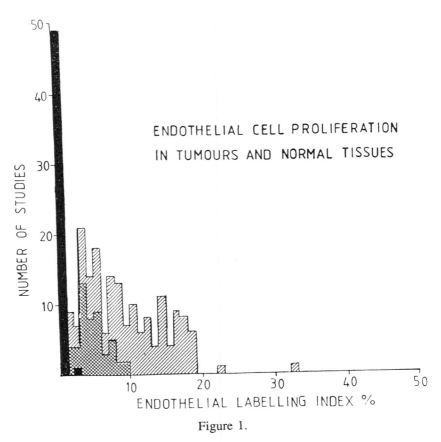

Figure 1.

Histogram summarising the studies of tritiated thymidine uptake into vascular endothelium. The solid black bar shows that almost all normal tissue values are below 1%. ☐ *= rodent tumours with volume doubling times of a few days.* ☐ *rodent tumours with volume doubling times of 10–14 days. Human tumour endothelial labelling indices range from 1–20%, i.e. similar to those in rodents.*

In 1982 we postulated that the massive difference in proliferation rates might allow antibody mediated therapies to be developed with antibodies raised against proliferating endothelial cell markers (Denekamp 1982). Subsequently we have realized that there may in addition be many other characteristics of the neovasculature that will allow agents to have a differential effect in tumours and normal tissues (Denekamp 1989). These are illustrated in Figure 2. It has long been realized that improving blood flow can increase radiosensitivity of tumours and can improve the delivery and effectiveness of cytotoxic drugs. However, decreasing the blood flow temporarily can enhance the response to hyperthermia, can trap drugs within tumours, can increase activation of drugs by reductive metabolism and can directly lead to ischaemic or haemorrhagic necrosis.

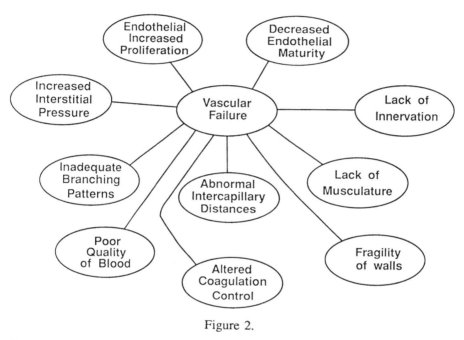

Figure 2.

In recent years, many agents have been shown to have a differential effect on tumour and normal vasculature, leading to a collapse of vessels or shut down of blood flow in tumours. This may be the main mechanism for a therapeutic benefit, rather than direct tumour cell kill, with many of these agents.

The newly formed vessels produce a chaotic disorganised network, with tortuous vessels, often sinusoidal in nature and always thin walled, traversing the tumour mass. It seems likely that the rare vessels seen in a tumour with a well developed multilayered wall are those that have been parasitized and engulfed by the expanding tumour mass. The branching pattern of the neovascular network is inadequate and intercapillary distances become so large that dead cells are sometimes visible between the sleeves or cords of tumour cells that surround each capillary. Figure 3 illustrates the diffusion gradient of nutrients. A rim of hypoxic (or anaerobic) cells is detectable with special stains at the boundary between the cord and the frankly necrotic debris (Hodgkiss et al. 1991). Such cords were demonstrated many years ago in human tumours before stains were available to identify the level of hypoxia (Thomlinson & Gray 1955).

VASCULAR INFLUENCES ON THE RESPONSE TO THERAPY

The nutrient gradient around each capillary has worried cancer therapists for many years because hypoxic cells are extremely resistant to radiation (requiring a threefold increase in dose to kill them). They would normally be doomed to die of their hypoxia if the tumour is left to grow. However if the tumour is treated, e.g. irradiated, they will receive an improved oxygen supply (i.e. be reoxygenated) as the sensitive cells near the vessels are killed. The distant cells are also resistant to cytotoxic drugs, for three reasons: 1) they are starved and therefore not actively engaged in proliferating and most chemotherapeutic drugs are preferentially toxic to cycling cells, 2) they are exposed to low drug doses because they are at such a great diffusion distance from the vessel, and 3) the interstitial pressure in tumours is high providing a further barrier to diffusion.

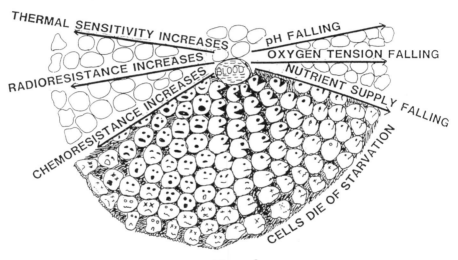

Figure 3.

In many tumours a cord of viable cells is detectable around each capillary, with necrosis at 6–10 cell diameters. Only when oxygen tensions fall to very low levels do the tumour cells become radioresistant. Nevertheless, 10% hypoxic cells can increase the dose needed to cure a tumour by a factor of 2.5. The same distant cells are resistant to chemotherapy, but may be sensitized to thermal therapy.

HYPERTHERMIA

Studies with a non conventional therapy, i.e. hyperthermia, in the late seventies and early eighties showed that surprisingly the hypoxic cells might be <u>more</u> sensitive to heat, both because of their nutrient deprivation and because of the lower pH that results from anaerobic glycolysis. However, a much more important step was the realization that the newly formed vessels were themselves much more prone to heat damage than normal vessels. In most animal tumours, vessel occlusion and/or a reduction in blood flow are seen after a 30 minute exposure to a moderate heat dose (42.5°–43°C) (Eddy 1980; Reinhold et al. 1990). By contrast, a much higher temperature of 46–47°C is needed to produce the same effect in normal vessels (Song 1984). Thus a large therapeutic window has been identified. Unfortunately this has not led to widespread clinical gain because the methods of administering deep heat are still a problem, and pain limits the temperature that can be achieved in practice in patients. It has also been shown in rodents that slow growing tumours need a higher heat dose to achieve a thermal response than fast growing tumours (Hill et al. 1989a), perhaps reflecting the proliferative state of the endothelial cells. The question therefore arises as to whether a higher temperature than 43°C will be needed in human tumours to lead to stasis.

The reason why vessel occlusion should occur at lower heat doses in tumours than in normal tissues is not clear, but rigidification of erythrocytes, adhesion of leukocytes to endothelium, and exceptional thermal sensitivity of proliferating endothelial cells have all been identified as possible mechanisms (Endrich et al. 1979; Fajardo et al. 1985).

The studies with heat were important because they illustrated that an agent may have been shown to cause direct tumour cell kill, and there may be a rationale for expecting it to be more effective on tumour than normal cells, but nevertheless the therapeutic benefit may result from something totally different, i.e. vascular-mediated cell starvation. Furthermore, it raises the question of whether the effectiveness of the agent might depend upon characteristics of rodent tumours which do not match those of humans, e.g. growth rate. A number of other forms of therapy, which were all developed as direct tumour cell killing agents have now also been shown to cause vascular shutdown or to induce the characteristic pattern of focal necrosis. Some of these are identified in Figure 2.

RADIOTHERAPY AND MODIFIERS OF RADIATION SENSITIVITY

Hypoxic cells, which exist because the excessive spacing of capillaries exceeds the oxygen diffusion gradient, give rise to the corded structure in Figure 3 with hypoxic cells dying at 100–150μm from the central vessels and appearing as necrotic islands (Thomlinson & Gray 1955). The emphasis in radiobiology has for decades been on the problems posed to the therapist by this inadequate vasculature. Little attention has been paid to the concept that vascular collapse would lead to an enhancement of the hypoxia-mediated killing. Indeed a process of "supervascularisation" has been described in shrinking tumours where the ratio of vessels to tumour cells appears to increase as the tumour cells are killed and resorbed. X-rays have, however, also been shown to cause vascular collapse, with vessel occlusion by thrombosis being described when the tumours started to regrow (Thomlinson & Craddock, 1967). Detailed morphometric studies have demonstrated that the occlusion is mainly in the narrower 10μm nutritive capillaries, with only small effects in vessels of 20–30 μm diameter (Solesvik et al. 1984). However, with radiation, the effect is small and unlikely to account for most of the observed tumour response to radiation.

Chemical radiosensitizers and radioprotectors have been developed which preferentially sensitize tumours because of their hypoxic cells or protect normal tissues because of differential uptake or critical levels of oxygenation. These are usually redox compounds, believed to interact with radiation-induced chemical species, i.e. free radicals. Misonidazole (a nitroimidazole) is the prime example of a tumour radiosensitizer whose radiosensitizing effect is clearly limited to hypoxic cells. In the presence of oxygen it completely loses its effect (Denekamp et al. 1974). In vitro studies with miso unexpectedly showed that it could be metabolised in hypoxic cells to a toxic product in the absence of radiation (Rose et al.

1980.). This was one of the earliest demonstrations that hypoxia could give the therapist an edge by converting a prodrug to a toxic metabolite. This has led to an expanding interest in the development of bioreductive drugs, which should be metabolised to a toxic product only within the hypoxic cells that exist in solid tumours. Mitomycin C is an existing cytotoxic drug in the clinic which also has such bioreductive activation. In spite of being developed to overcome the deficiencies of the tumour vasculature, the unexpected action of nitroimidazoles as bioreductive drugs has allowed a tumour-specific drug design approach to be developed based on the hypoxia resulting from the poor blood supply rather than on malignant versus normal cell differences.

A further unexpected "autocatalysis of bioreduction" was found when tumour blood flow reduction was observed in mice given miso alone, or in combination with the cytotoxic drug melphalan. An enhanced effectiveness of the two drugs had been reported, which had originally been attributed to independent subpopulations responding, i.e. the cycling cells close to vessels being killed by the antimetabolite melphalan and the distant hypoxic cells by the bioreductive activation of miso (Randhawa et al. 1984). However, miso was unexpectedly shown to cause a massive occlusion of the tumour vasculature, which lasted for 1–2 weeks (Murray et al. 1987). Thus it appears that the enhanced cytotoxicity of these agents when they are combined may not be simply due to direct cytotoxicity, but may also involve pharmacokinetic changes (trapping of the melphalan within the tumour when blood flow ceases), indirectly mediated ischaemic necrosis (resulting from starvation) and increased conversion of the prodrug miso to the toxic metabolite as the degree of hypoxia increases. These three vascular mediated effects were completely unexpected when miso was being developed as a radiosensitizer.

Thiol radioprotectors are assumed to protect by donating protons which inactivate damaging OH' radicals or restore damaged DNA. Ethiophos (WR 2721) is a phosphorothiol which was developed in the United States as a whole body radioprotective agent for military purposes. It was subsequently investigated as an adjunct to radiotherapy because it was shown to give much more of a radioprotective effect for bone marrow than for a rodent tumour (Yuhas & Storer, 1969). Radioprotection was subsequently observed in a range of normal tissues, particularly at short intervals after injection (Yuhas 1980). Studies using labelled Ethiophos showed rapid extensive uptake into normal tissues, and slower and lower uptake in tumours, but the cause of this difference was ill understood. Ethiophos is a phosphorylated prodrug, with the thiol group blocked by a phosphate. For activation, and indeed for transport into cells, the compound must be dephosphorylated. Yuhas suggested that the phosphatase activity might differ in tumour and normal cells. However, a reduced alkaline phosphatase activity in tumour endothelium now appears more likely to explain the effect. It could effectively prevent passage of ethiophos across the vessel wall, until a pool of the dephosphorylated compound, which is only generated in normal vessels, has saturated the normal tissues and provided a systemic reservoir of the diffusible product (Denekamp & Rojas 1988). Differential activation of other prodrugs based on differences in surface enzymes in mature and immature endothelium may be a productive path to investigate, e.g. for the activation of DNA repair enzymes in normal tissues after irradiation (Ward, 1986).

HAEMATOPORPHYRINS AND LIGHT

Photodynamic therapy depends upon the activation of a non toxic compound haematoporphyrin, by visible light to release toxic activated oxygen species which kill the cell in which it has been formed. These extremely reactive free radicals are unlikely to live long enough to diffuse to adjacent cells. It is evident under gross illumination that the compound, which is fluorescent, is retained preferentially in the tumour some 3–4 days after administration. At first, this observation was attributed to leaky tumour blood vessels, i.e. altered permeability (Bugelski et al. 1981), but recently it has been suggested that the retention is actually <u>within</u> the endothelial cells of the newly formed tumour blood vessels rather than extravascularly. Reduced blood flow and vascular destruction have both been demonstrated (Star et al. 1986) and delayed cell death consistent with ischaemia has been measured after photodynamic therapy (Henderson et al. 1985). The undesirable side effects of damage to normal tissues have also been shown to be vascular mediated. The reason for the putative endothelial localisation is totally unknown, but would indicate that differential endothelial retention must be an important parameter when screening for alternative compounds to replace the existing haematoporphyrin derivatives. At present the research emphasis in this field is almost entirely on development of more active compounds, or those responding to more penetrating wavelengths of light. Vascular endothelial cell uptake is unfortunately seldom monitored in the screening of such new agents.

CYTOTOXIC DRUGS

Most conventional cytotoxic drugs are believed to kill cells by causing damage to DNA, or by interfering with the normal processes of DNA replication or repair. Medicinal chemists aim to develop drugs which are freely diffusible through tissue to reach the tumour cells most distant from the capillaries. The transvascular and interstitial movement of cytotoxic agents may however be limited by the sinusoidal nature of many tumour blood vessels and the high interstitial pressures which result from poor lymphatic drainage (Jain 1991). If a drug is cycle specific a further problem is encountered because active proliferation is restricted to the first few cell layers around each vessel because of nutrient limitations. Furthermore flow through some vessels is arrested for many minutes or hours, which could lead to protected nests of surviving tumour cells. Thus the poor tumour vasculature creates problems in the effective use of cytotoxic drugs. Even so, significant tumour regression is seen in many solid tumours and it is pertinent to ask whether this is ever mediated by the drugs causing vascular mediated damage instead of direct tumour cell kill. There is a high rate of cell division of endothelial cells in all tumours. A tremendous avalanche of tumour cell death could occur for each occluded capillary resulting from cytotoxicity to the endothelial cells. Hence this could be an important target. However, very few studies of blood flow changes after chemotherapy have been performed, reflecting the paucity of interest in this possibility. In the study of melphalan and misonidazole described above (Murray et al. 1987) only a small effect on blood flow was seen when melphalan was used alone, compared with the much bigger effect from miso alone and an even bigger effect from the combination. Indirect evidence of a rapid, possibly host-mediated cytotoxicity is available from the studies of survival *in vitro* after excising tumours treated *in vivo* (Stephens & Peacock 1978). For most of the drugs tested, the yield of cells from tumours fell if they were harvested after 24 hours instead of immediately after drug administration. After the vinca alkaloids the cell yield fell to 25% of the original value. This is compatible with the rapid necrosis that is normally associated with vascular collapse.

An agent used to stabilize oxyhaemoglobin (BW12C) has also been shown to produce unexpected vascular shutdown in tumours (D. Honess, personal communication), as has Hoechst 33342, the bisbenzamide agent that is itself used as a functional vascular space marker (Smith et al. 1988). It is important to be aware of such effects as they may compromise the interpretation of tumour experiments. Fortunately they are usually limited to high drug doses and by carefully determining the safe dose, such agents can still be useful biological markers.

NOVEL CHEMOTHERAPY AGENTS

Two novel chemotherapeutic drugs have recently been taken into clinical trials, and have both proved disappointing compared with the dramatic responses that had been seen in a range of rodent tumours. These are the Lipha drug Flavone Acetic Acid and the Glaxo drug Phoquidone. Both of these have been shown to cause extensive haemorrhagic necrosis in mouse tumours, accompanied by a reduction in tumour blood flow.

Flavone acetic acid was developed originally as an anti-inflammatory agent, not as a cytotoxic drug. It has no antimetabolite activity and only weak DNA binding properties. Its action has many of the hallmarks of vascular mediated injury although these were not initially recognised as such. In NCI screens, it was found to be remarkably effective on all solid tumour models, but had no effect on lymphomas or leukaemias. It was ineffective when given immediately after tumour inoculation, but was most effective when given to large established (i.e. vascularised) solid tumours. It has little cytotoxic action *in vitro* and the effects *in vivo* are incompatible with this low *in vitro* toxicity, indicating that a host response must be involved. Tumour cell lines selected to be responsive or resistant to this agent in culture do not show a similar pattern *in vivo*. A rapid haemorrhagic necrosis is visible within a few hours of administering FAA and this is accompanied by, or preceded by a fall in tumour perfusion (Finlay et al. 1988). Decreased tumour perfusion has been detected within 15 minutes to 2 hours, depending on the assay technique and this persists for at least 24 hours. Hill et al. (1989b) have developed a panel of tumours showing a wide range of sensitivity to FAA. The FAA-sensitive tumours show extensive growth delay, or even local control, after single doses of the agent, whereas the unresponsive tumours show little delay even after double the drug dose. The extent of reduction in blood flow is correlated with the growth delay.

Recent studies by Zwi et al (1990) have demonstrated very convincingly that FAA action is largely vascular mediated but also has an immunological component. They have adapted an elegant model of intraperitoneal tumour spheroids. Dense aggregates of tumour cells are grown in culture to a size of 0.5 – 1.0 mm diameter. When injected intraperitoneally, these may exist as free floating non-vascularised spheroids or may attach to peritoneal surfaces and develop a vascular supply. Thus in the same animal the effect on similar sized tumours can be studied with and without a vascular network. The damage assessed histologically 18 hours after FAA was much more marked in the vascularised tumours, with a differential effect even being detectable in the vascularised and non-vascularised regions of the same spheroid. Even the non-vascularised spheroids showed much more effect than spheroids *in vitro* indicating the involvement of host cells.

An interesting feature of FAA is that it is much more toxic (by a factor of 2–3) when it is administered to tumour–bearing mice, in a manner that is related to tumour load and also to the tumour's sensitivity to FAA. The increased toxicity can be induced, though to a lesser extent, by other angiogenic stimuli, e.g. by surgery, or by the implantation of a surgical sponge. It therefore appears that a systemic effect is being induced both by the presence of a tumour, or by a simple network of immature blood vessels in granulation tissue. Changes in coagulation parameters have been demonstrated within a short time after drug injection. Initially the blood has a reduced clotting time, but after 24 hours this is reversed, perhaps indicating depletion of the clotting precursors (Murray et al. 1989). Detailed studies have indicated that the coagulation cascade is modulated by tissue factors produced by endothelial cells, and that these can be further altered by the presence of tumour cells, tipping the blood towards a procoagulant status. Evidence is also accumulating that this balance can be further tipped by the addition of TNF to the system. The effect of

FAA may even be mediated by immune effector cells, perhaps with leukotriene or TNF production forming part of a chain reaction (Fingar et al. 1991). The production of EDRF (nitrous oxide) is also implicated, since nitrates are detectable in the urine of patients treated with FAA (L. Thompsen - personal communication).

Flavone Acetic Acid should perhaps be regarded as a lead compound for future anti-vascular therapies. It has certainly indicated some potential new modes of vascular attack. However, it has been universally disappointing in the early clinical trials, in marked contrast to the dramatic effects seen in most rodent models. The early clinical studies showed that the main limiting side effect was severe hypotension, and also that decreased platelet aggregability was observed after 1–2 days. In order to minimise the troublesome hypotension, to maximise the chance of tumour cytotoxicity, and to avoid possible crystallization of FAA in the kidney tubules, a protracted clinical schedule was developed. This seemed to differ in only minor ways relative to the acute bolus administration in mice, but it may be the key to the disappointing clinical results. It consisted of slow infusion over 6–8 hours in patients pre-treated with bicarbonate to avoid potential crystallisation in the kidney tubules. If similar schedules are used in mice the dramatic tumour response is either abolished or grossly reduced.

BIOLOGICAL RESPONSE MODIFIERS

The cytokines themselves, especially interferon, the interleukins and TNF have also excited a lot of interest as cancer therapeutic agents. Again their use in rodent tumours gave great promise which has unfortunately not been realized with the schedules used in the clinic. Regional patchy necrosis was identified as the pattern of tumour response to endotoxin, TNF and interferon. A variety of studies have shown microvascular damage, changes in blood flow and/or altered responses to angiogenic stimuli with these and other cytokines (e.g. Bevilacqua et al. 1984; MacPherson & North 1986; Sidky & Borden 1987; Kotasek et al. 1988; Dvorak & Gresser 1989).

These effects are likely to be mediated by changes in endothelial <u>function</u>, e.g. in arachadonic acid metabolism, the control of coagulation by tissue factor and factor VIII production, or platelet or neutrophil adhesion, rather than by endothelial cell kill. It is clear that for a further understanding of these mechanisms a study of growth factors, cytokines, coagulation control and of adhesion molecule expression will be needed. Some of these features can be studied in single cell types grown in culture but the complex biochemical conversations between cells of different types, e.g. tumour cell, endothelial cell and macrophage, will need the *in vivo* situation for their complete expression.

FUTURE DIRECTIONS

The appeal of vascular-mediated ischaemic therapy is that it will be more effective on large tumours than small, but it will have the potential of being effective on any tumour with a newly evoked vascular network, i.e. of about 1mm in diameter. The studies with many different anticancer agents have illustrated the potential complexity of responses that can appear to cause tumour cell death by collapse or occlusion of the blood supply. They have also focused attention on features of disparate agents, e.g. TNF, FAA which may share similar pathways. No single feature of neovasculature can be highlighted as the route by which such antivascular therapy should be targeted (Figure 4). Rapid proliferation of the endothelial cells may prove to be a target, or may influence differentiation characteristics, so that the immature cells will function abnormally. The permeability of these poorly formed vessels may lead to extravasation of proteins (or erythrocytes), leading to increased interstitial pressures and by this means to an imbalance between intravascular and extra-vascular pressures and hence to collapse of the thin-walled vessels. Changes in systemic blood pressure, cardiac output, viscosity or coagulation and especially of regional perfusion would all have differential effects in tumours and normal vessels. Clearly both vascular patho-physiology and the complexity of endothelial cell function and its imbalance in neovasculature will be important in understanding the mechanism of action of anti–vascular strategies.

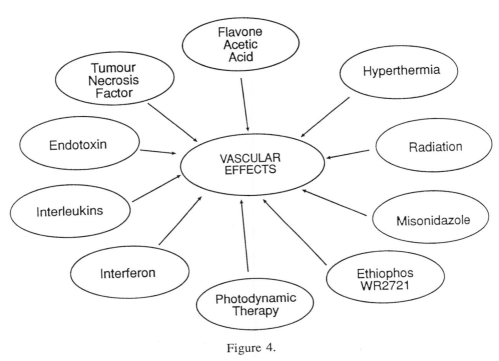

Figure 4.

Features of tumour vascular networks that differ from those in normal tissues, and which may lead to the observed differences in response to the therapeutic agents listed in Figure 2.

The body of evidence that vascular–mediated damage occurs after many existing forms of anti-tumour therapy is rapidly accumulating as illustrated at a recent Gray Conference (Moore & West 1991). Conventional screens of cells *in vitro* or using leukaemias or lymphomas will not detect this mode of action and may therefore miss effective agents. A change in the approach to experimental cancer therapy is needed to ensure that this important new avenue is fully investigated. The vascular patterns of different tumours may be critically important and these may change in different tumour sites, in transplanted versus spontaneous tumours and in xenografts versus established rodent models. The branching frequency of the vascular network will be critical in determining how many cells will die as a result of occlusion of a single capillary. The tumour cells' influence on endothelial cell function may be tumour specific. Absolute tumour size and the growth rate of tumour cells rather than micro–environmental parameters may be more important for these vascular mediated strategies than for direct cell kill. It is imperative that we assess whether the experimental models are appropriate in these respects. We have much to learn and it promises to be an exciting and challenging multidisciplinary frontier calling for cooperation and interchange between specialists from widely differing backgrounds. Conferences like this one are clearly very important steps in the right direction.

Acknowledgement: This work has been entirely supported by the Cancer Research Campaign.

REFERENCES

Belvilacqua, M.P., Pober, N.S., Majeau, G.R., Cotran, R.S., Gimbrone, M.A. Interleukin I (IL-I) induces biosynthesis and cell surface expression of procoagulant activity in human vascular endothelial cells. J.Exp.Med., 160, 618-623. 1984.

Bugelski, P.J., Porter, C.W., Dougherty, T.J. Autoradiographic distribution of hematoporphyrin derivative in normal and tumor tissue of the mouse. Cancer Res. 41, 4606-4612. 1981.

Clauss, M., Murray, J.C., Vianna, M., De Waal, R., Thurston, G., Nawroth, P., Gerlag, H., Gerlag, M., Bach, R., Familette, P. and Stern, D. A polypeptide factor produced by fibrosarcoma cells that induces tissue factor and enhances the procoagulant response of endothelium to tumour necrosis factor-cachectin. J.Biol.Chem. 265: 7078-7083. 1990.

Denekamp, J. Endothelial cell proliferation as a novel approach to targeting tumour therapy. Br.J.Cancer 45, 136-139. 1982.

Denekamp, J. Induced vascular collapse in tumours: a way of increasing the therapeutic gain in cancer therapy. Brit.J.Radiol. Report 19, 63-70. 1989.

Denekamp, J., Michael, B.D., Harris, S.R. Hypoxic cell radiosensitizers: comparative tests of some electron affinic compounds using epidermal cell survival in vivo. Radiat.Res. 60, 119-132. 1974.

Denekamp, J., Rojas, A. Radioprotection in vivo: cellular heterogeneity and fractionation. Anticarcinogenesis and Radiation Protection. Ed. P.A. Cerutti, O.F. Nygaard and M.G. Simic. Pub. Plenum, 421-430. 1988.

Dvorak, H.F., Gresser, I. Microvascular injury in pathogenesis of interferon-induced necrosis of subcutaneous tumors in mice. J.Nat.Canc.Inst. 81, 497-502. 1989.

Eddy, H.A. Alteration in tumour microcirculation during hyperthermia. Radiology 137: 515-521. 1980.

Endrich, B., Zweifach, B.W., Reinhold, H.S., Intaglietta, M. Quantitative studies of microcirculatory function in malignant tissue: Influence of temperature on microvascular hemodynamics during the early growth of the BA1112 rat sarcoma. Int.J.Radiat.Oncol.Biol.Phys. 5, 2021-2030. 1979.

Fajardo, L.F., Schreiber, A.B., Kelly, N.I., Hahn, G.M. et al. Thermal sensitivity of endothelial cells. Radiat.Res. 103, 276-285. 1985.

Fingar, V.H., Wieman, T.J., Doak, K.W. Mechanistic studies of PDT-induced vascular damage: evidence that ercosanoids mediate this process. Int.J.Radiat.Biol. 69: 303-309. 1991.

Finlay, G.J., Smith, G.P., Fray, L.M., Baguley, B.C. Effect of flavone acetic acid on Lewis Lung carcinoma: evidence for an indirect effect. J.N.C.I. 80, 241-245. 1988.

Folkman, J. Merler, E., Abernathy, C., Williams, G. Isolation of a tumour factor responsible for angiogenesis. J.Exp.Med. 133: 275-288. 1971.

Folkman, J. Tumor angiogenesis factor. Cancer Research 34: 2109-2113. 1974.

Folkman, J., Langer, R., Linhardt, R.J., Haudenschild, C., Taylor, S. Angiogenesis inhibition and tumor regression caused by heparin or a heparin fragment in the presence of cortisone. Science 221: 719-725. 1983.

Henderson, B.W., Waldow, S.M., Mang, T.S., Potter, W.R., Malone, P.P., Dougherty, T.J. Tumor destruction and kinetics of tumor cell death in two experimental mouse tumors following photodynamic therapy. Cancer Res. 45, 572-576. 1985.

Hill, S.A., Smith, K.A., Denekamp, J. Reduced thermal sensitivity of the vasculature in a slowly growing tumour. Int.J.Hyperth. 5, 359-370. 1989(a).

Hill, S.A., Williams, K.B., Denekamp, J. Vascular collapse after flavone acetic acid: a possible mechanism of its anti-tumour action. Europ.J.Cancer Clin.Oncol. 25l 1419-1424. 1989(b).

Hodgkiss, R.J., Jones, G., Long, A., Parrick, J., Smith, K.A., Stratford, M.R.L., Wilson, G.D. Flow cytometric evaluation of hypoxic cells in solid experimental tumours using fluorescence immuno detection. Brit.J.Cancer 63: 119-125. 1991.

Jain, R.K. Haemodynamic and transport barriers to the treatment of solid tumours. Int.J.Radiat.Biol. 60: 85-100. 1991.

Kotasek, D., Vercellotti, G.M., Ochoa, A.C., Bach, F.H., White, J.G., Jacob, H.S. Mechanism of cultured endothelial injury induced by lymphokine activated killer cells. Cancer Res. 48, 5528-5532. 1988.

Macpherson, G.G., North, R.J. Endotoxin mediated necrosis and regression of established tumours in the mouse: a correlative study of quantitative changes in blood flow and ultrastructural morphology. Cancer Immun.Immunother. 21, 209-216. 1986.

Moore, J.V., West, C.M.L. Vascular function and tissue injury in murine skin following hyperthermia and photodynamic therapy, alone and in combination. Radiat.Res. (submitted). 1990.

Moore, J.V., West, D.C. Vasculature as a target for anti-cancer therapy. Int.J.Radiat.Biol. 60: 1-421. (Proceedings of 16 L.H. Gray Conference). 1991.

Murray, J.C., Randhawa, V.S., Denekamp., J. The effects of melphalan and misonidazole on the vasculature of a murine sarcoma. Brit.J.Cancer 55, 233-238. 1987.

Murray, J.C. Flavone acetic acid induces a coagulopathy in mice. Brit.J.Cancer 60: 729-733. 1989.

Randhawa, V.S., Stewart, F.A., Denekamp, J., Stratford, M.R.L.S. Factors influencing the chemosensitization of melphalan by misonidazole. Brit.J.Cancer, 51: 219–228. 1984.

Reinhold, H.S., Zurcher, C., Van den Berg-Blok, A.E. Differential heat sensitivity of tumour microvasculature. Europ.J.Cancer 26: 541-543. 1990.

Rose, C.M., Millar, J.L., Peacock, J.H., Phelps, T.A., Stephens, T.C. Differential enhancement of melphalan cytotoxicity in tumor and normal tissue by misonidazole. Radiation Sensitizers; their use in teh clinical management of cancer. 250-257. Ed. L.W. Brady, Pub. Masson USA. 1980.

Sidky, Y.A., Borden, E.C. Inhibition of angiogenesis by interferons: effects on tumor-and lymphocyte-induced vascular responses. Cancer Res. 47, 5155-61. 1987.

Smith, K.A., Hill, S.A., Begg, A.C., Denekamp, J. Validation of fluorescent dye Hoechst 33342 as a vascular space marker in tumours. Brit.J.Cancer 57: 247-253. 1988.

Solsevik, O.V., Rofstad, E.K., Brustad, T. Vascular changes in a human malignant melanoma xenograft following single-dose irradiation. Radiation Research 98, 115-128. 1984.

Song, C.W. Effect of local hyperthermia on blood flow and microenvironment: a review. Cancer Res. 44, 4721-4730. 1984.

Star, W.M., Marijnissen, H.P.A., Van Den Berg Blok, A.E., Versteeg, J.A.C., Franken, K.A.P., Reinhold, H.S. Destruction of rat mammary tumor and normal tissue microcirculation by hematoporphyrin derivative photoradiation observed in vivo in sandwich observation chambers. Cancer Res. 46, 2532-2540. 1986.

Stephens, T.C., Peacock, J.H. Cell yield and cell survival following chemotherapy of the B16 melanoma. Br.J.Cancer 38, 591-598. 1978.

Thomlinson, R.H., Craddock, E.A. The gross response of an experimental tumour to single doses of X-rays. Br.J.Cancer 21, 108-125. 1967.

Thomlinson, R.H., Gray, L.H. The histological structure of some human lung cancers and the possible implications for radiotherapy. Br.J.Cancer 9, 539-547. 1955.

Ward, J.F. Mechanisms of DNA repair and their potential modification for radiotherapy. Int.J.Radiat.Oncol.Biol.Phys 12: 1027-1032. 1986.

Yuhas, J.M. and Storer, J.B. Differential chemoprotection of normal and malignant tissues. J.Natl.Cancer Inst. 42: 331-335. 1969.

Yuhas, J.M. Active versus passive absorption kinetics as the basis for selective protection of normal tissues by S-2-(3-Aminopropylamino)-ethylphosphorothioic acid. Cancer Res. 40, 1519-1524. 1980.

Zwi, L.J., Baguley, B.C., Gavin, J.B., Wilson, W.R. The use of vascularised spheroids to investigate the action of flavone acetic acid on tumour blood vessels. Brit.J.Cancer 62: 231-237. 1990.

THE DEVELOPMENT OF THE TUMOUR VASCULAR SYSTEM: 2-D AND 3-D APPROACHES TO NETWORK FORMATION IN HUMAN XENOGRAFTED TUMOURS

M.A. Konerding*, C. van Ackern*, F. Steinberg**, and C. Streffer**

*Institut für Anatomie, **Institut für Medizinische Strahlenbiologie,
Universitätsklinikum Essen, Hufelandstr. 55, D - 4300 Essen 1
Phone: 0201/723-4285; FAX: 0201/723-5916

INTRODUCTION

Angiogenesis plays a significant role during normal growth and pathological conditions such as inflammation, wound healing and tumour growth. However, our knowledge of angiogenesis is not so extensive as justified by the significance of this research front. Thus, for example, the "International Symposium on Angiogenesis" 1991 in St. Gallen as well as the meeting "Angiogenesis in Health and Disease" of the NATO Advanced Study Institute 1991 in Porto Hydra (Greece), where the latest experimental results were presented and the current interdisciplinary research on angiogenesis was summarised, showed that basic questions still remain unresolved despite extensive and detailed knowledge in various fields.

Among others, these are concerned with the morphology of angiogenesis, the detailed 3-dimensional structure of sprouts as well as the process of sprout fusion and regression during primary and secondary angiogenesis. Furthermore, comparisons between primary and secondary angiogenesis as well as the various secondary angiogenic processes are lacking. Knowledge of these events also appears to be of particular importance in the case of tumour angiogenesis, since the tumour vascular system plays a decisive role both in the genesis and growth as well as in metastasis formation and therapy. This becomes particularly apparent in connection with the concept of "Vascular endothelium as the vulnerable element in tumours" (1) or other strategies on antiangiogenesis (1-5).

However, since the initial extensive studies on capillary sprouting by THOMA in the last century (6), the prevailing view is that all morphologically tangible events of primary and secondary angiogenesis are similar even in such differing processes as normal growth, wound healing, and chronic inflammations as well as tumor angiogenesis (5, 7). Apart from common characteristics Hammersen et al. (8) showed that there exist grave differences between de novo vessel formation in malignomas and following tissue trauma. In addition, numerous studies on tumour vessels are concerned with pattern formations, course and density but not with the structure in the morphological sense (8). This can also be applied to tumour angiogenesis, since such important processes as sprout formation have only been investigated with respect to morphology in exceptional cases (5).

This is also reflected in the literature: over the period from 1983 to 1991, roughly 900 papers on angiogenesis were to be found in the Medline data bank, but of those, only 8 % or 70 reports were mainly or at least partially concerned with morphological aspects. Of these, only 21 publications reported on the morphology of tumor angiogenesis. In this respect, it is interesting that numerous studies on angiogenesis were carried out on in vitro and not on suitable in vivo tumor models. Studies on primary tumors, as has been frequently suggested in the past (9), are also extensively missing.

Angiogenesis in Health and Disease, Edited by M.E. Maragoudakis *et al.*
Plenum Press, New York, 1992

With respect to the significance of angiogenesis in general and the tumour biological and therapeutic relevance in the case of tumour angiogenesis in particular, we have studied the vascular architecture and structure of xenotransplanted human tumours on the nude mouse.

METHODOLOGY

Five human tumor lines, an undifferentiated squamous carcinoma 4197, an amelanotic melanoma MeWo as well as three soft tissue sarcomas were xenotransplanted on 84 thymus-aplastic immune deficient nude mice as described in detail previously (10-12). The melanoma was in the 2nd, the squamous carcinoma in the 8th, while the soft tissue sarcomas were in the 3rd, 11th or 42nd passage. After the occurrence of macroscopic tumor growth, the animals were perfused and the tumor material removed. The melanomas were sequentially investigated on day 8, 12, 18, 25 and 79 and those of the squamous carcinoma on day 4, 7, 10, 15 and 23. The sarcoma-bearing animals were perfused on day 6 to 49 at intervals of 5-7 days. In order to obtain detailed results on the structure and architecture of the vascular system with particular focus on sprout and network formation, apart from light microscopy both scanning and transmission electron microscopy as well as microvascular corrosion casting were used as complementary methods. The methods used for tissue preparation and animal maintenance were previously described in detail (11-13).

RESULTS

1. Characteristics of the Tumour Vascular System

A close side by side of differentiated, newly formed and destroyed vessels can be seen at all times after transplantation due to the high turnover rate of the tumour vascular system. However, in the case of differentiated vessels we can only speak of a differentiation in the sense of *structural stabilisation* since differentiation characteristics in the usual sense, e.g., the formation of a 3-layered vessel wall in arteries or veins or formation of a continuous basement membrane cannot be demonstrated in tumour vessels as a rule. Hence, in the following, we will refer to these structurally stabilised vessels as "established vessels".

Apart from an extensive lacking of differentiated arteries and veins, the vascular system of xenotransplanted human tumours on the nude mouse differs from "normal" tissues in several structural and architectonic respects, which will be briefly described. The most apparent architectonic characteristic is the *lack of a normal hierarchisation* of vessels. A tumour vascular envelope can regularly be seen around the tumours, supplied by several host vessels predominantly of the venous type. No normal branching pattern can be observed even in this vascular envelope (fig. 1). No characteristic pattern formations of the terminal tissue vessels around the transplantation site are to be found. Instead, one finds variations in calibre and atypical branching patterns. From this plexus-like network, the vessels either surround the tumour in a circular manner or enter these centripetally. Here, too, no hierarchical order can be observed but, instead, an exceptional heterogeneous distribution of the vessels is evident (fig. 2). The vascular density varies considerably in the individual regions of each tumour.

Apart from these characteristics, the frequent occurrence of sinusoidal vascular networks arising from veins and equally ending in venous vessels is of particular importance for the hemodynamics and thus the nutritive supply of the tissue (fig. 3). In this case, it is particularly apparent that a high vascular density does not guarantee a sufficient blood flow at all. Since these sinusoidal vessels with diameters of 5 - 50 µm are usually fed by venous vessels and rarely by arterial ones, a diminished perfusion can occur even at intercapillary distances of 15 µm (fig. 4).

The structural characteristics of the established tumour vascular system (see table 1), e.g., the lack of contractile wall elements or the alternatingly expressed discontinuities with poor structural stability also show that the tumour vascular system clearly distinguishes itself from all other vascular systems as a result of angiogenic activity.

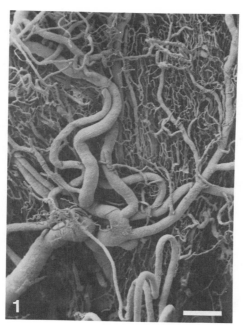

Fig. 1. Vascular corrosion cast specimen of a melanoma 12 days after transplantation. Peripheral vessels derived from subcutaneous vessels form a "tumour vascular envelope". Bar = 300 µm.

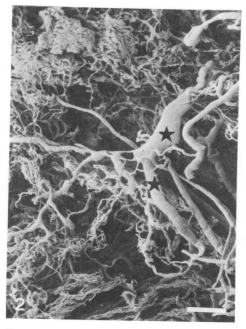

Fig. 2. Heterogeneous distribution of peripheral tumour vessels. Note the absence of vascular hierarchy, abnormal courses and flattening of main vessels (*). Bar = 300 µm.

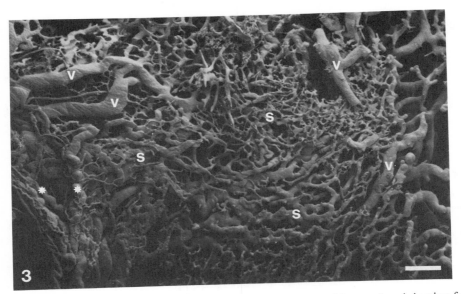

Fig. 3. Sinusoidal system (s) in a xenografted leiomyosarcoma with vessels originating from and draining into large calibre veins (v). Note the irregular vascular polymorphism, numerous blind ends, changes in diameter as well as tortuous vessel courses (*). Bar = 300 µm.

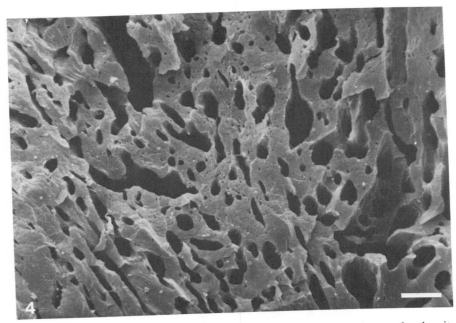

Fig. 4. SEM micrograph of a freeze-broken leiomyosarcoma with high vascular density in a necrotic area. The intercapillary distance is as low as 15 µm. Bar = 50 µm.

Table 1. Structural characteristics of the tumour vascular system

1.	capillary-like sinusoidal vessel wall construction
2.	flattening of the endothelium
3.	lack of differentiated arteries and veins
4.	low structural stability
5.	lack of contractile wall elements
6.	lack of pharmacological receptors
7.	varyingly expressed discontinuities
8.	low differentiation of contact structures
9.	incomplete formation of the basement membrane
10	varying formation of fenestrations
11.	irregular occurrence of pericytes
12.	lack of formation of lymphatic vessels.

2. Budding and Lumen Formation

The *sprout forming vessels*, without exception, are also representative of vessels with a capillary-like structure, independently of their diameter. The sprout forming endothelial cells as well as early forms are characterised by large amounts of endoplasmic reticulum, mitochondria and polyribosomes and the distinctly larger height of the endothelium. Two different types of de novo vessel formation can be distinguished by means of light and transmission electron microscopy and which occur independently of each other or in combination within a single sprout: sprout formation through intracellular or intercellular lumen formation.

In the first type, a primitive lumen in the form of a vacuole is formed intracellularly (fig. 5), which becomes enlarged and later makes contact with an established lumen or other endothelial cells. The vacuole, which forms the lumen later, stems from cisterna-like dilated sections of the endoplasmic reticulum or the Golgi-apparatus. However, it should be noted that not all intracellular vacuoles are indicative of an initial sprout formation; on the contrary, serial sections show that these vacuoles frequently only serve as lumen extensions. This form of sprout formation is frequently found in endothelial cells involved in the wall construction of established vessels thus facilitating the newly formed lumen to be connected to the already perfused blood vessel. However, it is also basically possible that intracellular lumina can arise in free endothelial cells which are not integrated into the wall of an established vessel and become associated with the perfused vascular network only after fusion with further vacuoles. In general this type of lumen formation is chiefly found in early tumour stages whereas the intercellular type occurs at all times.

Intercellular lumen formations arise through a protrusion or migration of neighbouring endothelial cells from early forms or established vessels. Here, more plump pseudopodia or the eccentrically situated perikaryon of the involved endothelial cells veer out from the endothelial structure leading to the formation of a slit like lumen (fig. 6). An intercellular lumen is thus already formed at the start of bud formation and not, as usually postulated, after formation of a solid endothelial cell cord. Initially, no continuous connection to the blood vessel can be seen (fig. 6). This can also be explained by the breadth of the lumina, often only 0.1 µm in size, which are frequently still attached to opposite cell membranes through point-like or cell-attached appositions and thus not yet traversable for cellular blood constituents.

An unusual type seen more rarely is the formation of lumina through protrusion of one or two cell ramifications with reflexive contact formation between these branches (fig. 5). 2) from the same cell that have developed contacts between themselves (small arrows). Fragments of the poorly developed basal lamina can be seen along the entire circumference (arrowheads). Pericytes are missing. Amelanotic melanoma MeWo 8 days after transplantation. Bar = 3 µm.

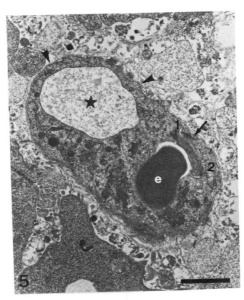

Fig. 5. Vascular sprout with advanced intracellular lumen formation (*). Note the high content of cell organelles and the height of the endothelium. The second lumen, which is occluded by an erythrocyte (e), was formed by an extensive reflexive overlapping of two ramifications (1,

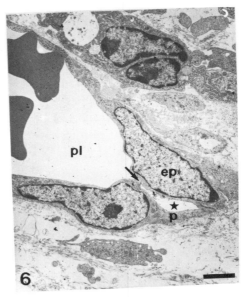

Fig. 6. Intercellular lumen formation and bud development through endothelial cell migration: the eccentrically situated perikaryon (ep) and a more plump pseudopodium (p) leading to the formation of a fine slit (*) veer out from the endothelial structure. Due to a punctiform contact (arrow) the newly formed lumen is not yet connected to the parent lumen (pl). Amelanotic melanoma MeWo 8 days after transplantation. Bar = 2 μm.

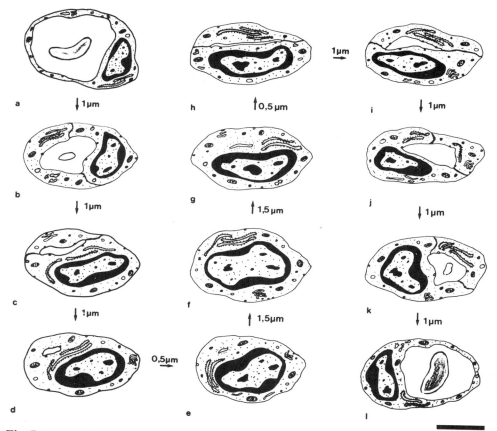

Fig. 7. Fusion of mixed sprouts with both inter- and intracellular lumen formation. Drawings of serial sections of a melanoma (MeWo) 8 days after transplantation. Numbers indicate the distance between the single section planes. **a:** already differentiated part of the sprout with flat endothelium. **b and c:** increased height of the endothelium in the mid part of the sprout. The lumen is confined by two cells. **c-d:** transition of intercellular lumen into an intracellular one. **e-f:** the intracellular vacuole becomes smaller until a lumen is no longer seen. **g:** again, formation of an intracellular vacuole. **g-h:** transition of an intracellular lumen into an intercellular one, confined by two endothelial cells. **h-k:** continuous decrease in endothelial height with simultaneous increase in lumen diameter. **l:** differentiated part of the sprout with flat endothelium. Compare the single sections with the reconstructions (fig. 8). Bar = 5 μm

Morphological characteristics of the newly formed vessels are the consistently higher and organelle-rich endothelium (fig. 5 und 6), which is already flattened in the early types but only shows the typical flattening of tumour vessels in the established forms. In general, the endothelium of sprouts and early forms is continuous and gaps only occur at later times. The source of cellular or non-cellular extravasates is thus seen more often in established or degenerated vessels. The cell contacts of the sprouts and early forms with vertical or obliquely angled appositions to the surface can be described as simply structured. Desmosomes or zonulae occludentes are not formed and overlapping of the cell boundaries or interlockings of the endothelia over longer stretches are only rarely found. These slight differentiation performances are also manifest in the construction of the basal lamina, which is seldom seen three-layered and continuous, in line with established vessels. Pericytic or subendothelial cells occur just as little which furthermore often cannot be seen to be representative of classical pericytes neither from their genesis nor their morphology.

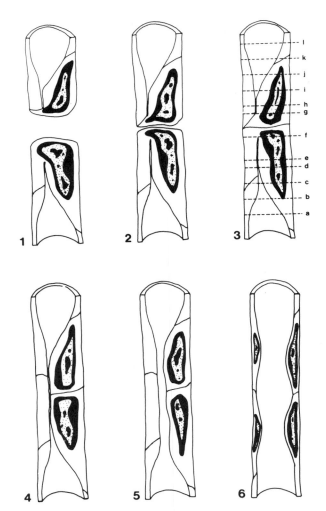

Fig. 8. Reconstruction of the course of sprout fusion. **1:** two sprouts grow towards each other. The lumen is restricted by two endothelial cells at the base of the bud while a single endothelial cell with intracellular lumen is located at the tip. After coming together (**2**) the tips meet vertically and fuse (**3**). The letters indicate the sectional heights from fig. 7. **4:** After fusion of the endothelial cells at the tip of the sprouts, the intracellular vacuoles form a uniform, continuous lumen. During the further course, this becomes wider with a decrease in endothelial height at the same time (**5, 6**).

3. Fusion of Vascular Sprouts

Vascular budding leads to the formation of new, nutritively effective vascular stretches only after connection to supplying vessels and perfusion. This connection can arise either through anastomosis with an established vessel or fusion with another vascular sprout. The latter type is more frequently found during tumour angiogenesis as observed in a series of semithin sections. However, no statistically evaluable data are available at present. This also applies to the frequency of connection of newly formed sprouts to the already perfused parts of the vascular system: the large number of degenerated sprouts early on, however, indicates that only a small fraction of vascular buds ever attaches to perfused stretches of vessels. Thus, a large "overproduction" of sprouts appears to exist.

Although convincing theories exist concerning the biochemical regulation of the coming together of sprouts for fusion purposes, the morphological "how" of fusion or anastomosis is still largely unresolved. Reconstructions from serial sections at the light and transmission electron microscopic level should yield information particularly on the mode of bud-fusion of sprouts with intracellularly formed lumina. It could be shown that endothelial cells with intracellular lumen are frequently to be found at the tip of multicelullar sprouts with already existing channeled intercellular lumen and function practically as "connective pieces" (fig. 7). Whereas several endothelial cell sections which form the lumen can be seen at the base of the sprout, only a solitary, cap-like endothelial cell with intracellular lumen is found at the tip. This can form contacts with the corresponding endothelial cell of the "partner" sprout after their coming together and thus facilitate a continuity of the lumen through fusion of the vacuoles (fig. 8). Whether a real fusion of the cell borders takes place or only contact structure formation occurs could not be convincingly demonstrated even in transmission electron microscopic sections up to now. Apart from that the described type of sprout construction and mode of fusion also explains why the sprouts even at the tip often show no signs of leakage.

The serial sections have shown that, on average, sprout fusion takes place over a distance of 10 - 15 µm (fig. 7, 8). This is, however, not to be confused with the length of newly formed stretches of vessels, which are considerably longer. However, since the endothelia involved in sprout formation differentiate very rapidly, only the cells at the tip of the sprout reveal their characteristic features: the endothelium becomes higher, and the slit-like lumina become more tighter.

4. Remodeling of the Tumour Vascular System

One major characteristic difference of tumour angiogenesis compared to other forms of secondary angiogenesis is that the reconstruction processes occur practically without higher differentiation of the vessels despite the high angiogenic activity. The basic pattern of the vascular architecture shows no great changes with time. Thus, the same architectonic characteristics can be observed at all times after transplantation, albeit to varying extents: glomeruloidal vascular convolutes in the periphery as a result of overshooting angiogenic activity ot tortuous courses of vessels are seen both at the 8th as well as 79th day after transplantation (figs. 9 a and b). This also applies to the architecture of sinusoids. Due to the tumour growth, there is a shift in the originally centrally located regions towards the periphery but the architecture remains unchanged (figs. 10 a and b). A genuine remodeling of the tumour vascular system does not take place, at least in xenotransplanted tumours.

DISCUSSION AND CONCLUSIONS

The central significance of angiogenesis in normal and pathological growth processes has called for numerous studies. In spite of this, however, there are comparatively few papers dealing with the morphology of tumor angiogenesis (e.g., 8, 14, 15, 16). Most have been concerned with the "why" of angiogenesis and paid less attention to the morphological "how". Literature reviews have also shown that numerous studies were carried out in vitro and not in vivo. This also applies to tumour angiogenesis. Investigations on human primary tumours are largely lacking. Since experimental tumours and, above all, endothelial cell cultures cannot be representative of the clinical situation, we have used the xenotransplanted human tumour on the nude mouse as the model in our studies. This model shows numerous parallelisms to the clinical primary tumour as described elsewhere (11, 13, 17, 18).

This also applies to the vascular system: Comparisons between xenografts and primary tumours have shown that most of the structural and architectonic characteristics of xenotransplants described by us also occur in primary tumours (19), although the stroma in the former stems from the host animal (20). Despite a tendency towards tumour-type specificity,

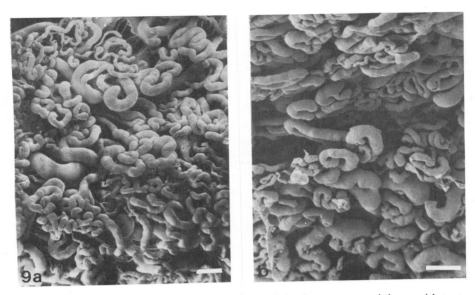

Fig. 9a and b. Glomeruloidal arrangement of vessels in the tumour periphery with tortuous courses and changes in diameter. Amelanotic melanoma MeWo, **a:** 8 days, **b:** 79 days after transplantation. There is no qualitative difference visible. Vascular corrosion casts. Bars = 100 μm

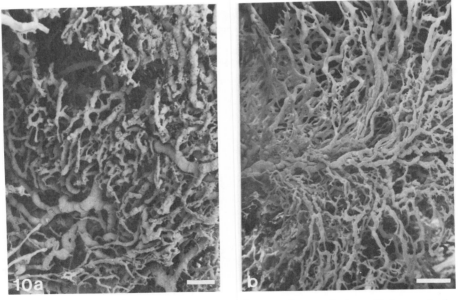

Fig. 10a and b. Vascular corrosion cast specimens of central sinusoidal vessels in an amelanotic melanoma on day 8 **(a)** and day 79 **(b)** after transplantation. Bar in a = 100μm, in b = 300 μm.

numerous characteristics of the tumour vascular system are found in all tumour entities (21). Even morphometric parameters such as vascular density, necrosis index or proliferation show parallelisms (22). The model used here can thus be taken to be valid. At the same time, further studies must show to what extent the described modes of angiogenesis can be determined in primary tumours.

Our studies have shown that there are different mechanisms of lumen formation. The commonly held view (e.g. 5) that migrated endothelial cells form a solid cord initially which later becomes channelised (6) does not hold for tumor angiogenesis in its present form. Apart from the fact that lumen formation in vascular sprouts is not dependent on the existence of two or more endothelial cells, according to our results at least a partial association with the primary lumen can certainly exist in the case of multicellular sprouts right from the beginning. Intracellular lumen formation in sprouts derived from a single endothelial cell was frequently found at least in the early stages in the xenotransplants studied. In standard text-books on normal and pathological secondary neovascularisation no allusions to intracellular lumen formation are yet to be found. This phenomenon, however, was already reported in the first in vitro study on angiogenesis in endothelial cell cultures by Folkman and Haudenschild (23): They found intracellular vacuoles both in endothelia from tumours as well as normal tissues, forming regular tubular systems. The role of these intracellular lumina has remained unclear up to now. Whether or not the "tube-like channels" described in cell cultures de facto show a continuous, uniform lumen is also not known. By the way, the formation of such lumina is not specific to endothelia but can also be observed in tumor cells (5, 12).

From our studies, the role of intracellular lumina is primarily to serve as "connective pieces" at the tip of the sprout; the fusion of sprouts exclusively with intracellular lumina over the entire length was not seen at all. In general, we consider the occurrence of the different forms of lumen formation as an answer to the enhanced metabolic requirements and the poor nutritive situation. However, an adequate nutrition is not at all assured despite the high angiogenic activity; an insufficient blood flow can occur even in the case of extremely high vascular densities with short intercapillary distances. On the one hand this is explained by the pathological vascular pattern formation with abnormal elongations of the vessels and, on the other, by an insufficient number of supplying arterial vessels. In this connection, we should remind ourselves of the frequent occurrence of veno-venous sinsusoidal systems. Since blood flow resistance is a function of vascular morphology (9), the architecture and structure of the vascular system determine the blood flow. Thus, morphometric studies have substantiated the finding that the number of vessels does not correlate with the frequency of necroses (22). The presence of vessels is only the pre-requisite, but not the guarantee for an efficient nutritive blood flow (24).

To what extent the primary and secondary angiogenesis is based on a general basic principle from the morphological point of view cannot be definitely assessed at the moment. Our own comparative studies on angiogenesis within the framework of repair processes, e.g., after epineural nerve suture (25) or subcutaneous transplantation of autologous tissue (Konerding et al., unpublished findings) have however shown that at the latest differences arise on further differentiation of newly formed vessels. Thus, we found regularly formed arteries with these types of secondary angiogenesis from day 21 on at the latest, which supply a differentiated capillary network and end in structurally normal veins. These results support the Hammersen's et al. (8) conclusion that "contrary to the current universal conceptions, tumor angiogenesis differs only to a minor extent from other neovascularisation processes ..., considerable structural differences ... occur".

REFERENCES

1. J. Denekamp, Vascular endothelium as the vulnerable element in tumors, Acta Radiol. Oncol., 23:217-225 (1984).
2. J. Folkman, Anti-angiogenesis: new concept for therapy of solid tumors, Ann. Surg., 175:409-416 (1972).
3. J. Folkman, R. Langer, R.J. Linhardt, C. Haudenschild, S. Taylor, Angiogenesis inhibi-

tion and tumor regression caused by heparin or a heparin fragment in the presence of cortisone, Sience, 221:719-725 (1983).

4. J. Denekamp, B. Hobson, Endothelial cell proliferation in experimental tumours, Br. J. Cancer, 46:711-720 (1982).

5. N. Paweletz, M. Knierim, Tumor-related angiogenesis, Crit. Rev. Oncol. Hematol., 9:197-242 (1989).

6. R. Thoma, Untersuchungen über die Histogenese und Histomechanik des Gefäß-Systems, Stuttgart, Enke (1893).

7. J.G. Simpson, R.A. Fraser, Angiogenese in malignen Tumoren, in: P. Vaupel, F. Hammersen eds., Mikrozirkulation in Forschung und Klinik. vol. 2, Karger, Basel-München-Paris-London-New York-Sydney, 1-14. (1989).

8. F. Hammersen, U. Osterkamp-Baust, B. Endrich, Ein Beitrag zum Feinbau terminaler Strombahnen und ihrer Entstehung in bösartigen Tumoren, in: P. Vaupel, F. Hammersen eds., Mikrozirkulation in Forschung und Klinik, vol. 2, Karger, Basel-München-Paris-London-New York-Tokyo-Sydney, 15-51 (1989).

9. R.K. Jain, Determinants of tumor blood flow: a review, Cancer Res., 48:2641-2658 (1988).

10. V. Budach, M. Bamberg, K. Donhuijsen, U. Schmidt, D. van Beuningen, M. Stuschke, Serial xenotransplantation of a human embryonal carcinoma in experimental urology, J. Urol.,136:1143-1147 (1986).

11. M.A. Konerding, F. Steinberg, C. Streffer, The vasculature of xenotransplanted human melanomas and sarcomas. 1. Vascular corrosion casting studies, Acta Anat., 136:21-26 (1989).

12. M.A. Konerding, F. Steinberg, C. Streffer, The vasculature of xenotransplanted melanomas and sarcomas. 2. Scanning and transmission electron microscopic studies, Acta Anat., 136:27-33 (1989).

13. M.A. Konerding, F. Steinberg, V. Budach, The vascular system of xenotransplanted tumors - scanning and light microscopic studies, Scanning Microsc., 3:327-336 (1989).

14. H.A. Eddy, G.W. Casarett, Development of the vascular system in the hamster malignant neurilemmoma, Microvasc Res., 6:63-82 (1973).

15. T.W. Grunt, A. Lametschwandtner, K. Karrer, The angioarchitecture of the Lewis lung carcinoma in laboratory mice, Scanning Microsc., II:557-573 (1986).

16. B.A. Warren, The ultrastructure of the microcirculation at the advancing edge of Walker 256 carcinoma, Microvasc. Res., 2:443-453 (1970).

17. K. Donhuijsen, D. van Beuningen, V. Budach, U. Schmidt, Instability of xenotransplanted soft tissue sarcomas: morphologic and flow cytometric results, Cancer, 61:68-75 (1988).

18. F. Steinberg, M.A. Konerding, C. Streffer, The vascular architecture of human xenotransplanted tumors: histological, morphometrical and ultrastructural studies, Int. J. Cancer Res., 116:517-524 (1991).

19. M.A. Konerding, F. Steinberg, U. Winkler, C. Streffer, Comparison of human primary and xenotransplanted tumours: the tumour vascular system as revealed by transmission and scanning electron microscopy, J. Cancer Res. Clin. Oncol., 116:514 (1990).

20. B.C. Giovanella, J. Fogh, Present and future trends in investigations with the nude mouse as a recipient of human tumor transplants, J. Fogh, B.C. Giovanella, eds., The Nude Mouse in Experimental and Clinical Research, Academic Press, New York, 281-297 (1978).

21. A. Bugajski, M. Nowogrodzka-Zagorska, J. Lenko, A.J. Miodonski, Angiomorphology of the humen renal clear cell carcinoma. A light and scanning electron microscopic study, Virchows Arch. (A), 415:103-113 (1989).

22. F. Steinberg, M.A. Konerding, A. Sander, C. Streffer, Vascularization, proliferation and necrosis in untreated human primary tumours and untreated human xenografts, Int. J. Radiat. Biol., 60:161-168 (1991).

23. J. Folkman, C. Haudenschild, Angiogenesis in vitro, Nature, 288:551-556 (1980).

24. P. Vaupel, H.P. Fortmeyer, S. Runkel, Blood flow, oxygen consumption, and tissue oxygenation of human breast cancer xenografts in nude rats, Cancer Res., 47:3496-3503 (1987).

25. M. Lehmann, M.A. Konerding, M. Blank, Vascularisation of the peripheral nerve after epineural suture, in: M. Samii, Peripheral Nerve Lesions, Springer, Berlin-Heidelberg-New York-London-Paris-Tokyo-Hongkong-Barcelona, 149-153 (1990)

TUMOR INDUCED ANGIOGENESIS: EFFECT OF PLATELET DERIVED GROWTH FACTOR (PDGF), PENTOXIFYLLINE, SODIUM DIETHYLDITHIOCARBAMATE, EPSILON AMINO CAPROIC ACID AND TRANEXAMIC ACID

J.L. Ambrus, C.M. Ambrus, C.A. Toumbis,
P. Forgach, C.P. Karakousis, and P. Niswander

Roswell Park Cancer Institute
State University of New York at Buffalo
Departments of Internal Medicine, Pediatrics,
Obstetrics-Gynecology, Ophthalmology, Experimental
Pathology, and Surgery

Most solid malignant tumors induce the formation of new blood vessels from the host into the tumor. Without this effect, most tumors are unable to propagate. This effect appears to be induced by tumor produced angiogenesis factors (Folkman et. al., 1971, 1974, 1974a, 1975; Brem et. al., 1975, 1976; Greenblatt et. al., 1968; Gullino, 1978, 1981; Langer et. al., 1976, 1981; Giambrone et. al., 1973; Eisenstein et. al., 1973). The extensive literature on angiogenesis has recently been reviewed by Klinworth (1991) and Rifkin and Klagsbrun (1987).

In the present study, we want to explore factors which may be involved in tumor induced angiogenesis and possible antagonists. In earlier studies (Ambrus et. al., 1976, 1978, 1978a, 1981, 1983) we found that in the tortuous A-V anastomoses rich tumor microvasculature there is sluggish flow and continuous aggregation of platelets which release various platelet factors including the platelet derived growth factors (PDGF). We have explored the angiogenic effect of crude platelet extracts and certain recombinant PDGFs. We also wanted to explore possible inhibitors of tumor induced angiogenesis. McAuslan and others (1979, 1980) reported that copper containing protein fractions and copper ions themselves had angiogenic activity. It was thought that copper prosthetic groups may be involved in angiogenesis. In earlier studies, we have found (Ambrus et. al., 1951) that the copper complexing agent, sodium diethyldithiocarbamate (NaDDTC) is relatively non-toxic in animals except for a mild anti-thyroid activity. Renoux et. al. (1977,1980, 1981) reported that this agent has immune modulatory activity. It increases cell mediated immunity, and it also recruits and activates T-cells. Others have found that this compound inhibits cis-platinum induced nephrotoxicity, acts as a radiation sensitizing agent and inhibits superoxide dismutase (Borch et. al., 1979; Risas et. al.,1980; Masos et. al.,1981; Guarnieri et. al., 1981). For these reasons, we wanted to explore this compound alone and in combination with other potential antiangiogenic agents.

Angiogenesis in Health and Disease, Edited by M.E. Maragoudakis *et al.*
Plenum Press, New York, 1992

In earlier studies, we found that the methylxanthine derivative pentoxifylline increases red cell membrane fluidity, inhibits platelet aggregation, and white cell adhesion, thus promoting the microcirculation (Ambrus et. al., 1990; Gastpar et. al., 1977, 1978). In earlier studies we have also investigated plasmin inhibitors (Ambrus et. al., 1968, 1968a, 1970, 1971). We hypothesized that the fibrinolysin system plays a role in keeping the abnormal microvasculature system of tumors open. For this reason, we thought that the fibrinolysin inhibitors such as epsilon amino caproic acid (EACA) or tranexamic acid (t-AMCHA) may reduce vascularization of neoplastic tissue. We also wanted to explore combinations of some of the above materials which presumably would work through different mechanisms producing an anti-angiogenic effect.

MATERIALS AND METHODS

In preliminary experiments, we explored angiogenic and anti-angiogenic effects in the dorsal air sac technique of Folkman et. al., (1971) modified by Phillips and Kumar (1979) using ICR/Ha male mice and in the chorioallantoic membrane assay of Gullino (1981) and Jacob et. al., (1977, 1978). On the basis of these results, experiments were planned using the corneal angiogenesis method of Folkman et. al., (1974, 1976, 1981; Langer et. al, 1981) in New Zealand white rabbits and Macaca arctoides monkeys. The materials to be studied were incorporated into purified Elvax-40 membranes inserted into the cornea. We found that Elvax-40 (ethylene-vinyl acetate copolymer, Dupont Co.) contained various interfering substances and for this reason it was necessary to purify it. It was washed with 2.5% ethanol at 37C for 8 hours. At least 50 washing cycles were employed. The material was then dissolved in methylene chloride at 37C to give a 10% solution (w/v). Material to be tested was added to this solution, thoroughly mixed, and then poured into a glass mold resting on a block of dry ice. After 10 minutes, the solidified membrane was transferred to a Nitex screen in a glass petri dish resting on a block of dry ice. After 30 minutes, the membrane was dried in a dessicator under vacuum at room temperature. These membranes were found not to be angiogenic by themselves. These membranes were inserted into the cornea. The operated area was inspected daily with a slit-lamp microscope using a green filter at 40X magnification. Final scoring was done 7 days after implantation by counting the vessels which appear to penetrate the operated area per high powered field. The enclosed figures show the actions scored:

> 1+ (5-20 ingrowing vessels)
> 2+ (20-50 ingrowing vessels)
> 3+ (>50 ingrowing vessels)

Materials studied include crude platelet derived growth factor prepared from normal human platelets. Extracts of 10^9 platelets/ml were incorporated into the purified Elvax-40/ml. The platelet extract was prepared by the method of Witkoski et. al.,(1982). Recombinant B/B homodimer of PDGF was kindly supplied by Drs. O. Friedman and P. Kelly of the Collaborative Research Laboratories, Boston, MA. Isolation, characterization and assay of platelet derived growth factor was reported by Ross et. al., (1974); Antoniades et. al., (1977, 1981); Westermark et. al., (1976) and Heldin et. al.,(1979).

Human malignant melanoma cell culture was established from biopsies, normal keratinocyte culture was established from the surrounding normal skin area of the same patients. NaDDTC

TABLE I

EFFECT OF HUMAN MELANOMA INDUCED ANGIOGENESIS

Group Corneal Implant	Mean degree of angiogenesis[x] +/- S.E. in RABBITS (n=4)	in Macaca arctoides MONKEYS (n=4)
Purified Elvax-40 (E)	0 +/- 0	0 +/- 0
E + Crude PDGF[xx] from 10^9 human platelets/ml	3 +/- 0	2 +/- 0
E + rPDGF B/B homodimer 50 half max. units/ml	2 +/- 0	2 +/- 0.82
E + 10^9 human malignant melanoma cells (M)/ml	2.25 +/- 0.96	2.25 +/- 0.96
E + 10^9 human keratinocytes/ml	0 +/- 0	0 +/- 0
E+M+NaDDTC[xxx]25mg/ml	2 +/- 0.82	1 +/- 0.82
E+M+Pentox.(10mg/ml)	2 +/- 0.82	2 +/- 0.82
E+M+Pentox.+NaDDTC	1.75 +/-0.96	0 +/- 0
E+M+EACA[xxxx]20mg/ml	2 +/- 0.82	2 +/- 0.82
E+t-AMCHA[xxxxx]20mg/ml	2 +/- 0.82	2 +/- 0.82
TOTAL	*1.70+/-0.18*	*1.33 +/- 0.17*

**Between groups, F=12.1, p<0.00005
Cochran's C-test for homogeniety of variance,p=0.161

[x]Number of Capillaries Designation

HPF(40X)	
0-4	0+
5-20	1+
21-50	2+
>50	3+

[xx]PDGF = Platelet derived growth factor

[xxx]NaDDTC = Sodium diethyldithiocarcamate

[xxxx]EACA = Epsilon aminocaproic acid (Amicar)

[xxxxx]t-AMCHA = Tranexamic acid (Cyclokapron)

TABLE II

SIGNIFICANCE OF DIFFERENCES BETWEEN GROUPS

Elvax control vs. Malignant melanoma cells (M) $p < 0.05$
Normal keratinocytes vs. Malignant melanoma cells (M) $p < 0.05$
Elvax control vs. a) E+Crude PDGF $p < 0.05$
                       b) rPDGF B/B homodimer $p < 0.05$
E+malignant melanoma cells vs. E+M+NaDDTC $p < 0.05$
E+malignant melanoma cells vs. E+M+Pentoxifylline $p < 0.05$
E+malignant melanoma cells vs. E+M+EACA $p < 0.05$
E+malignant melanoma cells vs. E+t-AMCHA $p < 0.05$

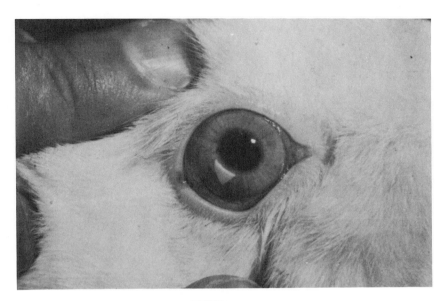

FIGURE 1A

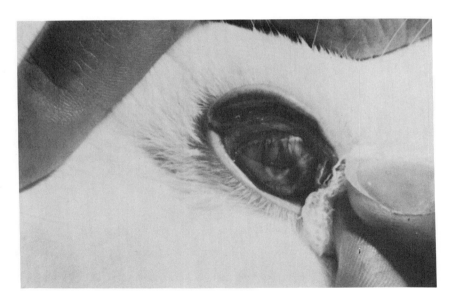

FIGURE 1B

FIGURE 2.

1+ REACTION

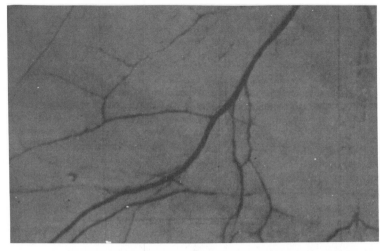

2+ REACTION

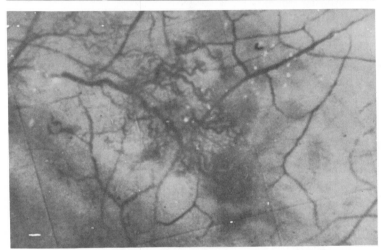

3+ REACTION

was obtained from the Sigma Chemical Co., (St. Louis, MO). Pentoxifylline (Trental) was kindly supplied by Hoechst Roussel Pharmaceutical Co., (Somerville, NJ). Epsilon aminocaproic acid (Amicar) was kindly supplied by the Lederle Co.. Aminocaproic acid (Cyclokapron, Kabi Vitrum) was kindly supplied by Kabi Pharmaceutical Co., (Stockholm, Sweden).

RESULTS AND DISCUSSION

Tables I and II summarize the results in both rabbit and monkey corneas. Purified Elvax-40 had no angiogenic effect but when crude PDGF was incorporated, significant angiogenesis developed more so in rabbits than in monkeys. Recombinant
E. Coli derived PDGF, B/B homodimer was angiogenic in both species. Human malignant melanoma cells were highly angiogenic but keratinocytes cultured from surrounding normal skin had no effect. When melanoma cells were combined with NaDDTC, angiogenesis was inhibited more in monkeys than in rabbits. The hemorrheologic agent pentoxifylline also had inhibitory activity. When pentoxifylline and NaDDTC were combined there was a significant potentiation of anti-angiogenic activity by these two agents, which presumably work through different mechanisms. The two anti-fibrinolytic agents epsilon aminocaproic acid and tranexamic acid were both effective in slightly reducing tumor induced angiogenesis. Figure 1A shows a rabbit with an implanted Elvax membrane and Figure 1B shows the same eye one week later with massive corneal angiogenesis developing. Figure 2 shows high powered fields of the operating microscope through which the degree of angiogenesis was scored. It shows the 1+, 2+, and 3+ reactions.

It appears from this data that platelet aggregation and the release of PDGF may contribute to the development of angiogenesis. Human malignant cells appear to be more angiogenic than normal cells in both rabbits and monkeys. This angiogenic effect is inhibited by the copper complexing agent NaDDTC and the hemorrheologic agent pentoxifylline. The two work presumably through different mechanisms and synergize with each other. The two antifibrinolytic agents epsilon aminocaproic acid and tranexamic acid were also shown to have some anti-angiogenic effect. This suggests that the fibrinolysin system may play a role in keeping newly formed capillary networks open. Anti-angiogenic agents may have therapeutic potential in preventing growth of newly established metastatic tumors. New studies with systemic therapy are being planned.

REFERENCES

Ambrus, CM, Ambrus, JL, Harrisson, JWE: Effect of sodium diethyl-dithio-carbamate
    on thyroid activity. Am. J.Pharm. 123: 129-30, 1951.
Ambrus, CM, Ambrus, JL, Lassman, HB, Mink, IB: Studies on the mechanism of action of
    inhibitors of the fibrinolysin system. Ann. N.Y. Acad. Sci. 146: 430-447, 1968.
Ambrus, JL, Ambrus, CM, Stutzman, L, Schimert, G, Niswander, KR, Woodruff, MW, Magoss,
    IV: Treatment of fibrinolytic hemorrhage with proteinase inhibitors: A preliminary report.
    Ann. N.Y. Acad. Sci. 146: 625-641, 1968a.
Ambrus, CM, Ambrus, JL, Lassman, HB, Mink, IB: On the heterogeneity of activity
    some plasmin inhibitors. Res.Comm. Chem. Pathol. Pharmacl. 1: 67-85, 1970.

Ambrus, JL, Schimert, G, Lajos, TZ, Ambrus, CM, Mink, IB, Lassman, HB, Moore, RH, Melzer, J: Effect of antifibrinolytic agents and estrogens on blood loss and blood coagulation factors during open heart surgery. J. Med. 2: 65-81, 1971.

Ambrus, JL, Ambrus, CM: Blood coagulation in neoplastic disease. In: Gastpar, H (ed.): Onkohamostaseologie (Hematologic Problems in Cancer). F. K. Schattauer Verlag, Stuttgart-New York, pp. 167-193, 1976.

Ambrus, JL, Ambrus, CM, Gastpar, H: Studies on platelet aggregation and platelet interaction with tumor cells. In: Platelets: A Multidisciplinary Approach. De Gaetano, C, and Garattini, S (eds.). Raven Press, New York, NY, pp.　467-480, 1978.Ambrus, JL, Ambrus, CM: Anticoagulants, fibrinolytic enzymes and platelets interaction with tumor cells. In: Cancer Chemotherapy III (The 46th Hahnemann Symposium) Brodsky, I, Kahn, SB, Conroy, JF (eds.), Grune and Stratton, Inc., New York, NY, pp. 97-109, 1978a.

Ambrus, JL, Ambrus, CM, Gastpar, H: The role of platelet aggregation in metastatic dissemination and thromboembolic complications in cancer patients. Prevention with pentoxifylline. In: Disorders of Blood Flow: New Therapeutic Aspects. Manrique, RV, Muller, R (eds.), Excerpta Medica, Amsterdam-Oxford Princeton, pp. 29-42, 1981.

Ambrus, Jl, Gastpar, H: Plattchenaggregationschemmer in der krebsforschung (Platelet aggregation inhibitors in oncology). In: Onkologie und Blutgerinnung (Oncology and Blood Coagulation). Kostering, H (ed.), Roche Editions, Basel, Switzerland, pp. 201-214, 1983.

Ambrus, JL: Microvascular changes in diabetes. In: Diabetes and Vascular Disease. Kerstein, MD (ed.), J.B. Lippincott Co., Phil., PA., pp. 9-39, 1990.

Antoniades, HN, Scher, CD: Radioimmunoassay of a human serum growth factor for Balb/c-3 T3 cells: Derivation from platelets. Proc. Natl. Acad. Sci. 74: 1973-1977, 1977.

Antoniades, HN: Human platelet-derived growth factor (PDGF): Purification of PDGF-II and serparation of their subunits. Proc. Natl. Acad. Sci. 78: 7314-7317, 1981.

Borch, RF, Plasants, ME: Inhibition of cis-platinum nephrotoxicity diethyldithiocarbamate rescue in a rat model. Proc. Natl. Acad. Sci. 76: 6611-6614, 1979.

Brem, H, Folkman, J: Inhibition of tumor angiogenesis mediated by cartilage. J. Exp. Med., 141: 427-439, 1975.

Brem, S, Brem, H, Folkman, J, Finkelstein, D, Patz, A: Prolonged tumor dormancy by prevention of neovascularization in the vitreous. Cancer Res. 36: 2807-2812, 1976.

Eisensten, R, Sorgente, N, Soble, LW, Miller, A, Kuettner, KE: The resistance of certain tissues to invasion. Am. J. Pathol. 73: 765-772, 1973.

Folkman, J, Merler, E, Abernathy, C, Williams, G: Isolation of a tumor responsible for angiogenesis. J. Exp. Med. 133: 275-288, 1971.

Folkman, J: Tumor angiogenesis factor. Advanced Cancer Research, 19: 331-358, 1974.

Folkman, J: Tumor angiogenesis factor. Cancer Res. 34: 2109-2113, 1974a.

Folkman, J: Tumor angiogenesis. In: Biology of Tumors. Becker, FF (ed.), Plenum Press, NY, NY, pp. 355-388, 1975.

Gastpar, H, Ambrus, Jl, Ambrus, CM, Spavento, P, Weber, FJ, Thurber, LE: Study of platelet aggregation in vivo III. Effect of pentoxifylline. J. Med. 8: 191-197, 1977.

Giambrone, MA, Jr, Leapman, SB, Cotran, RS, Folkman, J: Tumor angiogenesis: Iris neovascularization at a distance from experimental intraocular tumors. J. Natl. Cancer Inst. 50: 219-228, 1973.

Greenblatt, M, Shubik, P: Tumor angiogenesis: Transfilter diffusion studies in the hamster by the treatment by the transparant chamber technique. J. Natl. Cancer Inst. 61: 639-643, 1968.

Guarnieri, C, Flamisni, F, Ventura, C, Rossoni-Caldarera, C: Inhibition of rat heart superoxidase dismutase activity by dietyldithiocarbamate and its effect on mitochondrial function. Biochem. Pharmacol. 30: 2174-2176, 1981.

Gullino, PM: Angiogenesis and oncogenesis. J. Natl, Cancer Inst. 61: 639-643, 1978.

Gullino, PM: Angiogenesis factor(s). In: Tissue Growth Factors. Baserga, R (ed.), Springer-Verlag, pp. 428-449, 1981.

Heldin, CH, Westermark, B, Wasteson, A: Purification platelet derived growth factor and partial characterization. Proc. Natl. Acad. Sci. 76: 3722-3726, 1979.

Jakob, W, Jentzsch, KD, Mauersberger, B, Oehme, P: Demonstration of angiogenesis activity in the corpus luteum of cattle. Exp. Pathol. 13: 231-236, 1977.

Klinworth, GK: Corneal Angiogenesis: A Comprehensive Critical Review. Springer Verlag, NY, NY, 1991.

Langer, J, Brem, H, Flaterman, H, Klein, M, Folkman, J: Isolation of a cartilage factor that inhibits tumor neovascularization. Science. 193L 70-71, 1976.

Langer, R, Folkman, J: Angiogenesis inhibitors. In: Molecular Actions and Targets for Cancer Chemotherapy Agents. Sartorelli, AC, Lazo, JS, Bertino, J, (eds.) Academic Press, NY, NY, pp. 511-525, 1981.

Masos, L, Jarvis, JA: The paradox effect of carbon disulfide and diethyldithiocarbamate of adrenal dopamine-beta-hydroxylase. G. Ital. Med. Lav. 3: 107-108, 1981.

McAuslan, BR, Hoffman, H: Endothelium stimulating factor from Walker carcinoma cells. Relation to tumor angiogenic factor. Exp. Cell Res. 119: 181-190, 1979.

McAuslan, BR: A new theory of neovascularization based on identification of an angiogenic factor and its effect on cultured endothelial cells. In: Control Mechanisms in Animal Cells. Jimenez de Asua, L, Shields, R, Levi-Montalcini, R, Iacobelli, S, (eds.), Raven Press, NY, NY, pp. 285-292, 1980.

Phillips, P, Kumar, S: Tumour angiogenesis factor (TAF) and its neutralization by a xenogeneic antiserum. Int. J. Cancer. 23: 82-88, 1979.

Renoux, G, Renoux, M: Thymus-like activities of sulfer derivatives on T-cell differentiation. J. Exper. Med. 145: 466-471, 1977.

Renoux, G, Renoux, M: The effects of sodium diethyldithiocarbamate, azathioprine, cyclophosphamide, or hydrocortisone acetate administered alone or in association for 4 weeks on the immune responces of BALB/c mice. Cloin. Immunol. Immunopath. 15: 23-32, 1980.

Renoux, G: Levamisole and sodium diethyldithiocarbamate. Trends Pharmacol. Sci. 2: 248-249, 1981.

Rifkin, DB, and Klagsbrun, M: Angiogenesis: Mechanisms and Pathobiology. Cold Spring Harbor Laboratory Press, 1987.

Risas, DA, Esinitis-Risas, C, Bisl, RH, Stankova, L, Head, C: Biphasic radio-sensitive
of human lymphocytes by diethyldithiocarbamate: possible involvement of superoxide
dismutase. Int. J. Radiat. Biol. 38: 257-266, 1980.

Ross, R, Glomset, J, Kariya, B, Harker, L: A platelet dependent serum factor
that stimulates the proliferation of arterial smooth muscle cells in vitro. Proc. Natl.
Acad. Sci. 71: 1207-1210, 1974.

Westermark, B, Wasteson, A: A platelet factor stimulating human normal glial cells. Exp.
Cell Res. 98: 170-174, 1976.

VASCULAR GROWTH IN THE INTERMITTENTLY ISCHEMIC HEART:

A STUDY ON GROWTH FACTORS EXPRESSION

Hari S. Sharma, Martin Wünsch, Stefan Sack,
Wolfgang Schaper

Max-Planck-Institute for Physiological and Clinical
Research, 6350-Bad Nauheim, Germany

INTRODUCTION

The formation of blood vessels, i.e. angiogenesis, is an important component of various normal and pathological conditions including wound healing, fracture repair and in females during follicle development, ovulation and pregnancy. Under normal physiological conditions, angiogenesis is tightly regulated, whereas during pathological conditions such as tumor growth and retinopathies it seems to be uncontrolled. Various methods including corneal micropocket and chick embryo chorioallantoic membrane assay have been described to study angiogenesis in vivo. New capillaries originated mainly from sprouting of small venules and localized degradation of basement membrane of the parent venule was followed by movement of endothelial cells towards the angiogenic stimulus[1]. Angiogenesis involves the orderly migration, proliferation and differentiation of vascular cells[2]. The mechanism of angiogenesis and endothelial cell growth control, especially with respect to collateralization (collateral vessel development) is an important research area of cardiovascular physiology.

Poly-peptide growth factors probably contribute in the regulation of new blood vessel growth during myocardial collateralization. Poly-peptide growth factors are hormone like modulators of cell proliferation and differentiation. They bind to the specific cell surface receptors to exhibit their functions for subsequent alterations in gene expression in responsive cells[3]. We have observed that during a progressive coronary artery stenosis, growth of new blood vessels and enlargement of existing ones (sprouting and non sprouting angiogenesis) occur in the porcine heart. This growth adaptation can be very efficient to prevent any myocardial infarction even during chronic coronary artery occlusions. Therefore, if we can understand the process of coronary angiogenesis better, we might be able to design drugs for regulated myocardial angiogenesis to help patients with coronary heart disease. At the cellular level, it is not yet

Angiogenesis in Health and Disease, Edited by M.E. Maragoudakis *et al.*
Plenum Press, New York, 1992

clear how this adaptive mechanism of ischemic defence is regulated. To understand the molecular basis of ischemia induced myocardial angiogenesis, we examined the expression of some angiogenic poly-peptides like endothelial cell growth factor-β (β-ECGF), vascular endothelial growth factor (VEGF) and transforming growth factor-β1 (TGF-β1) using polymerase chain reaction (PCR), Northern and in situ hybridization techniques in a porcine model of progredient chronic coronary artery occlusion.

Myocardial Collateralization

The progredient stenosis of a major epicardial coronary artery, (as in human atherosclerotic heart disease) is a potent mitogenic stimulus for mesodermal derived cells such as fibroblasts, smooth muscle cells and endothelial cells. The result of this stimulation is the adaptive growth of new and enlargement of existing vessels in the region at risk of infarction and at its border[4-6]. This compensatory growth seems to be the most important event in the prevention of myocardial infarction. The formation of new blood vessels in response to ischemic stimulus in the heart can be referred as collateralization. Collateral vessels may simply be defined as routes of blood supply to the heart that are not functional or present under normal circumstances. These alternate routes of blood supply to the jeopardized myocardium arise both from preformed and newly formed collateral vessels[1]. In past years, the process governing collateralization in the adult heart has been studied extensively both in experimental animals and in living human patients. In symptomatic patients transmural infarction leading to death is often avoided by timely enlargement and formation of new collaterals[7]. In experimental animals various laboratories have shown the growth of new arterioles and capillaries in response to slowly progressing coronary stenosis and in a significant percentage, this growth adaptation of collaterals salvage the ischemic myocardium[5,6,8,9].

Models of Myocardial Collateralization

Based on the coronary artery occlusions, various methods have been described in the literature to induce myocardial collateralization in experimental animals. These methods usually interfere with the coronary blood flow. Several models studied so far, include acute coronary artery occlusion, stenosis by fixed known degree, progredient stenosis until complete occlusion, repeated short occlusions, microembolization, chronic coronary vasodilation by severe anemia and drugs and combination of above methods[8-12]. The diagnostic and experimental techniques used to study collateral vessel growth include angiography, histology, ultrastructure and cell biology. Canine heart was mostly used to develop these method but pigs were also used often. Species differences exist within mammals and from an anatomic point of view, the pig coronary and collateral circulation is closest to that of the human heart[8].

We opted for a porcine model of myocardial angiogenesis by progressive coronary artery occlusion due to some obvious reasons. The ameroid model is an established method to induce coronary artery stenosis slowly by hygroscopic swelling[5]. The principle of this method is that the hygroscopic plastic material on the basis of formalin-treated casein swells by absorbing the tissue fluid and compresses the artery. In the normal pig heart, there are practically no anatomically demonstrable collaterals[8]. As it was demonstrated recently by Görge and co-workers[13], coronary blood flow was reduced to about 20% of normal and collateral flow rose to 60% of maximal normal flow after 2-4 weeks of ameroid implantation in pigs. Numerous vessels developed throughout the area at risk in a dense "blush" like fashion. For these characteristics the pig model approximates the situation in human heart better than other animal models, having either a large number of naturally occuring collaterals as in dog or no possiblity for collateral opening and formation as in rat[8,14]. However, the ameroid dependent stenosis, as a biological process differs widely. It is conceivable that the complete arterial occlusion that mostly occurs, is due to the pronounced foreign body reaction against the ameroid material, shrinkage of the fibrous tissue and arterial thrombosis. The mortality of ameroid occlusion of the left circumflex coronary artery in pigs is about 30% which is largely due to ventricular fibrillation[8].

Experimental Procedure for Myocardial Angiogenesis in Pigs

Details of the surgical procedures have been described elsewhere[13]. Several domestic pigs, weighing 18-20 kg were intubated endotracheally with room air enriched in oxygen using a pressure respirator under general anaesthesia. After a lateral thoracotomy a hygroscopic ameroid constrictor was implanted around the left circumflex coronary artery (LCx). Stenosis of the coronary artery was determined by angiography after 2-3 weeks of ameroid implantation. At more than 90% of LCx stenosis and in some cases with collateral growth from right coronary artery, pigs were sacrificed, the heart was removed quickly and rinsed in ice-cold saline. Small tissue pieces from the LCx region (macroscopically examined to exclude any infarction) and normally perfused interventricular septum were excised and snap frozen in liquid nitrogen and stored at -80° C and analysed for the expression of various angiogenic growth factors.

Angiogenic Growth Factors

In the last decade, a number of angiogenic poly-peptide growth factors have been purified and characterized (for reviews see ref.[1,3,17,18]). These poly-peptides include vascular endothelial growth factor (VEGF), acidic and basic fibroblast growth factor (FGF), angiogenin, transforming growth factor α and β (TGF) platelet derived endothelial cell growth factor (PD-ECGF), and tumor necrosis factor-α (TNF-α)[15-18]. Analysis of these factors strongly suggests that there must be several different mechanisms for stimulating angiogenesis[17]. As the endothelial cell is considered to be the primary cell involved in all forms of angiogenesis, efforts have been concentrated to identify and characterize poly-peptide growth factors that control endothelial cell motility and proliferation. Endothelial cells form the inner surface of all functional vessels and under normal conditions they are remarkably quiescent cells. However, in a number of pathological conditions endothelial cells rapidly proliferate and thereby forming new blood vessels[19]. On the basis of their action on either motility or proliferation of endothelial cells, these poly-peptide growth factors can be grouped as shown in Table 1. The role of various growth factors have been studied with cell culture systems, relatively little is known about the physiological significance of these molecules in vivo. Recently, we and others have isolated and characterized heparin binding growth promoting poly-peptides from hearts of various species including pigs[20-22]. In 1971 Schaper et al.[7] discovered that the cell proliferation and DNA synthesis was the basis of collateral enlargement. These findings encouraged for further studies exploring the stimulus for vascular growth.

Table 1. Effects of angiogenic peptides on endothelial cells

GF	MW (kD)	Angiogenesis	EC Proliferation	EC Migration
VEGF	46	Yes	Yes	Yes
FGF	18	Yes	Yes	Yes
PD-ECGF	46	Yes	Yes	Yes
TGF-β1	25	Yes	Inhibition	Inhibition
TGF-α	6	Yes	Yes	Yes
TNF-α	17	Yes	Inhibition	Yes
Angiogenin	14	Yes	No	No
Angiotropin	5	Yes	No	Yes

GF = growth factor, EC = endothelial cell, MW = molecular weight

Angiogenic Growth Factors Expression During Collateralization in Pigs

The present study was conducted to elucidate whether angiogenic growth factors like β-ECGF, VEGF and TGF-β1 could play a role in the ischemia induced myocardial angiogenesis and if their mRNAs' expression is altered during collateralization in pigs. It is now well established that the pig heart is a much better model for the clinical situation and the effect of chronic coronary occlusion on the collateralization has been reported by cell proliferation assays in vivo[6,7]. Our results on the expression of angiogenic peptides in the collateralized myocardium are presented in following headings.

Endothelial Cell Growth Factor-β

Endothelial cell growth factor-β, a heparin binding growth factor is an anionic polypeptide mitogen of 20 kd with nearly 55% amino acid identity to basic fibroblast growth factor[3]. Amino acid sequence analysis revealed that the β-ECGF is a precursor of α-ECGF and acidic fibroblast growth factor[23]. It is well documented that both acidic and basic fibroblast growth factors are mitogenic in vitro for cells of vascular, nervous, connective and muscle tissue origin[18]. Unlike other angiogenic peptides an intriguing feature of aFGF and bFGF is that they lack a signal sequence to direct secretion from the cell and they appears to be cell associated[24].

We searched for the mRNAs encoding β-ECGF in the porcine myocardium employing PCR and Northern hybridization techniques. Frozen tissue was pulverized in the presence of liquid nitrogen and total cellular RNA was isolated according to the method of Chomczynski and Sacchi[25]. RNA concentration was measured by optical density measurement and the quality of RNA was tested on a denatured formaldehyde agarose gel[26]. To perform PCR, single stranded cDNA was synthesized from total RNA derived from normal and collateralized myocardia. An aliquot of above cDNA was used for PCR amplification. For each cDNA synthesis a blank was prepared using all the reagents except the RNA (volume substituted by water). β-ECGF specific primers were designed from the published bovine and human cDNA sequences[27,28]. Highly conserved regions were chosen to design sense and anti-sense primers spanning over the whole coding region of the mRNA. Primers for 18S ribosomal RNA[29] were also synthesized to use as a standard amplification reaction in PCR. cDNA products were amplified using each sense and anti-sense primers, in an automatic DNA thermal cycler (Perkin-Elmer Cetus, USA) for 30 cycles[30]. The PCR products were analysed by poly-acrylamide gel electrophoresis, stained with ethidium bromide and visualized under UV light. For Southern hybridization, PCR products were electrophoresed on agarose gels, blotted on a nylon membrane and hybridized with a radiolabeled 2.1 kb β-ECGF cDNA insert from a clone pDH15 isolated from human brain stem cDNA library[27] (kindly provided by Dr. T. Maciag). Filters were washed under stringent conditions and auto-radiographed at -70° C for 1-2 hrs.

In designing β-ECGF specific primers, inter-species cDNA sequence differences were taken into account and 3' end base of both sense and anti-sense primers was taken strictly either C or G for better anealing. These primers were hoped to generate a 480 bp DNA fragment spanning over the whole coding region of β-ECGF. PCR amplification of cDNAs derived from collateralized as well as normal myocardium yielded one major product of expected size. We have established the cDNA sequence of β-ECGF from porcine heart generated by asymmetric amplification of cDNAs derived from the normal and/or collateralized myocardium[31]. The specific amplification of β-ECGF was verified by Southern blot analysis of PCR products, where a major band of 480 bp hybridized with a radiolabeled cDNA insert while marker lane showed no hybridization. By ethidium bromide staining of PCR products on the agarose gel, we observed amplification only in one out of three preparations from the normal myocardium. However, a DNA fragment of 405 bp specific for 18S rRNA was amplified by PCR in all preparations, confirming the intactness of mRNAs and efficiency of reverse transcription.

To visualize the differences in β-ECGF expression in the collateralized and normal myocardia, Northern blot analysis was performed. For Northern hybridization, total RNA from each tissue was size fractionated on a formaldehyde containing agarose gel in the presence of ethidium bromide as described previously[26]. RNA was transferred to hybond-N membrane by vacuum blotting, UV crosslinked and ribosomal RNA bands were marked under UV light. Blots were hybridized with a radiolabeled β-ECGF cDNA probe at 42°C in a buffer containing 50% formamide and washed under stringent conditions and subsequently, exposed to Kodak XAR films at-70° C for 2-4 days. Using the radiolabeled human cDNA probe encoding β-ECGF, we detected a mRNA species of 4.6 kb after an overnight exposure in the collateralized myocardium. However, a similar band also appeared in the normal myocardium after 48-72 hrs of exposure indicating enhanced expression of β-ECGF in the collateralized tissue as compared to the normal. This is for the first time that by Northern hybridization β-ECGF mRNA is detected in the myocardial tissue[32]. There are several reports showing the presence of β-ECGF protein in the normal myocardium[20-22] and the mRNA expression only in the primary cultures of isolated adult and neonatal rat myocytes[33,34]. The size of the mRNA for β-ECGF is almost identical to other β-ECGF mRNA transcripts previously detected in human brain[27], bovine retina[28].

It is clear that β-ECGF is present in the porcine myocardium but the cell type(s) producing this factor under normal and experimental conditions such as collateralization are virtually not known. To localize the β-ECGF transcripts, we performed in situ hybridization. For in situ hybridization porcine normal and collateralized tissue sections were hybridized with ^{35}S labeled nick translated cDNA insert encoding β-ECGF. In one of the several pigs studied, β-ECGF transcripts were mainly seen in the vascular wall presumably in the endothelial and smooth muscle cells. However, using a monoclonal antibody for β-ECGF, Bernotat-Danielowski et al.[35] localized protein by immunoflourescence technique as a focal distribution in cardiomyocytes boardering fibrotic tissue. When total protein extract of the collateralized myocardium was tested in immuno-blotting using a monoclonal antibody, no β-ECGF specific protein band could be seen probably due to the lower sensitivity of the method. Weiner and Swain[33] have recently reported that cultured neonatal myocytes produce aFGF (short form of β-ECGF) as they could detect the mRNA and protein in the extracts of myocytes. To compare the in vitro situation to the in vivo is difficult and secondly neonatal rat myocytes could still be transcriptionally active to make β-ECGF.

Transforming Growth Factor-β1

Transforming growth factor-β1 (TGF-β1) is a 25 kDa homodimeric polypeptide differentiation factor which modulates the growth of normal and neoplastic cells[36]. Although, TGF-β1 is an inhibitor of endothelial cell proliferation, it has been found to elicit a potent angiogenic reaction when injected subcutaneously into new born mice[37] or when applied locally in wound healing experiments[38]. An interesting property of TGF-β is that it is secreted in a biologically inactive latent form which can be activated by heat, acidification and proteases[39]. TGF-β1 is largely found in platelets and most organs, including heart[40]. The role(s) of TGF-β1 in the heart is not yet clear, its properties in vitro suggests that it might be important in various myocardial situations such as cardiac embryogenesis, hypertrophy, atherogenesis, healing of myocardial infarction and development of coronary collaterals[41].

TGF-β1 specific oligonucleotide primers were designed from the published porcine cDNA sequence[42]. Employing reverse transcriptase-PCR, we amplified a DNA product of 423 bp in cDNA preparations derived from normal and collateralized myocardia. PCR product was verified for TGF-β1 specificity by Southern hybridization with a human cDNA insert encoding TGF-β1 (kindly provided by Dr. R. Derynck) and reamplification using internal primers. In a semi-quantitative approach, higher amplification of a TGF-β1 specific DNA fragment of 257 bp was observed after 15 cycles of PCR in collateralized as compared to the normal myocardium. In all experiments, a blank was included (cDNA substituted by water) to check the contamination of reagents used in PCR and no amplification in the blank tube was observed. By PCR amplification of cDNAs, we learnt that TGF-β1 transcripts are present in the swine myocardium and probably this growth factor is produced more during collateral growth. This difference became more apparent when TGF-β1 expression was evaluated by Northern hybridization. Employing Northern hybridization we found a significantly enhanced expression of a 2.4 kb TGF-β1 mRNA in collateralized myocardium as compared to control in pigs with critical stenosis without major infarction[43]. The presence of TGF-β1 mRNA in the normal pig heart suggests that there is ongoing transcription, translation and utilization of this peptide.

Furthermore, we attempted to localize the TGF-β1 transcripts by in situ hybridization. ^{35}S labeled an anti-sense cRNA probe for TGF-β1 was used to hybridize myocardial preparations from normal and collateralized tissue. We found TGF-β1 specific mRNA activity predominantly in the cardiac myocytes in histologially intact myocardium. We performed immunoblotting to detect the protein in the porcine myocardial extracts and found TGF-β1 specific protein band of 25 kDa. No difference in the intensity of signal between normal and collateralized myocardium was found. This could be attributed to the higher turnover of TGF-β1 and probably upregulation of TGF-β1 receptors would also lead to the enhanced internalization and degradation of TGF-β1 in the collateralized tissue. Similar to in situ localization of TGF-β1 mRNA, immunocytochemically we observed the staining in the cardiac myocytes and surprisingly dense staining in the conducting Purkinje cells[43]. Connective tissue, especially around blood vessels, did not show more than background activity. There was no difference in the TGF-β1 mRNA expression in normal and the tissue derived from pigs with complete LCX occlusion and a major infarction. Although TGF-β1 plays an important role in tissue repair and is secreted by fibroblasts and activated macrophages[38,40], after acute myocardial infarction at least in pigs it does not seem to play a significant role. In contrast, Thompson et al.[44] reported that TGF-β1 transcripts were markedly increased in the infarcted rat heart but TGF-β1 immunoreactivity was lost from the cardiac myocytes during infarction.

Vascular Endothelial Growth Factor

Recently described, vascular endothelial growth factor (VEGF) is a highly glycosylated cationic 46-48 kD dimeric heparin binding protein. VEGF is very similar to vascular permeability factor which contain 24 amino acid insertion[44]. VEGF cDNA has been cloned and analysis of its nucleotide sequence suggests that it is a member of platelet derived growth factor (PDGF) family with 21-24% amino acid homology[45]. VEGF is a secretary protein and has been isolated from a number of normal and tumor cells[17]. It is a highly specific peptide mitogen for endothelial cells derived from small and large vessels and is a potent angiogenic factor in vivo[16].

As described above, using similar molecular biological methods, we examined the expression of VEGF in the normal and collateralized myocardium. As a diagnostic search, we performed PCR on cDNA templates prepared from extracted total RNAs from normal and collateralized tissue. We amplified a DNA product of 319 bp by PCR in all cDNA preparations i.e. cDNAs derived from the normal and collateralized myocardium. This PCR product was subjected to Southern hybridization using a cDNA insert of 930 bp encoding human VEGF (kindly provided by Dr. N. Ferrara). A clear dominant band of 319 bp hybridized to the human VEGF insert while the marker lane showed no hybridization confirming the specific amplification of the VEGF in the swine myocardium. Furthermore, the PCR product was cloned into a plasmid vector and recombinant double stranded plasmid DNA was sequenced. The nucleotide sequence analysis of the PCR fragment cloned in the plasmid revealed a homology of 89% with the published human cDNA sequence. Using the PCR product and the human cDNA insert encoding VEGF we detected two mRNAs of 3.9 and 1.7 kb in the normal and collateralized myocardium, however, the expression of VEGF in the collateralized tissue was enhanced[46]. It is clear that VEGF is also expressed in the myocardium, and also probably contribute in the collateral vessel growth in response to ischemia but the onset of its expression and its contributing relationship to the other growth factors is not clear.

SUMMARY

In this study, we examined the expression of angiogenic growth factors β-ECGF, TGF-β1 and VEGF in a porcine model of coronary vascular growth as a reaction to progressive coronary stenosis leading to myocardial ischemia. Myocardial angiogenesis (collateralization) was induced by progredient coronary occlusion of left circumflex coronary artery (LCx) with the help of a hygroscopic constrictor and verified angiographically. By reverse transcriptase PCR, RNA from collateralized as well as from normal myocardium yielded DNA bands corresponding in size spanned by the two primers for β-ECGF, VEGF and TGF-β1. Southern hybridization of the amplified DNA fragments with respective human cDNA probes confirmed the specificity of amplification. By Northern hybridization, we obsereved enhanced expression of β-ECGF, TGF-β1 and VEGF in the collateralized myocardium as compared to the normal myocardium. These results are summarized in Table 2. Furthermore, β-ECGF transcripts were localized in the

Table 2.   Detection and localization of β-ECGF, TGF-β1 and
            VEGF in the porcine myocardium.

GF	Normal tissue (mRNA)	(protein)	Collateralized tissue (mRNA)	(protein)	Localization
β-ECGF	+	-	++	+	Blood vessel wall, myocytes
TGF-β	+	+	++	+	Purkinje cells, myocytes
VEGF	+	ND	+++	ND	ND

- = Not detectable by conventional methods; + = Detectable
signal; ++/+++ = Enhanced expression; ND = Not determined.

blood vessel wall, whereas TGF-β1 was seen in myocytes and
Purkinje cells by in situ hybridization. From these findings,
we conclude that β-ECGF, VEGF and TGF-β1 are expressed in the
heart and they are upregulated during collateralization
suggesting their vital role in the ischemia induced
angiogenesis and in the development of neovascular collateral
vessels. The existence of these angiogenic molecules in the
normal adult heart where virtually no vascular proliferation
is supposed to take place raises questions of further control
mechanisms for their functions.

REFERENCES

1. J. Folkman and M. Klagsbrun, Angiogenic factors. Science 235:442-447 (1987).
2. P. A. D'Amore and R. W. Thompson, Mechanism of angiogenesis. Ann. Rev. Physiol. 49:453-464 (1987).
3. W. H. Burgess and T. Maciag, The heparin binding (Fibroblast growth factor family proteins. Ann. Rev. Biochem. 58:575-606 (1989).
4. W. Schaper and J. Schaper, Adaptation to and defense against myocardial ischemia. Cardiology 77:367-372. (1990).
5. W. Schaper, "The Collateral Circulation of the Heart" North Holland Publishing Co., Amsterdam (1971).
6. S. Pasyk, W. Schaper, J. Schaper, K. Pasyk, G. Miskiewicz, B. Steinseifer, DNA synthesis in coronary collaterals after coronary artery occlusion in concious dog, Am. J. Physiol. 242:H1031-H1037 (1982).
7. W. Schaper, M. De Brabander, P. Lewi, DNA synthesis and mitosis in coronary collateral vessels of the dog, Circ. Res. 28:671-679 (1971).
8. W. Schaper, G. Görge, B. Winkler, J. Schaper, The collateral circulation of the heart, Prog. Cardiovasc. Dis. 31:57-77 (1988).
9. J. Litvak, L. E. Siderides, A. M. Vineberg, The experimental production of coronary artery insufficiency and occlusion, Am. Heart J. 53:505-518 (1957).

10. M. Mohri, H. Tomoike, M. Noma, T. Inone, K. Hisano, M. Nakamura, Duration of ischemia is vital for for collateral development: Repeated brief coronary artery occlusions in conscious dogs, Circ. Res. 64:287-296 (1988).

11. F. C. White, D. M. Roth, C. M. Bloor, Coronary collateral reserve during exercise induced ischemia in swine. Basic Res. Cardiol. 84:42-54 (1989).

12. W. M. Chilian, H. J. Mass, S. E. Williams, S. M. Layne, E. S. Smith, K. W. Scheel, Microvascular occlusions promote coronary collateral growth, Am. J. Physiol. 258:H1103-H111 (1990).

13. G. Görge, T. Schmidt, B. R. Ito, G. A. Pantely, W. Schaper, Microvascular and collateral adaptation in swine hearts following progressive coronary artery stenosis. Basic Res. Cardiol. 84:524-535 (1989).

14. G. A. Pantely and H. H. Kerr, Failure to reduce infarct size in the rat, Am. J. Cardiol. 47;443 (1981).

15. F. Ishikawa, K. Miyazono, U. Hellman, H. Drexler, C. Wernstedt, K. Usuki, F. Takaku, W. Risau, C. H. Heldin, Identification of biologic activity and the cloning and expression of platelet-derived endothelial cell growth factor, Nature 338:557-562 (1989).

16. D. W. Leung, G. Cachianes, W. J. Kuang, D. V. Goeddel, N. Ferrara, Vascular endothelial growth factor is a secreted angiogenic mitogen, Science 246:1306-1309 (1989).

17. M. Klagsbrun, P. A. D' Amore, Regulators of angiogenesis. Ann. Rev. Physiol. 53:217-239 (1991).

18. D. Gospodarowicz, N. Ferrara, L. Schweigerer, G. Neufeld, Structural characterization and biological functions of fibroblast growth factor, Endocrine Reviews 8:95-109 (1987).

19. W. Risau and H. G. Zerwes HG, Role of growth factors in the formation of blood vessels, Z. Kardiol. 78:9-11 (1989).

20. W. Quinckler, M. Maasberg, S. Bernotat-Danielowski, N. Lüthe, H. S. Sharma, W. Schaper, Isolation of heparin-binding growth factors from bovine, porcine and canine hearts, Eur. J. Biochem. 181:67-73 (1989).

21. W. Cassells, E. Spier, J. Sasse, M. Klagsbrun, P. Allen, M. Lee, B. Calvo, M. Chiba, L. Haggroth, J. Folkman, S. Epstein, Isolation characterization, and localization of heparin-binding growth factors in the heart, J. Clin. Inv. 85:433-441 (1990).

22. H. Sasaki, H. Hoshi, Y. M. Hong, T. Suzuki, T. Kato, H. Sasaki, M. Saito, H. Youki, K. Karube, S. Kono, M. Onodera, S. Aoyagi, Purification of acidic fibroblast growth factor from bovine heart and its localization in the cardiac myocytes, J. Biol. Chem. 264:17606-17612 (1989).

23. W. H. Burgess, T. Mehlman, D. R. Marshak, B. A. Fraser, T. Maciag, Structural evidence that endothelial cell growth factor factor-β is the precursor of both endothelial cell growth factor a and acidic fibroblast growth factor, Proc. Natl. Acad. Sci. USA 83:7216-7220 (1986).

24. I. Vlodosky, R. Friedman, R. Sullivan, J. Sasse, M. Klagsbrun, Aortic endothelial cells synthesize basic fibroblast growth factor which remains cell associated and platelet derived growth factor like protein which is secreted, J. Cell. Physiol, 131:402-408 (1987).

25. P. Chomczynski and N. Sachhi, Single-step method of RNA isolation by acid guanidinium thiocyanate-phenol-chloroform extraction, Anal. Biochem. 162:156-159 (1987).

26. K. M. Rosen, E. D. Lamperti, L. Villa-´Komaroff, Optimizing the Northern blot procedure, Biotechniques 8:398-403 (1990).

27. M. Jaye, R. Howk, W. Burgess, G. A. Ricca, I.-M. Chiu, M. W. Ravera, S. J. O'Brien, W. S. Modi, T. Maciag, W. N. Drohan, Human endothelial cell growth factor: Cloning, nucleotide sequence and chromosome localization, Science 233:541-545 (1986).

28. F. Esch, N. Uenao, A. Baird, F. Hill, L. Denoroy, N. Ling, D. Gospodarowicz, R. Guillemin, Primary structure of bovine brain acidic fibroblast growth factor (FGF), Biochem. Biophys. Res. Comm. 133:554-562 (1985).

29. F. S. McCallum and B. E. H. Maden, Human 18S ribosomal RNA sequence infered from DNA sequence. Biochem. J. 232:725-733 (1985).

30. R. K. Saiki, D. H. Gelfand, S. Stoffel, S. J. Scharf, R. G. Higuchi, G. T. Horn, K. B. Mullis, H. A. Erlich, Primer directed enzymatic amplification of DNA with a thermostable DNA polymerase, Science 239:487-491 (1988).

31. M. Schmidt, H. S. Sharma, W. Schaper, Amplification and sequencing of an mRNA encoding acidic fibroblast growth factor from porcine heart, communicated.

32. H. S. Sharma, R. Kandolf, T. Markert, W. Schaper, Localization of endothelial cell growth factor-β mRNA in the pig heart during collateralization, Circulation 80:II-453 (1989).

33. H. L. Weiner HL and J. L. Swain, Acidic fibroblast growth factor mRNA is expressed by cardiac myocytes in culture and the protein is localized to the extracellular matrix, Proc. Natl. Acad. Sci. (USA) 86:2683-2687 (1989).

34. E. Spier, Z. Yi-Fu, M. Lee, S. Shrivastva, W. Casscells, Fibroblast growth factors are present in adult cardiac myocytes in vivo, Biochem. Biophys. Res. Comm. 157:1336-1340 (1988).

35. S. Bernotat-Danielowski, R. J. Schott, H. S. Sharma, P. Kremer, W. Schaper, Fibroblast growth factor (FGF), an endothelial mitogen, is localized in cardiomyocytes of the ischemic collateralized pig heart, Circulation 82:III-377 (1990).

36. M. B. Sporn, A. B. Roberts, L. M. Wakefield, B. Crombrugghe, Some recent advances in the chemistry and the biology of TGF-β1, J. Cell. Biol, 105:1039-1045 (1987).

37. A. B. Roberts, M. B. Sporn, R. K. Assoian, J. M. Smith, N. S. Roche, L. M. Wakefield, U. I. Heine, L. A. Liotta, V. Falanga, J. H. Kehrl, A. S. Fauci, TGF-β1: Rapid induction of fibrosis and angiogenesis in vivo and stimulation of collagen formation in vitro, Proc. Natl. Acad. Sci. (USA) 83:4167-4171 (1986).

38. T. A. Mustoe, G. F. Pierce, A. Thomason, P. Gramates, M. B. Sporn, T. F. Deuel, Accelerated healing of incisional wounds in rats induced by TGF-β1, _Science_ 237:1333-1335 (1987).

39. D. A. Lawrence, P. Pircher, P. Jullien, Conversion of a high molecular weight latent beta-TGF from chicken embryo fibroblasts into a low molecular weight active beta-TGF under acidic conditions, _Biochem. Biophys. Res. Commun._ 133;1026-1034 (1985).

40. A. B. Roberts, M. A. Anzano, L. C. Lamb, J. M. Smith, M. B. Sporn, New class of transforming growth factors potentiated by epidermal growth factor: Isolation from non neoplastic tissues, _Proc. Natl. Acad. Sci. (USA)_ 78:5339-5343 (1981).

41. W. Schaper, H. S. Sharma, W. Quinkler, T. Markert, M. Wünsch, J. Schaper, Molecular biologic concepts of coronary anastomases, _J. Am. Coll. Cardiol._ 15:513-518 (1990).

42. R. Derynck and L. Rhee, Sequence of the porcine TGF-β1 precursor, _Nucleic Acid. Res._ 15:3187 (1987).

43. M. Wünsch, H. S. Sharma, T. Markert, S. Bernotat-Danielowski, R. J. Schott, P. Kremer, N. Bleese, W. Schaper, In situ localization of transforming growth factor-β1 in porcine heart: Enhanced expression after chronic coronary artery occlusion, _J. Mol. Cell. Cardiol._, in press.

44. P. J. Keck, S. D. Hauser, G. Krivi, K. Sanzo, T. Warren J. Feder, D. T. Connolly, Vascular permeability factor, an endothelial cell mitogen related to PDGF, _Science_ 246:1309-1312 (1989).

45. E. Tischer, D. Gospodarowicz, R. Mitchell, M. Silva, J. Schilling, K. Lau, T. Crisp, J. C. Fiddes, J. A. Abraham, Vascular endothelial growth factor: A new member of the platelet derived growth factor gene family, _Biochem. Biophys. Res. Commun._ 165:1198-1206 (1989).

46. H. S. Sharma, S. Sack, W. Schaper, Vascular endothelial growth factor in the porcine heart: Enhanced expression during ischemia induced collateralization, _in preparation._

ROLE OF MECHANICAL FACTORS IN ANGIOGENESIS UNDER

PHYSIOLOGICAL AND PATHOPHYSIOLOGICAL CIRCUMSTANCES

O.Hudlicka

Department of Physiology
University of Birmingham Medical School
Birmingham B15 2TJ, UK

INTRODUCTION

The role of growth factors in angiogenesis has been extensively
studied under pathological circumstances such as wound healing,
regeneration, tumour growth, diabetic microangiopathy, retrolental
fibroplasia or development of collateral circulation. Their importance
has been recently revied by Folkman & Klagsbrun (1987), D'Amore &
Thompson (1987) or Hudlicka & Tyler (1986). However, it has become
increasingly obvious that mechanical factors, such as increased shear
stress or stretch can either induce proliferation of endothelial cells in
vitro on their own, or can modify the sensitivity of cells – or vessels –
to growth factors. Thus Folkman & Moscona (1978) showed that
incorporation of ^{3}H thymidine into endothelial cells was the greater the
flatter the cells. Dewey et al (1981) and Ando et al (1987) reported
proliferation of endothelial cell exposed to shear stress in cultures.
Increased tension (ensured by small intraluminal springs) stimulated
uptake of labelled thymidine and proline into the walls of arterial
segments in vitro (Hume, 1980). Ingber et al (1987) demonstrated higher
^{3}H thymidine incorporation into stretched endothelial cells with larger
area particularly in the presence of fibroblast growth factor.

Support for the role of the mechanical factors in growth of vessels
in vivo comes mainly from studies during development. It was suggested
many decades ago (Thoma 1893) that vessel growth is the result of
combined effects of blood flow and tension exerted by the surrounding
growing tissue. Role of blood flow was quite clearly demonstrated during
development and regeneration by Clark (1918) and Clark & Clark (1940).
High blood flow is also found in many of the above mentioned pathological
circumstances such as inflammation, healing wounds or diabetic
microangiopathy (see Hudlicka & Tyler,1986). This paper brings some
evidence how increased blood flow, and factors connected with it,such as
high shear stress or capillary wall tension, can explain capillary
proliferation in skeletal and cardiac muscle. In addition, the effect of
mechanical stretch exerted on capillarie s in skeletal and cardiac muscle
is also considered and the role of increased pressure in the alteration
of the vascular bed is briefly discussed.

Angiogenesis in Health and Disease, Edited by M.E. Maragoudakis *et al.*
Plenum Press, New York, 1992

CAPILLARY GROWTH IN SKELETAL MUSCLES

Growth of capillaries in skeletal muscles has been described mainly as a result of endurance training or exposure to low temperatures (for review see Hudlicka et al, 1991) – situations where blood flow is increased either as a result of repeated muscle contractions or in response to increased demands for oxygen. Much greater increase in capillary growth – demonstrated either as a presence of sprouts which usually appeared at a bending of an existing capillary (Myrhage & Hudlicka, 1978) or as an increase in capillary density (number of capillaries/mm^2,CD), capillary/fibre ratio (C/F) or capillary length based on estimation of all anatomically present capillaries by counting the profiles in low–power electron micrographs (Hudlicka et al, 1987) or by staining of capillary endothelium for alkaline phosphatase (Brown et al, 1976) was shown in chronically stimulated muscles that were contracting at 10 Hz for 8 h/day. This activity resulted in capillary proliferation within 4 days (Brown et al, 1976) and the size of capillary bed doubled within 4 weeks. In spite of a great increase in capillary supply, a low molecular angiogenic factor was found in the same proportion (25%) of extracts from normal and stimulated muscles (Hudlicka et al,1989). Stimulated muscles had also increased activity of oxidative enzymes (Pette et al,1973) which could be taken as an indicator of hypoxia. However, when blood flow was limited by ligation of the main supplying (iliac) artery, there was no increase in capillarization although hypoxia – which is sometimes considered as an important stimulus for endothelial cell proliferation – was even greater than in stimulated muscles with intact vascular supply (Hudlicka & Price, 1990). Further evidence for the role of blood flow in capillary proliferation was provided by experiments where capillary growth was induced by a long–term administration of vasodilating drugs:Tornling et al (1980) induced capillary growth by application of dipyridamole, Ziada et al (1984) by adenosine or xanthine derivative HWA 285 and Dawson & Hudlicka (1989) by prazosin.

By what mechanism could increased blood flow stimulate capillary growth? Capillary growth in vivo – as well as in confluent endothelial cell cultures in vitro – can only start after the removal of contact inhibition (see D'Amore & Thompson, 1987). This can result from cell wounding which, however, can take place even when the endothelial cell surface is only slightly damaged. Disturbance of the endothelial cell results either in release of proteases that would break down the basement membrane and thus enable migration and consecutive mitosis of endothelial cells resulting eventually in the formation of sprouts, or it can release growth factors, particularly bFGF, from the basement membrane (D'Amore & Orlidge, 1988).

There are very few direct measurements in vivo which could elucidate which factors connected with increased blood flow could initiate angiogenesis. High blood flow can lead to increased intravascular pressure and/or increased vessel diameter, and thus increased wall tension. It can also result in increased velocity of flow and thus increased shear stress and/or increased blood cell – endothelial cell interaction. In order to differentiate which of these factors is crucial, we measured velocity of flow and capillary diameters (using intravital observations, Dawson et al, 1987) in experiments where capillary growth was induced either by chronic stimulation or by long–term administration of prazosin. In addition, increased red blood cell – capillary endothelial cell interaction was brought about by a drastic increase in haematocrit (from 45 to 60%) by long–term administration of $CoCl_2$ (Miller & Hale, 1970). Administration of prazosin resulted in a 76% higher velocity of flow than in control animals; chronic electrical stimulation increased red blood cell velocity by 55% while administration of $CoCl_2$ was without effect. Capillary diameters were decreased by 10% after prazosin treatment and increased by 6% in both chronically stimulated and CoCl2 treated animals. Capillary/fibre ratio was increased by 20% in stimulated and prazosin treated animals, but there was no change in C/F after $CoCl_2$ treatment. Calculated shear stress was 12 dyn/cm^2 in capillaries in control muscles, 20 in prazosin treated and 19 dyn/cm^2 in stimulated muscles. Calculated capillary wall tension, assuming capillary pressure of 25 mmHG (Fronek & Zweifach, 1975) was increased by 20% in stimulated muscles and decreased by 8% after prazosin treatment.

How increased shear stress and/or increased capillary wall tension can induce endothelial cell proliferation is still open to speculation. They may brake down gap junction and thus disturb the contact inhibition (Davies et al, 1988). Shear stress increases permeability for Ca^{++} and hence intracellular calcium concentration (Ando et al, 1988) – a prerequisite for cell division (Berridge, 1975). Since Ca^{++} channels can also be opened by stretch (Lansman & Hallam, 1987), increased wall tension found in capillaries in chronically stimulated muscles could initiate capillary growth together with increased shear stress while shear stress would be the main factor in capillary growth in prazosin treated animals.

Prazosin treatment resulted in a 20% increase in capillary supply after 5 weeks of administration. Similar increase was achieved by chronic stimulation applied for only 7 days, and 4 weeks stimulation caused a 100% increase in both CD and C/F. Acute administration of the dose of prazosin that was given chronically caused a greater increase in blood flow than acute muscle stimulation (Hudlicka et al, 1984). Thus muscle stimulation obviously exterts some other influence in addition to increased flow such as stretching and shortening of muscle fibres and their supplying capillaries during contractions and relaxation. If stretch is important in the initiation of capillary growth, increased capillarisation should appear in muscles exposed to stretch only. Extirpation of muscle agonists (eg tibialis anterior) overloads the remaining muscles (extensor digitorum longus) and causes a permanent stretch and fibre hypertrophy (eg Saleo et al, 1980). This situation indeed resulted in an increase in C/F ratio as well as CD from control values of 1.4 and 540/mm^2 to 2.1 and 960/mm^2 respectively. Stimulation of stretched muscles caused a similar increase to that found in overloaded muscles alone. Thus capillary growth in chronically stimulated muscles can be iniated by increased shear stress, increased capillary wall tension and stretch of the capillary wall in conjunction with stretch of the muscle fibres, while increased shear stress is very probably the sole factor initiating capillary proliferation during administration of vasodilating drugs where the growth is less extensive than in stimulated muscles.

CAPILLARY GROWTH IN THE HEART

Unlike in skeletal muscles, endurance training elicited capillary growth in the hearts of only young animals (Tomanek, 1970, Ljungqvist et al, 1984). However, hearts of so called athletic animals (hare, wild rat) are known to have a much higher capillary density than those of similar sedentary species (rabbit, laboratory rat, Poupa et al, 1970). Athletic animals also have a considerably lower heart rate. When the heart rate in rabbits was artificially lowered to about half of the original value by chronic bradycardial pacing, capillary density increased within 2-4 weeks up to 70 % over that found in hearts of comparable size with normal heart rate (Wright & Hudlicka, 1981) without any sign of heart hypertrophy (Hudlicka et al, 1988). Since the extent of capillary supply was estimated either by counting of all capillary profiles based either on low-power electron micrographs or staining of capillary endothelium for alkaline phosphatase, this finding indicates a considerable capillary growth. In contrast, heart hypertrophy due to either pressure or volume overload is linked with a decrease in capillary density because capillaries do not grow while myocytes increase in size. Neither training nor administration of drugs succeeded in stimulating capillary growth in pathological heart hypertrophy (Rakusan, 1987). However, when heart hypertrophy was elicited by aortic valve lesion and the hearts started to be bradycardially paced one month later, capillary density increased by almost 60% within a month of pacing (Wright et al, 1989). Bradycardial pacing was also beneficial in cases of subendocardial necrosis elicitied by high doses of noradrenaline (Brown & Hudlicka, 1988).

What factors could initiate capillary growth in the heart? In parallel with skeletal muscles, the most obvious one could be increased blood flow: bradycardial pacing prolongs the duration of diastole during which blood flow is much higher than during systole. Moreover, long-term administration of vasodilators caused capillary growth in the heart: Tornling (1982) found capillary proliferation after treatment with dipyridamole, Mall et al (1980) after long-term administration of ethanol and Ziada et al (1984) in hearts of animals infused for 3-5 weeks with either adenosine or a xanthine derivative HWA 285. Prazosin treatment which did not increase coronary blood flow was ineffective in the heart - in contrast to skeletal muscle (Ziada et al, 1989). Maximal coronary blood flow is lower in hypertrophic hearts (Brown & Wright, 1991) with lower capillary density. However, resting or maximal blood flow was not higher in bradycardially paced hearts (Hudlicka et al, 1989), so it is unlikely that increased shear stress resulting from high blood flow could be the stimulus for capillary growth. Increased capillary wall tension is more likely involved: Tillmans et al (1974) described greater capillary diameters during diastole than during systole which implies greater capillary wall tension. Since the duration of diastole is considerably longer during bradycardial pacing, stretching of capillary wall due to increased tension might be important. In addition, increased end-diastolic volume stretches the myocytes, and hence stroke volume and stroke work are increased (Wright & Hudlicka, 1981). Thus stretch of the capillary wall due to stretch of muscle fibres could affect the capillary basement membrane and possibly lead to its damage, release of growth factors and capillary growth. This hypothesis was strenghtened by the finding of low molecular angiogenic factor in all paced hearts (Hudlicka et al, 1989) and by experiments with long-term administration of a drug with a positively inotropic effect - dobutamine. Two weeks administration of dobutamnine increased capillary density by 30% in the absence of a significant increase in coronary blood flow while the force of contraction assessed by dp/dt index and maximal cardiac work were significantly increased (Brown & Hudlicka, 1991).

CHANGES IN THE VASCULAR BED IN HYPERTENSION

Increased blood pressure has been long recognised as a stimulus for proliferation and hypertrophy of smooth muscle cells. However, there is no evidence that hypertension is linked with capillary growth. Gray (1984) found no difference in C/F ratio or capillary density between normotensive and spontaneously hypertensive (SHR) animals and Henrich described even lower than control values in muscles from SHR (1979) and from hypertensive patients (1988). This can be explained by the fact that capillary pressure is not different in hypertensive and control animals (Bohlen et al, 1977). However, there are also fewer arterioles (Prewitt et al, 1982) and small arteries (Struyker-Boudier et al, 1988) and there is no good explanation for this phenomenon.

There is no doubt that high pressure induces hypertrophy and hyperplasia in the media of large arteries (Bevan et al, 1976), and Gabbiani et al (1979) described even endothelial thickening in the aorta in hypertensive animals. However, the data on the thickening of the wall in smaller arteries and arterioles are very controversial. Folkow (1982) described increased wall/lumen ratio in small precapillary resistance vessels in the hind limbs of SHR and this was recently confirmed in the heart in rats with renal hypertension by Tomanek et al (1990). In their experiments larger arterioles (25µm) and small arteries (50-60µm) were not different from those in normotensive animals. In contrast, Aalkjaer et al (1987) did not find any difference in wall/lumen ratio in small (100µm) arteries from subcutaneous tissue of hypertensive patients, but reported a considerable increase in wall thickness in arteries twice that size. Mulvany & Aalkjaer (1990) concluded that although these vessels probably contribute to the increase in the peripheral resistance less than previously assumed, thickening of the wall and a consecutive decrease in the lumen may diminish the maximal dilatation in individual organs and thus cause ischaemia.

CONCLUSIONS

Mechanical factors involved in the remodeling of the vascular tree include blood pressure and blood flow as well as stretch by surrounding tissues. While blood pressure is most important in the growth of vascular smooth muscle during hypertension, its importance as a stimulus for growth of capillaries, if any, is questionable.

Our current hypothesis regarding the role of mechanical versus growth factors in skeletal and cardiac muscle can be summarized as follows. In skeletal muscle, increase in blood flow causes increased velocity of red blood cells and hence increased shear stress; in chronically stimulated muscles, the capillary wall tension is also increased due to wider capillaries. These changes are likely to disturb the luminal side of capillary endothelial cell and consequently release bound proteases which would disrupt the intergrity of the basement membrane. Externally imposed stretch could provide additional factor. The disruption of the basement membrane can lead to a release of growth factors although there is so far little or no experimental evidence for this. Migration of endothelial cells outside, mitosis and formation of sprouts then follows. In the heart, capillary growth may be elicited by procedures which do (vasodilators) or do not (bradycardia, dobutamine) increase blood flow. Increased velocity of flow and high shear stress therefore do not seem to play an important role. Bradycardia results in increased wall tension due to wider capillaries in diastole. In addition, increased force of contraction can stretch the capillary wall tension from outside and possibly disturb the basement membrane and relase growth factors.

REFERENCES

Aalkjaer, C., Heagerty, A.M., Petersen, K.K., Swales, J.D. & Mulvany, M.J. Evidence for increased media thickness, increased neuronal amine uptake and depressed excitation-contraction coupling in isolated resistance vessels from essential hypertensives. Circ. Res. 61, 181-186, 1987.

Ando, J., Nomura, H., and Kamiya, A. The effect of fluid shear stress on the migration and proliferation of cultured endothelial cells. Microvasc. Res. 33, 62-70, 1987.

Ando, J., Komatsuda, T., and Kamiya, A. Calcium response to fluid shear stress in cultured vascular endothelial cells. In vitro, 24, 871-877, 1988.

Berridge, M.J. The interaction of cyclic nucleotides and calcium in the control of cellular activity. Adv. Cyclic Nucleotide Res. **6**, 1–98, 1975.

Bevan, R.D., Van Marthens, E., and Bevan, J.A. Hyperplasia of vascular smooth muscle in experimental hypertension in the rabbit. Circ. Res. Suppl. II **38**, 58–62, 1976.

Bohlen, H.G., Gore, R.W., and Hutchings, P.M. Comparison of microvascular pressures in normal and spontaneously hypertensive rats. Microvasc. Res. **13**, 125–130, 1977.

Brown, M.D. and Hudlicka, O. Protective effects of long–term bradycardial pacing against catecholamine–induced myocardial damage in rabbit hearts. Circ. Res. **62**, 965–974, 1988.

Brown, M.D. & Hudlicka, O. Capillary growth in the heart. In Angiogenesis, eds Weisz PB, Langer R & Steiner, R. Karger, Basel, in press, 1991.

Brown, M.D. & Wright. A.J.A. The relation of anatomical capillary supply to coronary blood flow and cardiac performance in anaesthetised rabbits. J. Physiol. **435**, 17P, 1991.

Brown, M.D., Cotter, M.A., Hudlicka, O., and Vrbova, G. The effects of different patterns of muscle activity on capillary density, mechanical properties and structure of slow and fast rabbit muscles. Pflgers Arch., **361**, 241–250, 1976.

Clark, E.R. Studies on the growth of blood vessels in the tail of the frog. Am. J. Anat. **23**, 37–88, 1918.

Clark, E.R. & Clark, E.L. Microscopic observation on the extraendothelial cells of living mammalian blood vessels. Am. J. Anat. **66**, 1–49, 1940.

D'Amore, P.A. and Orlidge, A. Growth factors and pericytes in microangiography. Diabete & Metabolism **14**, 495–504, 1988.

D'Amore, P.A. and Thompson, R.W. Mechanisms of angiogenesis. Ann. Rev. Physiol. **49**, 453–464, 1987.

Davies, P.F., Olensen, S.P., Clapham, D.E., Morrell, E.M. and Schoen, F.J. Endothelial communication: State of the art lecture. Hypertension **11**, 563–572, 1988.

Dawson, J.D. & Hudlicka, O. The effects of long–term administration of prazosin on the microcirculation in skeletal muscles. Cardiovasc. Res. **23**, 913–920, 1989.

Dawson, J.D., Tyler, K.R., and Hudlicka, O. A comparison of the micro–circulation in rat fast glycolytic and slow oxidative muscles at rest and during contractions. Microvasc. Res. **33**, 167–182, 1987.

Dewey, C.F. Jr., Bussolari, S.R., Gimbrone, M.A. Jr., and Davies, P.F. The dynamic response of vascular endothelial cells to fluid shear stress. J. Biomech. Eng., **103**, 177–185, 1981.

Folkman, J. and Klagsbrun, M. Angiogenic factors. Science **235**, 442–446, 1987a.

Folkman, J. and Moscona, A. Role of cell shape in growth control. Nature, Lond. **273**, 245–349, 1978.

212

Folklow, B. Physiological aspects of primary hypertension. Physiol. Rev. **62,** 347–504, 1982.

Fronek, K. and Zweifach, B. Microvascular pressure distribution in skeletal muscle and the effect of vasodilatation. Am. J. Physiol. **228,** 791–796, 1975.

Gabbiani, G., Elemer, G., Guelpa, C.H., Vallotton, M.B., Badonnel, M.C. & Huttner, I. Morphologic and functional changers of the aortic intima during experimental hypertension. Am. J. Pathol. **96,** 399–442, 1979.

Gray, S. Morphometric analysis of skeletal muscle capillaries in early spontaneous hypertension. Microvasc. Res. **27,** 39–50, 1984.

Hudlicka, O., Brown, M.D. & Eggington, S. Angiogenesis in skeletal and cardiac muscle. Physiol. Rev. in press.

Hudlicka, O. and Price, S. The role of blood flow and or muscle hypoxia in capillary growth in chronically stimulated fast muscles. Pflugers Arch. **417,** 67–72, 1990.

Hudlicka, O., West, D., Kumar, S., El Khelly, F., and Wright, A.J.A. Can growth of capillaries in the heart and skeletal muscle be explained by the presence of an angiogenic factor? Br. J. Exp. Path. **70,** 237–246, 1989.

Hudlicka, O., Wright, A.J.A., Hoppeler, H., and Uhlmann, E. The effect of chronic bradycardial pacing on the oxidative capacity in rabbit hearts. Resp. Physiol. **72,** 1–12, 1988.

Hudlicka, O. and Tyler, K.R. Angiogenesis. The growth of the vascular system. London: Academic Press, P. 221, 1986.

Hudlicka, O., Tyler, K.R., Wright, A.J.A., and Ziada, A.M.A.R. Growth of capillaries in skeletal muscles. Prog. Appl. Microcirc. **5,** 44–61, 1984.

Hume, W.R. Proline and thymidine uptake in rabbit ear artery segments in vitro increased by chronic tangential load. Hypertension **2,** 738–743, 1980.

Ingber, D.E., Madri, J.A., and Folkman, J. Endothelial growth factors and extracellular matrix regulates DNA synthesis through modulation of cell and nuclear expansion. In vitro **23,** 387–394, 1987.

Lansman, J.B. and Hallam, T.J. Single stretch–activated ion channels in vascular endothelial cells as mechanotransducers? Nature **325,** 811–812, 1987.

Ljungqvist, A., Tornling, G., Unge, G., Jurdahl, E.B., and Larsson, B. Capillary growth in the heart and skeletal muscle during dipyridamole treatment and exercise. Prog. Appl. Microcirc. **4,** 9–15, 1984.

Mall, G., Mattfeldt, T., and Volck, B. Ultrastructural morphometric study on the rat heart after chronic ethanol feeding. Virchow's Arch. Abt. A Path. Anat. **389,** 59–77, 1980.

Miller, A.T. and Hale, D.M. Increased vascularity of brain, heart and skeletal muscle of polycythemic rats. Am. J. Physiol. **219,** 702–704, 1970.

Mulvany, M.J., Aalkjaer, C. Structure and function in small arteries. Physiol. Rev. **70**, 921–962, 1990.

Myrhage, R. and Hudlicka, O.   Capillary growth in chronically stimulated adult skeletal muscle as studied by intravital microscopy and histological methods in rabbits and rats. Microvasc. Res. **16**, 73–90, 1978.

Pette, D., Smith, M.E., Staudte, H.W., and Vrbova, G.   Effect of long–term electrical stimulation on some contractile metabolic characteristics of fast rabbit muscles. Pflgers Arch. **338**, 257–272, 1973.

Poupa, O., Rakusan, K., and Ostadal, B.   The effect of physical activity upon the heart of the vertebrates. Medicine Sport. **4**, 202–235, 1970.

Prewitt, R.L., Chen, I.I.H., and Dowell, R.   Development of microvascular rarefaction in the spontaneously hypertensive rat. Am. J. Physiol. **243**, H243–H251, 1982.

Rakusan, K. Microcirculation in the stressed heart. In: The Stressed Heart, edited by M.J. Legato. Boston: Martinus Nijhoff Publishing, p. 107–123, 1987.

Saleo, A., Anastasi, G., La Spada, G., Falzea, G. & Denaro, M.G. New muscle fibre production during compensatory hypertrophy. Med. Sci. Sports. Exerc. **12**, 268–273, 1980.

Struyker–Boudier, H.A.J., Le Noble, J.L.M.L., Slaaf, D.W., Smits, J.F.M., and Tangelder, G.J.   Microcirculatory changes in cremaster muscle during early spontaneous hypertension in the rat. J. Hypertension **6**, S185–S188, 1988.

Thoma, R. Untersuchungen uber die Histogenese und Histomechanik des Gefssystems. Enkeverlag, Stuttgart, 1893.

Tillmans, T.H., Ikeda, S., Hansen, H., Sarma, J.S., Fauvel, J.H., and Bing, R.J.   Microcirculation in the ventricle of the dog and turtle. Circ. Res. **34**, 561–569, 1974.

Tomanek, R.J.   Effects of age on the extent of the myocardial capillary bed. Anat. Rec. **167**, 55–62, 1970.

Tomanek, R.J., Aydelotte, M.R. & Butters, C.A., Late onset renal hypertension in old rats alters myocardial microvessels. Am. J. Physiol. **259**, H1681–H1687, 1990.

Tornling, G. Capillary neoformation in the heart of dipyridamole treated rats. Acta Pathol. Microbiol. Scand. sec A. **90**, 269–271, 1982.

Tornling, G., Unge, G., Adolfsson, J., Ljungovist, A. and Carlsson, S. Proliferative activity of capillary wall cells in skeletal muscle in rats during long–term treatment with dipyridamole. Arzneimittel–Forsch. **30**, 622–623, 1980.

Wright, A.J.A. and Hudlicka, O.   Capillary growth and changes in heart performance induced by chronic bradycardial pacing in the rabbit. Circ. Res. **49**, 469–478, 1981.

Wright, A.J.A., Hudlicka, O., and Brown, M.D.   Beneficial effect of chronic bradycardial pacing on capillary growth and heart performance in volume overload heart hypertrophy.   Cric. Res. **64**, 1205–1212, 1989.

Ziada, A.M.A.R., Hudlicka, O., Tyler, K.R., and Wright, A.J.A.   The effect of long–term vasodilatation on capillary growth and performance in rabbit heart and skeletal muscle.   Cardiovasc. Res. **18**, 724–732, 1984.

Ziada, A.M.A.R., Hudlicka, O., and Tyler, K.R.   The effect of long–term administration of '1–blocker prazosin on capillary density in cardiac and skeletal muscle.   Pflgers Arch. **415**, 355–360, 1989.

LOW MOLECULAR MASS NON-PROTEIN ANGIOGENIC FACTORS AND THEIR ROLE IN
THE DISEASE PROCESSES

Jacqueline B. Weiss, and Raj Odedra

Wolfson Angiogenesis Unit,
Rheumatic Diseases Centre,
University of Manchester, Clinical Sciences Building
Hope Hospital Salford M6 8HD, UK

Introduction

In current times when molecular biology is King attention is
generally directed towards protein factors at the expense of perhaps
equally important low molecular weight non-protein factors. To a large
extent this has been true for angiogenesis. Protein angiogenic factors
such as fibroblast growth factors (FGFs) have received dispro-
portionate attention despite being notoriously non-specific for the
microvasculature where angiogenesis occurs.

However, many low molecular weight angiogenic factors, some of
them totally specific to the microvasculature receive less attention.
These small (not lesser) molecules may have roles either as initiators
or as co-factors in the neovascularisation process or may act
synergistically with protein initiators. Weiss et al (1979) isolated a
low molecular weight compound (Mr $\pm$ 400) which gave a positive result
on the chick chorioallantoic membrane (CAM) test, from the so called
tumour angiogenic factor (TAF) which was a crude extract of both
intracellular material and extracellular matrix rat Walker tumours,
(Folkman et al 1971). After extraction of this factor the angiogenic
activity of TAF was very significantly reduced. In fact, it was not
possible to obtain a positive result on the CAM with the same amount
of TAF after the extraction process. Subsequently, Weiss et al (1983)

Angiogenesis in Health and Disease, Edited by M.E. Maragoudakis *et al.*
Plenum Press, New York, 1992

showed that the small molecule extracted from TAF not only gave an angiogenic response on the chick chorioallantoic membrane but also stimulated microvessel endothelial growth in culture. For this latter reason they call the factor endothelial cell stimulating angiogenesis factor (ESAF).

At the same time that ESAF was extracted from TAF, workers in Australia (McAuslan and Hoffman 1979) described a factor from extra-cellular matrix of rat Walker tumour which they extracted via a method which was not the same as that of Folkman and which had a molecular mass of the order of 200Da. They called their factor endothelial stimulating factor (ESF). Although the molecular weights did not agree, they are almost certainly the same molecule. In both cases the original molecular mass determination was by gel filtration which is notoriously inaccurate. However, although the definite structure of ESAF is not yet known, it is now clear from mass spectrometric analysis that the molecular weight is less than 600Da. Since this time ESAF has been shown to have a variety of interesting biochemical properties. Perhaps the most useful property is its ability to activate pro and latent MMP-1 (classical collagenase) since this enables a method for its quantification to be designed. The manner and mechanism by which it does this will be discussed in a separate paper.

This discussion of low molecular mass factors is not meant to be all encompassing. For a comprehensive revue the reader is referred to Odedra and Weiss (1991).

What is an Angiogenic Factor?

A major problem in defining an angiogenic factor is whether it gives a positive test in more than one type of _in vivo_ system, whether it can stimulate microvessel cells in culture and whether it is specific for these cells, and again is it chemotactic for microvessel endothelial cells. Since many accepted high molecular mass angiogenic factors do not, as has been mentioned earlier, fulfill all these criteria it may seem hard to be restrictive for the non-protein factors but the fact is that many low molecular mass factors may induce angiogenesis _in vivo_ by indirect means. In other words they may act as initial metabolic signals for release or synthesis of angiogenic factors from tissues stressed perhaps by low oxygen tension, or as cofactors for enzymes involved in the synthesis of an angiogenic factor or activation of receptors.

Lactic Acid

Some low molecular mass factors are either co-factors for anaerobic metabolism or actual products of such metabolism. One such metabolite of the anaerobic metabolism of glucose is lactic acid. This has variously been described as an angiogenic molecule in the cornea (Imre 1984) and to have no effect on endothelial cells _in vitro_ (Atherton, 1977) or _in vivo_ on CAM (Barnhill and Ryan 1983). Such contradictory evidence can possibly be explained by the observation that macrophages may be stimulated to produce angiogenic activity if they are incubated _in vitro_ with lactate (Jensen et al 1986). This is of great interest as it suggests that lactate may influence angiogenesis indirectly by an accumulation or activation of macrophages. This is an excellent example of a relationship between a small molecule influencing the production of a protein angiogenic factor - in this case from macrophages, Leibovich et al (1987) have identified the angiogenic factor in macrophages as tumour necrosis factor (TNF). In wound healing where some degree of anoxia has been linked with angiogenesis (Knighton et al 1984, Hunt et al 1984) such a system could very easily be seen to operate.

Metal Ions as Potentiators of Angiogenesis

Both copper and selenium have been implicated in angiogenesis. A considerable body of work has been carried out on copper compounds (Ziche et al 1982, Alessandri et al 1983, 84) and Alessandri and co-workers have demonstrated that heparin complexed with copper was chemotactic for microvessel endothelial cells. This chemotactic effect was absolutely specific to microvessel endothelial cells and had no effect whatsoever on any other cell types. A low molecular weight growth factor Gly-His-Lys showed similar properties and all the copper complexed proteins examined gave positive results _in vivo_ in the corneal assay system. Thus it would appear that copper or copper complexes are in some way involved in angiogenesis in the cornea.

Perhaps by complexing copper with protein or peptide it is possible that the copper is maintained locally in high concentrations where angiogenesis in the cornea is appearing. It is quite possible that copper in this instance is acting as a co-factor for an enzyme which may be essential for inducing angiogenesis in the cornea. It is quite important to stress that effects _in vivo_ in the cornea only have been noted and that no activity was obtained on the CAM. On the other hand selenium in the form of the sodium salt or the selenous acid did give a positive response both in the corneal pocket assay and on the CAM (McAuslan and Reilly 1986) However, these results conflict with the observations of the same workers and of Jacobs et al (1980) that selenite actually retarded angiogenesis in rat tumours.

Lipid Factors

Several lipids with angiogenic properties have been described as possessing an angiogenic action and perhaps the gangliosides described by Niinikoski et al 1986 are the most interesting. They were able to accelerate capillary endothelial cell migration when complexed with fibronectin (Alessandri et al 1986). Another lipid faction, hexosyl ceramide obtained from Solcoseryl, which is a non-protein faction of calf blood marketed by Selco Basle Ltd., caused rapid vascularisation of the rabbit ear chamber. However, in our hands we were unable to find any activity of this fraction in the chick chorioallantoic membrane.

Nucleotides and Nucleotide Metabolites

Although adenosine and adenosine diphosphate have been suggested as angiogenic factors by several workers (Dusseau et al 1986, Dusseau and Hutchins 1988, Meininger et al 1987, Adair et al 1989) the amounts needed to produce an angiogenic effect are rather high and indeed are suggestive of a pharmacological rather than a physiological response. However, Meininger et al (1987) have shown that an adenosine receptor blocker is able to prevent endothelial cell proliferation _in vitro_ in response to anoxia. This might indicate that the release or synthesis of mitogenic activity is mediated by adenosine binding to it specific receptor. A metabolic derivative of adenosine has also been implicated in neovascularisation. Fenselau (1984) reported stimulation of

220

angiogenesis in the cornea with implants of inosine (2mg/pellet) and
Adair et al (1989) were able to induce neovascularisation in the CAM
after repeated applications of 1mg inosine/day. A single application
of 3mg gave a negative response (Dusseau et al 1986, Morris et al
1989). There are no reports of the effects of these two factors on
endothelial cells.

Nicotinamide has also been put forward as a possible signal for
angiogenesis (Kull et al 1987). However, although a positive response
could be obtained on the CAM, nicotinamide was not able to elicit any
response from endothelial cells in culture. Morris et al (1989)
concluded from these observations that the _in vivo_ effect was
indirect.

Hyaluronic Acid

Very few factors so far described have been shown to have any
relationship to disease with the exception of ESAF and hyaluronic acid
fragments. The hyaluronic acid fragments were a low molecular weight
with a size varying between 5 and 25 disaccharide units (West et al
1985, West and Kumar 1989a). This material gave a positive response on
the CAM and this response could be obtained with fragments between 3
and 16 disaccharide units. A six disaccharide unit could elicit
proliferation of microvessel endothelial cells and also stimulate
angiogenesis on the CAM (West and Kumar 1989b). Although these
fragments are not specific to microvessel endothelial cells but are
also active on aortic endothelial cells they are not active on any
non-endothelial cell type. Clinically, evidence for the presence of
hyaluronic acid in the serum of children with bone metastasising renal
tumours and Wilm's tumours has been reported (Kumar et al 1989).
Removal of the tumour resulted in drastic reduction in hyaluronic acid
levels. It is not yet clear whether these findings relate fundamentally to angiogenesis or are simply a reflection of matrix
degradation which one would expect in a growing tumour.

ESAF - Involvement in Pathological Conditions and in Embryonic Development

ESAF has been implicated in a number of disease states where angiogesis is involved and also in non-disease systems such as embryonic growth. In early work using a kitten model of retrolental fibroplasia as a model for angiogenesis, Taylor et al (1986) measured ESAF in both the retina and the vitreous of kittens who were either exposed to very high oxygen tension at birth for 10 days followed by 10 day return to normal oxygen which resulted in retinal anoxia and angiogenesis compared to a control group of kittens who had not been exposed to high oxygen levels. The levels of ESAF in the retina were approximately five times greater in the oxygen exposed kittens than in those in normal ambient air. Subsequently Taylor et al (1988) investigated the levels in the vitreous using the same experimental model and showed an increase of approximately three times in the levels of ESAF in the oxygen exposed animals than in the control ones. An interesting side line of this research was that in the normal kittens the levels of ESAF in the vascular retina and avascular vitreous were the same, suggesting that control of angiogenesis in the retinal vitreal system was probably through an inhibitor. An inhibitor of angiogenesis has indeed been isolated from vitreous (Taylor et al 1985, Thompson and Weiss 1989) and shown to act biochemically in a manner which is the opposite of that of ESAF.

Diabetic Retinopathy

Subsequent work by the same group (Taylor et al 1989) focussed on patients with diabetic retinopathy and diabetes without retino-pathy. Circulating levels of ESAF were measured in these patients and a statistically highly significant difference was found between the levels of circulating ESAF in patients with diabetes without proliferative retinopathic complications and patients with such complications.

New Bone Growth

A study of ESAF levels in synovial fluid of patients with rheumatoid arthritis and osteoarthritis gave what seemed at the time a rather remarkable result namely that levels in patients with osteoarthritis were much higher in general than those in patients with rheumatoid arthritis although with some exceptions. This led to the hypothesis that perhaps in osteoarthritis where new bone was forming (ie osteophytes) an angiogenic factor such as ESAF would be needed in order to vascularise the cartilage. Examination of bovine growth plate confirmed this hypothesis since very high levels were found in this tissue. Perhaps more strikingly the levels of ESAF in the growth plate were highest in the early foetus where growth was more rapid and fell towards maturity of the foetus. In a more recent study levels of ESAF in the serum of patients with osteoarthritis and rheumatoid arthritis have been determined and circulating levels of ESAF shown to be higher in osteoarthritis than in rheumatoid arthritis (Brown and Weiss 1988). Patients with rheumatoid arthritis and with higher levels of ESAF generally showed the presence of osteophytes. These results were interesting because it suggested a divergence of pathways between inflammatory angiogenesis as in rheumatoid arthritis and non-inflammatory as in osteoarthritis where the angiogenesis was indicative of a situation in which the growth plate re-opened and new blood vessel formation occurred. The highest levels of circulating ESAF in rheumatic disease have been found in ankylosing spondylitis, perhaps because here there is continuous endochondral ossification taking place (Weiss and Murphy personal communication).

Inflammatory Versus Non-Inflammatory Angiogenesis

As has been mentioned previously, inflammatory angiogenesis may be controlled by a totally different pathway from "normal" angio-genesis such as in embryonic growth and probably menstruation. ESAF has certainly been extracted from human uterine angiogenic factor and from human and bovine decidua (Taylor et al, 1991a). Levels of ESAF in human uterine angiogenic factor were high and it is interesting that not only ESAF but fibroblast growth factor could be extracted from this preparation. This observation will be discussed later.

Malignant and Benign Tumours

ESAF was originally isolated from a tumour source (Weiss et al 1979) and recently a study of patients undergoing surgery for malignant or benign brain tumours has been undertaken. A highly significant correlation was shown between ESAF levels in the malignant tumours and those in benign ones (Table 1). There was also a difference between patients with rapidly growing and slowly growing tumours (Taylor et al 1991b).

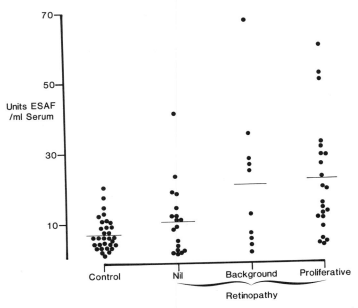

Fig. 1. ESAF levels in patients with diabetes with and
without proliferative retinopathic complications
(with permission of the Lancet)

Table 1

ESAF levels in intracranial tumours

CONTROL	MALIGNANT	BENIGN
0.34	4.23	1.45
$\pm$ 0.33	$\pm$2.7	$\pm$1.35

ESAF levels in human brain tumour samples compared with levels in control brains expressed as units of ESAF where 1 unit = the amount activating sufficient procollagenase to degrade lug collagen/hr (control samples were either patients with head injuries or brain haemorrhage or post mortem from patients dying from myocardial infarction). Malignant and benign tumours were both significantly different from control brain p=0.03 and 0.04 respectively.

Codistribution and synergism of ESAF and FGF

As has been mentioned earlier ESAF was extracted from the original tumour angiogenesis factor of Folkman et al (1971) and subsequently fibroblast growth factor has also been extracted from this source (Shing et al 1985). The same is true for an angiogenic extract derived from decidua obtained from first trimester legal termination of preganancy and called human uterine angiogenic factor (HUAF) (Maroudas et al 1987). Both FGF and ESAF have been found in this highly angiogenic material. It has been shown that these two factors act synergistically. Evidence for such a synergism between ESAF and basic FGF in the proliferative response of microvessel endothelial cells in culture was observed by Odedra and Weiss (1987) where the mitogenic effect of FGF was potentiated threefold by an amount of ESAF not in itself sufficient to stimulate proliferation of these cells. The effect was not additive nor was it observed with other cell types such as skin fibroblasts. Subsequently Thompson (1989) showed a strong synergism between these two molecules in the chick yolk sac membrane (YSM) test. Table 2. The synergism of ESAF and FGF is maybe of more than passing importance since ESAF is present in quite high levels in foetal calf serum as in many embryonic tissues and since the majority of _in vitro_ assays involving FGF or other protein angiogenic factors are carried out in the presence of 10% foetal calf serum then they will always be in the presence of an appreciable amount of ESAF.

Table 2

Synergism of ESAF and FGF on chick Yolk Sac Membrane

ESAF (Units)	FGF (ng)	Angiogenic Response(24hrs)
2.4	0	Positive
0	2.0	Positive
1.5	0	Negative
0	1.5	Negative
0.83	0.05	Positive

Methyl cellulose pellets containing either ESAF or FGF were applied to 4 day fertile chick yolk sac membranes (Thompson 1989) with permission of the author.

Many tissues in the body have been investigated for the levels of ESAF and the pineal gland has been shown to have an order of magnitude more ESAF than other tissues. This is presumably not due to the fact that it is a highly vascular tissue (Hodde, 1979) since neither kidney nor liver both of which are also highly vascular tissues have appreciable levels of ESAF. Studies have shown that the pineal gland is remarkably resistant to tumour growth and pineal tumours are extremely rare (Lapin 1975, Tapp 1979, Blask 1984). One of the products of the pineal gland, serotonin has anti-tumour effects and other so far unspecified anti-tumour factors have been extracted from the pineal gland. If the anti-tumour properties of serotonin and other pineal factors is on solid tumours and not on ascites tumour as has been suggested by Huxley and Tapp 1972 then the levels of ESAF present in the pineal may be necessary to over-ride the vascular inhibitory activities and maintain the highly vascular bed which the pineal contains.

To date only ESAF of all putative angiogenic factors has been shown to have any correlation with disease activity. However, we do not claim ESAF as the only angiogenic factor but merely as perhaps an important low molecular weight adjunct to other angiogenic factors and growth factors.

References

Adair, T.H., Montani, J.P. Strick, D.M. and Ruyton, A.C. (1989) Vascular development in chick embryos: a possible role for adenosine. Am. J. Physiol. 256: h240-h246

Alessandri, G., Raju, K. and Gullino, P.M. (1984) Angiogenesis in vivo and selective mobilisation of capillary endothelium in vitro by a heparin-copper complex. Microcirculation, Endothelium, Lymphatics 1(3): 329-346

Alessandri, G., Raju, K. and Gullino, P.M. (1983) Mobilisation of capillary endothelium in vitro induced by effectors of angiogenesis in vitro. Cancer Res. 43: 1790-1797

Alessandri, G., Raju, K.S. and Gullino, P.M. (1986) Interaction of gangliosides with fibronectin in the mobilisation of capillary endothelium. Possible influence on the growth of metastasis. Invasion Metastasis 6(3): 145-165

Atherton, A. (1977) Growth stimulation of endothelial cells by simultaneous culture with sarcoma 180 cells in diffusion chambers. Cancer Res. 37: 3619-3622

Barnhill, R.L. and Ryan, T.J. (1983) Biochemical modulation of angiogenesis in the chorioallantoic membrane of the chick embryo. J. Invest. Derm. 81(6): 485-488

Blask, D.E (1984) The pineal gland: an oncostatic gland? In: The Pineal Gland, pp253-284, Reiter R.J. ed. Raven Press, New York.

Brown, R.A. and Weiss, J.B. (1988) Neovascularisation and its role in the osteoarthritic process. Ann. Rheum, Dis. 47(11): 881-885

Dusseau, J.W. and Hutchins, P.M. (1958). Hypoxia induced angiogenesis in chick chorioallantoic membranes: a role for adenosine. Resp. Physio 71(1): 33-44

Dusseau, J.W., Hutchins, P.M. and Malbasa, D.S. (1986) Stimulation of angiogenesis by adenosine on the chick chorioallantoic membrane. Circ. Res. 59(2): 163-170

Fenselau, A. (1984) An angiogenic role for adenine nucleotide. Fedn. Proc. 43: 587

Folkman, J., Merler, E., Abernathy, C. and Williams, G. (1971) Isolation of a tumour factor responsible for angiogenesis. J. Exp. Med 133: 275-289

Heininger, C.J., Shelling, M.E. and Granger, J.H. (1987) The proliferation of cultured aortic endothelial cells is stimulated by hypoxia or adenosine. Fedn. Pro. 46: 1535

Hodde, K.C. (1979) The vascularisation of the rat pineal organ. In: Progress in Brain Research, Vol 52. pp 39-44, Kapper J.A. and Pevet P. (eds) Elsevier/North Holland, Amsterdam

Hunt, T.K., Knighton, D.R., Thakri, K.K., Goodson, W.H. and Andrews, W.S. (1984) Studies on inflammation and wound healing: angiogenesis and collagen synthesis stimulated in vivo by resident and activated would macrophages. Surgery, St. Louis 9b(1): 48-54

Huxley, M. and Tapp, E. (1972). The histological measurement of the lipid content in pineals of rats suffering from malignancy. Brain Res. 39: 123-131

Imre, G. (1984) The role of increased lactic acid concentration in neovascularisations. Acta. Morph. Hung. 32(2): 97-103

Jensen, J.A., Hunt, T.K., Schuenstuhl, H. and Banda, M.J. (1986) Effect of lactate, pyruvate and pH on secretion of angiogenesis and mitogenesis factors by macrophages. Lab. Invest. 54(5): 574-578

Knighton, D.R., Oredsson, S., Banda, M. and Hunt, T.K. (1984) Regulation of repair: hypoxic control of macrophage mediated angiogenesis. Surg. Sci. Ser. 2: 41-49

Kull, F.C., Brent, D.A., Parikh, I. and Cuatrecasas, P. (1987) Chemical identification of tumour derived angiogenic factor. Science 236: 843-845

Kumar, S., West, D.C., Ponting, J.M. and Galtemine, H.R. (1989) Sera of children with renal tumours contain low molecular mass hyaluronic acid. Int. J. Cancer 44: 445-448

Lapin, V. (1975) The pineal gland and neoplasia. Lancet 1: 341-344

Leibovich, S.J., Polverini, P.J., Shepard, H.M., Wiseman, D.M., Shively, V. and Nuseir, N. (1987) Macrophage induced angiogenesis is mediated by tumour necrosis factor. Nature 239: 630-632

Maroudas, N.G., Fuchs, A. Lindenbaum, E.S. (1987) The non-tumour perspective geometry in growth control. In "Ocular Circulation and Neovascularisation" (Ben Ezra D., Ryan J. Eds) Dordrech, Martinus Nijhoff. pp. 499-504

McAuslan, B.R. and Hoffman, H. (1979) Endothelium stimulating factor from Walker carcinoma cells Exp. Cell Res. 119: 181-190

McAuslan, B.R. and Reilly, W. (1986) Selenium induced cell migration and proliferation: relevance to angiogenesis and microangiopathy. Microvasc. Res. 32: 111-120

Morris, B.P., Ellis, M.N. and Swain, J.L. (1989) Angiogenic potency of nucleotide metabolites: potential role in ischaemia-induced vascular growth. J. Molec. Cell. Cardiol. 21: 351-358

Odedra, R. and Weiss, J.B. (1987) A synergistic effect on microvessel cell proliferation between basic fibroblast growth factor (bFGF) and endothelial cell stimulating angiogenesis factor (ESAF) Biochem. Biophys Res. Commun. 143(3): 947-953

Odedra, R. and Weiss, J.B. (1991) Low molecular weight angiogenesis factors. Pharmac. Ther. 49: 111-124

Shing, Y., Folkman, J., Haudenschild, C., Lund, D., Grun, R. and Klagsbrun, M. (1985) Angiogenesis is stimulated by a tunour-derived endothelial cell growth factor. J. Cell Biochem. 29(4): 275-287

Tapp, E. (1979) The histology and pathology of the pineal gland. In: Progress in Brain Research, Vol 52: The Pineal Gland of Vertebrates including Man. pp. 481-500

Taylor, C.M. and Weiss, J.B. (1985) Partial purification of a 5.7K glycoprotein from bovine vitreous which inhibits both angiogenesis and collagenase activity. Biochem. Biophys. Re. Commun. 133(3): 911-916

Taylor, C.M., Weiss, J.B., Kissun, R.D. and Garner, A. (1986) Effect of oxygen tension on the quantities of procollagenase activating angiogenic factor present in the developing kitten retina. Br. J. Ophthalmol. 70: 162-165

Taylor, C.M., Weiss, J.B., McLaughlin, B. and Kissun, R.D. (1988) Increased procollagenase activating angiogenic factor in the vitreous humour of oxygen treated kittens. Br. J. Ophthalmol. 72(1): 2-4

Taylor, C.M., Kissun, R.D., Schor, A., McLeod, D., Garner, A. and Weiss, J.B. (1989) Endothelial cell stimulating angiogenesis factor in vitreous from extra-retinal neovascularisations. Invest. Ophthal. Vis. Sci. 30(10): 2174-2178

Taylor, C.M., McLaughlin,B., Weiss, J.B. and Maroudas, N. (1991a) Concentrations of endothelial cell stimulating angiogenesis factor, a major component of human uterine angiogenesis factor, in embryonic tissues and decidua. J. Reprod. and Fertility (In press)

Taylor, C.M., Weiss, J.B. and Lye, R.H. (1991) Raised levels of latent collagenase activating angiogenesis factor (ESAF) are present in actively growing human intracranial tumours. Br. J. Cancer 63: (In press)

Thompson, J. (1989) Ph.D. Thesis Manchester University

Thompson, J. and Weiss, J.B. (1989) Isolation of an inhibitor of collagenase from bovine vitreous. The possible inhibitor of angiogenesis. Int. J. Microcirc. 8:

Weiss, J.B., Brown, R.A., Kumar, S. and Phillips, P. (1979) An angiogenic factor isolated from tumours: a potent low molecular-weight compound. Br. J. Cancer 40(3): 493-496

Weiss, J.B., Hill, C.R., Davis, R.J., McLaughlin, B., Sedowofia, K.A. and Brown, R.A. (1983) Activation of a procollagenase by low molecular weight angiogenesis factor. Biosci. Rep. 3(2): 171-177

West, D.C., Hampson, I., Arnold, F. and Kumar, S. (1985) Angiogenesis induced by degradation products of hyaluronic acid. Science 228: 1324-1326

West, D.C. and Kumar, S. (1989a) The effect of hyaluronate and its oligosaccharides on endothelial cell proliferation and monolayer integrity. Expl. Cell Res. 183(1): 179-196

West, D.C. and Kumar, S. (1989b) Hyaluronan and angiogenesis. <u>Ciba Found. Symp.</u> 43: 187-201

Ziche, M., Jones, J. and Gullino, P.M. (1982) Role of prostaglandin E1 and copper in angiogenesis. <u>J. Natn.Cancer Inst.</u> 69(2): 475-482

CALCIFICATION-LINKED ANGIOGENESIS AND BONE GROWTH

Robert A.Brown, [2]Clive D.McFarland, Michael Kayser,
S.Yousuf Ali, [1]Barry McLaughlin, [1]Jacqueline B.Weiss

Institute of Orthopaedics, Stanmore, Middx., HA7 4LP U.K.
[1]Rheumatic Diseases Centre, Hope Hospital, Salford, M6 8HD. U.K.
[2]Biomolecular Engineering, CSIRO, North Ryde, N.S.W., Australia

Adult cartilage is normally considered to be an avascular tissue.
Certainly, it is a rich source of anti-angiogenic factors and a number of
studies have identified anti-invasion factors which may act by inhibition
of matrix proteases[1,2,3]. Despite this, cartilage neovascularisation does
occur - most commonly during calcification and ossification associated with
long bone growth and fracture repair. Paradoxically, it is its resistance
to angiogenesis which makes cartilage so important in the study of its
control. Where it does occur, the control of cartilage neovascularisation
is both stark and rigid, for example in the normal physeal growth plate
(fig. 1). Invasion is through the non-mineralised transverse cartilage
septa which separate each chondrocyte. Progress is synchronised to produce
an invasive 'front' of vessels which resorbs the cartilage at a rate just
equal to its synthesis. This can be rapid, averaging 0.3mm per 24 hour, or
one chondrocyte plus its matrix every 3 hours, in young rats[4]. We must ask
ourselves, therefore, what is the mechanism by which microvessels are
suddenly able to invade and erode cartilage. Inhibitors of matrix
degradation[1] have been implicated in the restraint of invasion and the
present thesis is that activation (ie. angiogenesis) is regulated through
localised removal of that inhibition.

MORPHOLOGY OF CALCIFICATION-LINKED NEOVASCULARISATION

Figure 1 illustrates growth plate location and structure, with its
columns of chondrocytes progressing through their maturation and synthetic
stages[5]. This ends abruptly with capillary invasion and resorption of the
cartilage. There is also thought to be active remodelling and lysis of the
matrix around expanding chondrocytes of the hypertrophic cell zone[6]
(fig.1). The concept expanded here, then, of calcification-linked
angiogenesis has its basis in the general pattern of skeletal growth.
However, experimental and pathological observations corroborate the idea.
In particular, both rickets (impaired calcification) and periosteal
stripping[7] (producing impaired capillary invasion) result in growth plate
thickening through a failure to resorb the hypertrophic cartilage zone.

There are good grounds to suspect that cellular control of
cartilage resorption at the chondro-osseous junction is far more
sophisticated than is presently understood. At the interface between
capillary and chondrocyte, cells assume a distinct pattern of behaviour.

Angiogenesis in Health and Disease, Edited by M.E. Maragoudakis *et al.*
Plenum Press, New York, 1992

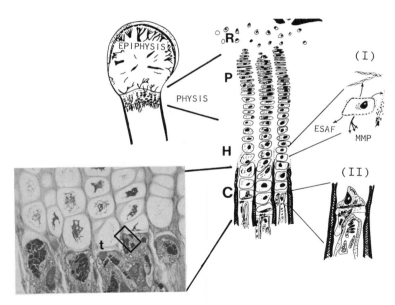

FIG. 1

Figure 1. Schematic relationship of the physeal growth plate and epiphysis in a mammalian long bone. Structure of the growth plate (right) with columns of developing chondrocytes (reserve zone-R, proliferative zone-P, hypertrophic zone-H, calcification/neovascularisation-C). [I] hypertrophic chondrocytes produce matrix metalloproteinases (MMP), ESAF and calcify their matrix. [II] perivascular cells breach the last transverse cartilage septum (t). Micrograph (bottom left) shows the front of microvascular invasion: box indicates the area shown in figs. 2 and 3.

Much of the microvascular penetration seems to be mediated through 'perivascular' cells[8,9] at the tip of the capillary sprout[10]. Since they resemble migrating endothelial cells during angiogenesis [11], we have proposed that they are invading cells which have left the vessel wall[9]. These cells form close contacts with the microvascular endothelium (fig. 2) whilst pushing cell processes up into the non-mineralised cartilage septum. The delicate appearance of these processes, surrounded by cartilage, suggests a controlled, local lysis. At the same time, the non-fibrillar, vacuolated appearance of this cartilage (fig 3 and ref.[12]) suggests that the collagen has been degraded. It seems likely that this is achieved by a localised activation of collagenolytic enzymes notably collagenase/matrix metalloproteinases -MMP-1. It is important, then, to determine whether all the elements of this system come from the perivascular endothelial cells or are already in the cartilage and only require activation.

On the other side of the cartilage septum, the last hypertrophic chondrocyte does not appear to be a passive by-stander. Farnum and Wilsman [12] have identified a dramatic final stage in which this cell condenses and attaches, always onto the last transverse septum (fig. 3). Such condensed cells are a natural feature, rather than a product of in-process shrinkage [12]. Our studies on rabbit growth plate have located the same form of chondrocyte (fig.1 -box- and fig. 3), also attached to the junctional septum, which is being degraded. Two additional features can be identified; (i) a polarised change in the cartilage matrix and (ii) a perforation

through the septum associated with a chondrocyte process (fig. 3B). (i)
Cartilage adjacent to the chondrocyte was packed with collagen fibrils, yet
matrix on the endothelial side of the same septum was amorphous and non-
fibrillar. Disappearance of the collagen component was abrupt, producing a
line in the matrix. This double layer was not seen in septa above the
chondro-endothelial junction. In addition (ii), the chondrocyte in fig 3B
appeared to have pushed a cell process through the junctional septum,
towards the capillary. The precise source of this perforation must remain
uncertain, though there was no evidence, in this or serial sections, of
endothelial cell involvement. Furthermore, matrix disruption was quite
different to that around perivascular cells. Cartilage at the rupture site
was rich in fibrillar collagen and a stream of fibrils had been disaggregated and freed from the matrix. Indeed, some collagen fibrils
remained attached to the chondrocyte process. There would, then, appear to
be two mechanisms for cartilage breakdown at this site. The first (mediated
by perivascular endothelial cells) produces complete lysis of the collagen.
The second (putatively a chondrocyte mechanism) involves disaggregation of
the collagen matrix. This is consistent with a considerable degree of
cooperation between the two cell types and a greater participation by the
chondrocyte (perhaps in matrix preparation) than previously envisaged.

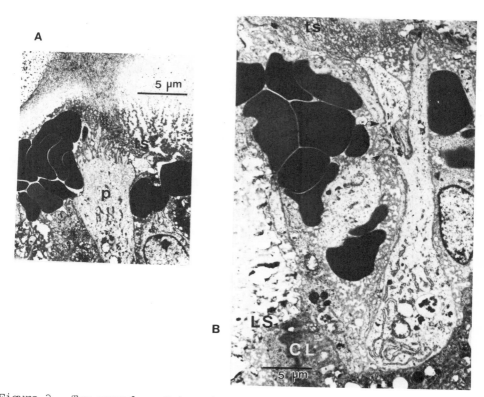

Figure 2. Two examples of invasive perivascular endothelial cells (see
fig.1 box). Microvascular invasion is from bottom to top. [A] perivascular
cell and processes inserted into a transverse septum (ts). [B] perivascular
cell in intimate contact (note interdigitations, arrows) with a lumenal
endothelial cell; below is an osteoclast (CL) resorbing a calcified
longitudinal septum (LS). **From reference [9], with permission.**

REGULATION OF COLLAGENOLYSIS DURING CARTILAGE CALCIFICATION

We have previously investigated the role of a low molecular weight activator of latent collagenase[13], Endothelial cell Stimulating Angiogenesis Factor (ESAF) in chondrocyte regulation of cartilage neovascularisation[14,15] (see Weiss et al, this volume, for review of ESAF). Foetal bovine growth plate was found to be a particularly rich source of ESAF[14] and rat growth plate also releases substantial quantities[16] (McFarland, Brown, Dean and Howell, unpublished data). ESAF is able to stimulate matrix degradation in vitro, by activation of latent collagenase [13,17,18] and to attract microvascular invasion. The observation that growth plate is rich in this compound, capable of activating both angiogenesis and collagenolysis, led us to develop the hypothesis that ESAF is pivotal to cartilage neovascularisation through its activation of matrix lysis[14,19].

The matrix breakdown which is most obviously important to angiogenesis occurs around the tip of invading capillaries (fig 2). Yet in order to grow, hypertrophic chondrocytes also produce collagenase and degrade their matrix[4,6,20,21] (Fig. 1). It then becomes important to question whether these two foci of matrix resorption are independent or interdependent. Both are clearly localised and closely regulated.

We proposed and tested the concept of CALCIFICATION-LINKED ANGIOGENESIS using an in vitro chondrocyte calcification model[14,15]. This model consists of rabbit growth plate chondrocytes in high density cultures

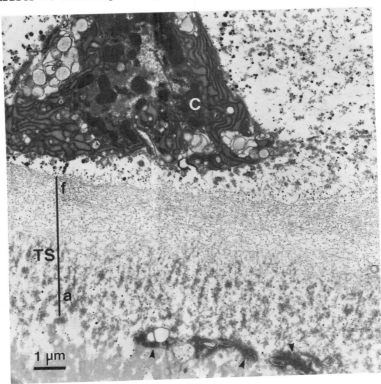

Figure 3. (A) A condensed chondrocyte (C) attached to the last transverse septum (TS), within a few microns of the microvasculature. The septum has a fibrous appearance (f) on the chondrocyte side and an amorphous (a) nonfibrillar structure on the microvascular side. Cell processes, similar in appearance to the chondrocyte, appear on the vascular side (arrows).

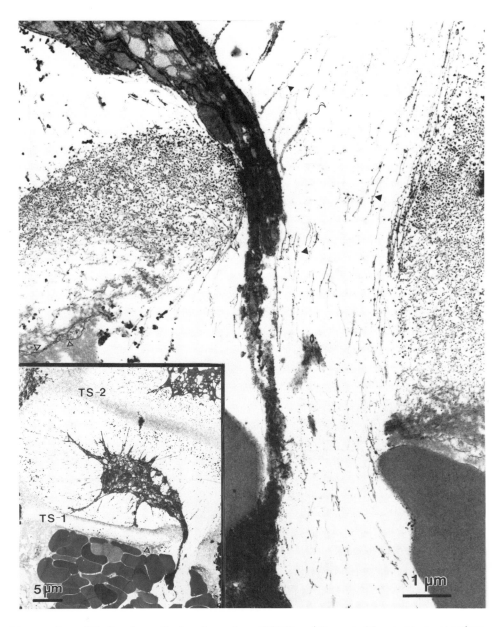

Figure 3. (B) Condensed chondrocyte, INSET, with a cell process passing
through the septum to the endothelium (open arrow heads). Contrast the
last septum (TS-1) with TS-2 which has a homogeneous fibrillar appearance.
MAIN PICTURE: detail of the perforated septum showing the smoothed
fibrillar edge and stream of disaggregated collagen fibrils, (filled arrow
heads), some of which remain attached to the chondrocyte.

(micro or megamass: 2×10^5 and 10^7 cells respectively). Cultures rapidly
produce a cartilage matrix which calcifies on addition of sodium or calcium
beta glycerophosphate (NaBGP and CaBGP)[14,16,22].

Figure 4 shows the appearance of micromass cultures with and without calcification. Deposition of mineral is visible in the matrix surrounding the deepest chondrocytes (i.e. those closest to the culture dish). Treatment of cultures with levamisole (an inhibitor of alkaline phosphatase) together with CaBGP reduced the formation of mineral deposits. Figure 4B showns the level of calcium accumulated, together with matching alkaline phosphatase activities in cell pellets after 48 hours. Both CaBGP and Ca^{++} supplemented NaBGP produced calcification, together with a corresponding fall in residual alkaline phosphatase levels.

Stimulation of ESAF production following in vitro calcification has been demonstrated using larger megamass scale cultures, treated with 3 forms of beta glycerophosphate reagent (fig. 5). The most effective was $CaCl_2$ supplemented NaBGP which produced similar calcification to CaBGP and stimulated ESAF by 4.7 fold, relative to untreated cultures or controls given levamisole or NaCl[15]. The peak of ESAF appeared after 48 hours, when production returned to basal, even in the presence of fresh organic phosphate. Inhibition of protein synthesis, with cycloheximide abolished ESAF stimulation. Since ESAF is not a peptide this suggests that induction of an enzyme may be necessary for its production[15].

Effects of ESAF on cartilage breakdown were monitored using the release rate of ^3H-protein from cultures labelled with ^3H-proline. Cultures received either exogenous, purified ESAF or were stimulated to calcify (hence produce their own, endogenous ESAF). Release of ^3H-proline labelled protein was stimulated to varying extents, depending on the age of the

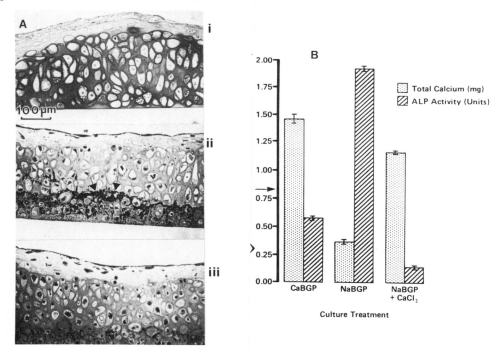

Figure 4. (A) Sections of micromass cultures (i) untreated: (ii) 48h post calcification + 4mM CaBGP (iii) as (ii) but +2mM levamisole. Calcification (arrowed in ii) was predominantly adjacent to the culture dish (^{H}umphrey's trichrome). [B] Three forms of beta glycerophosphate treatment and their effects on calcification and alkaline phosphatase activity in megamass cell pellets (48h post treatment: pre-treatment levels of Ca^{++}, arrow-head: alkaline phosphatase, arrow). **From reference 15 with permission.**

culture (from 30% to 240% above controls)[18]. 2.4 fold activation produced complete matrix lysis by day 6. Figure 6 shows the 30% level of activation, together with its inhibition by simultaneous addition of TIMP[18]. Inhibition by TIMP indicates that degradation is by an activation of matrix metalloproteinases and supports the findings of McLaughlin et al[17] that ESAF is an antagonist of TIMP. In contrast, stimulation of endogenous ESAF production, by initiation of calcification, did not increase matrix breakdown at any stage over a 9 day culture period (data not shown)[18]. Consequently, whilst ESAF can activate dramatic breakdown of cartilage, it does not seem to be produced in sufficient quantities to do this during calcification of micromass cultures.

Recent studies have focused on the production of matrix metallo-proteinases, principally collagenase, by calcifying chondrocyte cultures. Enzyme release was measured in conditioned media, over a time course post-mineralisation, using native type I and type II collagen and gelatin substrates (fig. 7). Proteinases were measured both in the active form and as total enzyme (ie. including latent, activatable enzyme). Latent enzyme was activated with an organo-mercurial (amino phenylmercuric acetate), used widely for matrix metalloproteinase activation[6,21,23]. Production of enzymes degrading types I and II collagen was almost the same (type I only shown). Pre-calcification release of total collagenase was negligible and calcification did not increase the level of active enzyme. However, by 48 hours post-mineralisation substantial amounts of activatable, latent enzyme were present. Stimulation of latent collagenase was 4 fold on type I collagen and 10 fold on type II, relative to the levamisole control. Stimulation relative to pre-mineralisation levels was even greater. In contrast, gelatinase was only weakly responsive to calcification. It was produced in both active and latent forms, before and after treatment[18].

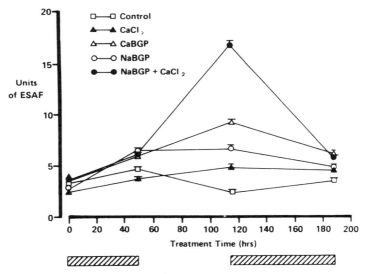

Figure 5. Effects of calcification (sodium or calcium beta glycero-phosphates) on ESAF production, as activation of procollagenase (megamass scale cultures). Media were supplemented with $CaCl_2$, CaBGP, NaBGP all at final concentrations of 4mM, or with no addition. Start and duration of treatments are shown by the bars, with unsupplemented medium between treatments (1 unit of ESAF stimulates breakdown of 1μg/h of ^{3}H-collagen by otherwise inactive procollagenase). **From reference 15 with permission.**

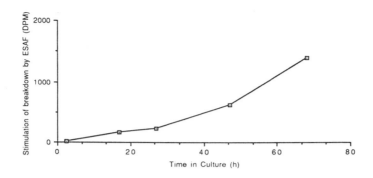

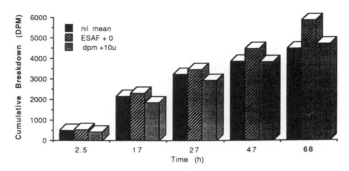

Figure 6. (A) Effect of ESAF on the breakdown of tritiated micromass cultures. There was a significant increase in release of matrix-incorporated label over the time course, relative to untreated controls (means of duplicate cultures: $p < 0.05$ by student's t-test). (B) Histogram showing inhibition of <u>total</u> release of tritiated matrix by TIMP (generously provided by Dr. Gillian Murphy, Strangeways Laboratories, Cambridge). Inhibition was seen at all concentrations between 1 and 20 units[18].

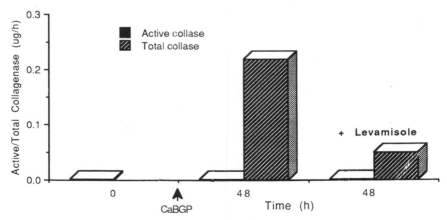

Figure 7. Effect of calcification (CaBGP; arrowed) on collagenase released to the medium by micromass cultures. Levels of active collagenase were negligible in all cases. However, latent collagenase, activatable by amino phenylmercuric acetate, rose dramatically by 48h post mineralisation. This increase was reduced by addition of 2mM levamisole together with CaBGP.

SUMMARY AND PATHOLOGICAL IMPORTANCE

The investigation of calcification-linked angiogenesis has a twofold purpose. Firstly, it provides an excellent model for investigating the control of microvascular invasion of connective tissues. In addition, the interlinking of microvascular invasion and collagenase activation provides an opportunity to study a potential pathogenic mechanism in joint disease.

Although the primary growth plates disappear by adulthood, latent growth plates can persist[24], as vestiges of the secondary centres, in the deep zone of articular cartilage. Re-initiation of calcification at these sites is thought to be important in the remodelling of joints, osteophyte formation and subchondral changes associated with osteoarthritis [19,24,25,26]. The importance of controlled collagenolysis in angiogenesis and the apparent failure of that control in ageing and osteoarthritis [27,28,29,30] is consistent with the idea that angiogenesis can contribute to degeneration of articular cartilage. This idea was supported by the identification of accumulations of ESAF in synovial fluids of patients with osteoarthritis[31]. It is tempting to suggest that these accumulations are a result of calcification-linked angiogenesis associated with osteophytic bone formation, though it is uncertain if synovial fluid-ESAF contributes directly to the pathogenesis.

The initial suggestion (14) that ESAF from calcifying chondrocytes acts directly and alone on microvascular invasion by activating the collagenase produced by endothelial cells (32,33), now seems too simple. The dramatic increase in release of latent collagenase following calcification suggests that the surrounding cartilage is undergoing preparation for breakdown. Yet almost all of the enzyme was inactive and calcification alone did not stimulate cartilage lysis. It may be that production of ESAF and latent collagenase by chondrocytes provides the means for cartilage degradation, but that a further activation step is normally required. Microvascular endothelial cells could provide such an activation, giving a localised and restricted pattern of collagenolysis. The potential for inappropriate activation of this system is clear and could provide a framework for interpretion of early osteoarthritic changes.

REFERENCES

1. K.E. Kuettner, L. Soble, R.L. Croxen, B. Marczynska, Tumour cell collagenase and its inhibition by a cartilage-derived protease inhibitor, Science 196:653.
2. R. Eisenstein, K.E. Kuettner, C. Neapolitan, W. Soble, N. Sorgente, The resistance of certain tissues to invasion, Am. J. Pathol. 81:337 (1975)
3. N. Sorgente, C.K. Dorey, Inhibition of endothelial cell growth by a factor isolated from cartilage, Exp. Cell Res. 128:63 (1980).
4. E.B. Hunziker, R.B. Schenk, L.M. Cruz-Orive, Quantitation of Chondrocyte performance in growth-plate cartilage during longitudinal bone growth, J.Bone Joint Surg. 69A: 162 (1987).
5. C.T. Brighton, The growth plate Orthop. Clin. North Am. 15:571 (1984).
6. D.D. Dean, O.E. Muniz, D.S. Howell, Association of collagenase and tissue inhibitor of metalloproteinase (TIMP) with hypertrophic cell enlargement in the growth plate, Matrix 9:366 (1989).
7. R.H. Yabsley, W.R. Harris, The effect of shaft fractures and periosteal stripping on the vascular supply to epiphyseal plates, J. Bone Joint Surg. 47A:551 (1965).
8. R.J. Schenk, J. Wiener, D. Spiro, Fine structural aspects of vascular invasion of the tibial epiphyseal plate of growing rats, Acta Anat. 68:1 (1968).
9. R.A. Brown, J.A. Rees, C.D. McFarland, D. Lewinson, S.Y. Ali,

Microvascular invasion of rabbit growth plate cartilage and the influence of dexamethasone, Bone Mineral, 9:35 (1990).

10. A.L. Arsenault, Microvascular organisation at the epiphyseal-metaphyseal junction of growing rats, J. Bone Min. Res. 2:143 (1987).

11. D.H. Ausprunk, J. Folkman, Migration and proliferation of endothelial cells in preformed and newly formed blood vessels during tumour angiogenesis, Microvasc. Res. 14:53 (1977).

12. C.E. Farnum, N.J. Wilsman, Morphologic stages of the terminal hypertrophic chondrocyte of growth plate cartilage, Anat. Rec. 219:221 (1987).

13. J.B. Weiss, C.R. Hill, R.J. Davis, B. McLaughlin, K.A. Sedowofia, R.A. Brown, Activation of procollagenase by low molecular weight angiogenesis factor, Bioscience Rep. 3:171 (1983).

14. R.A. Brown, C.M. Taylor, B. McLaughlin, C.D. McFarland, J.B. Weiss, S.Y. Ali, Epiphyseal growth plate cartilage and chondrocytes in mineralising cultures produce a low molecular mass angiogenic procollagenase activator, Bone Mineral 3:143 (1987).

15. C.D. McFarland, R.A. Brown, B. McLaughlin, S.Y. Ali, J.B. Weiss, Production of endothelial cell stimulating angiogenesis factor (ESAF) by chondrocytes during in vitro cartilage calcification, Bone Mineral 11:319 (1990).

16. C.D. McFarland, An Investigation into the mechanism linking cartilage calcification and neovascularisation, Ph.D. Thesis, University of London (1990).

17. B. McLaughlin, T. Cawston, J.B. Weiss, Actin of the matrix metalloproteinase inhibitor complex by a low molecular weight angiogenic factor, Biochem. Bio. Phys. Acta. 1073:295 (1991).

18. R.A. Brown, C.D. McFarland, Regulation of growth plate cartilage degradation in vitro: effects of calcification and a low molecular weight angiogenic factor (ESAF). (manuscript submitted).

19. R.A. Brown, J.B. Weiss, Neovascularisation and its role in the osteoarthritic process, Ann. Rheum. Dis. 47:887 (1988).

20. C.C. Brown, R.M. Hembry, J.J. Reynolds, Immunolocalisation of metalloproteinases and their inhibitor in the rabbit growth plate, J. Bone Joint Surg. 71A:580 (1989).

21. D.D. Dean, O.E. Muniz, J.F. Woessner, D.S. Howell, Production of collagenase and tissue inhibitor of metalloproteinase (TIMP) by rat growth plates in culture, Matrix 10:320 (1990).

22. J.R. Ralphs, S.Y. Ali, Induction of calcification in cultures of epiphyseal chondrocytes with calcium beta glycerophosphate, In: "Proceedings of the IV International Conference on cell mediated calcification and matrix vesicles", ed. S.Y. Ali, Elsevier, Amsterdam (1986).

23. D.D. Dean, O.E. Muniz, I. Berman, J.C. Pita, M.R. Carreno, J.F. Woessner, D.S. Howell, Localisation of Collagenase in growth plate of rachitic rats, J. Clin. Invest. 76:716 (1985).

24. S.Y. Ali, New knowledge of Osteoarthrosis, J. Clin. Path. suppl 12:191 (1978).

25. S.Y. Ali, The distribution of alkaline phosphatase activity and calcific changes in articular cartilage and in osteoarthritis, in: "Pendragon Papers 2" eds. A.K. Thould, A. St.J.Dixon, P.A. Dieppe, Boots Company, Nottingham.

26. P.G. Bullough, The geometry of diathrodial joints, its physiological maintenance and the possible significance of age related changes in geometry to load distribution and the development of osteoarthritis, Clin. orthop 156:61 (1987).

27. J. Trueta, Degenerative Arthritis in:'Studies of the development and decay of the human frame, Heinemann, London (1968).

28. L.B.Lane, A. Villacin, P.G. Bullough, The vascularity and remodelling of subchondral bone and calcified cartilage in adult human femoral and

humeral heads, J. Bone Joint. Surg. 59B:272 (1977).

29. M.G. Ehrlicn, H.J. Mankin, H. Jones, R. Wright, C. Crispen, G. Vigliani, Collagenase and collagenase innibitors in osteoarthritic and normal human cartilage, J. Clin. Invest. 59:226 (1977).

30. D.D. Dean, Azzow, J. Martel-Pelletier, J.R. Pelletier, J.F. Wassner, Levels of metalloprotease and tissue inhibitor of metalloprotease in human osteoarthritic cartilage, J. Rheumatol. 14(suppl. 14):43 (1987).

31. R.A. Brown, I.W. Tomlinson, C.R. Hill, J.B. Weiss, P. Phillips, S. Kumar, Relationship of angiogenesis factor in synovial fluid to various joint diseases, Ann. Rheum. Dis. 42:301 (1983).

32. D.A. Moscatelli, D.B. Rifkin, E.A. Jaffe, Production of latent collagenase by human umbilical vein endothelial cells in response to angiogenic preparations, Exp. Cell Res. 156:379.

33. G.S. Herron, Z. Werb, K. Dwyer, M.J. Banda, Secretion of metalloproteinases by stimulated capillary endothelial cells, J. Biol. Chem. 261:2810 (1986).

INVOLVEMENT OF A LOW MOLECULAR MASS ANGIOGENIC FACTOR (ESAF) IN THE
ACTIVATION OF LATENT MATRIX METALLOPROTEINASES

Jacqueline B. Weiss and Barry McLaughlin

Wolfson Angiogenesis Unit
Rheumatic Diseases Centre
University of Manchester, Clinical Sciences Building
Hope Hospital, Salford M6 8HD, UK

Introduction

There is no doubt that during angiogenesis connective tissue
breakdown occurs. Ausprunk, 1979 showed that in the early stages of
vascularisation of the cornea after implantation of a tumour transplant,
capillary sprouts from limbal venules and capillaries began to grow
through the connective tissue towards the implant. On day one after the
implant she observed that the basal lamina which separates the microvessel
cells from their perivascular connective tissue began to fragment. By the
second day she observed that cells extended large pseudopods into the
connective tissue and began to migrate towards the implant.

It is noteworthy that the cornea is a tissue consisting of densely
packed collagen fibre bundles lying in close apposition to each other in
both vertical and horizontal planes (Fig 1). Penetration through such a
matrix is not easily envisaged in the absence of some mechanism for
degradation of the connective tissue. Such degradation would necessarily
have to be strictly limited to the environs of the advancing capillary
tip.

Extracellular matrix degrading enzymes (MMPs)

The turnover of extracellular matrix and basement membrane is under
the control of a group of neutral zinc containing proteinases known as the
matrix metalloproteinases (MMPs) (Alexander and Werb 1990, Matrisian and

Angiogenesis in Health and Disease, Edited by M.E. Maragoudakis *et al.*
Plenum Press, New York, 1992

Hogan, 1990). There are at least seven of these but the major enzymes synthesised by connective tissue cells such as fibroblasts and chondrocytes and even from microvessel endothelial cells are MMP-1, MMP-2 and MMP-3. (see table 1). These enzymes are secreted as proenzymes or zymogens and pro MMP-1 and 3 can be activated by plasmin (Alexander and Werb 1990). ProMMP-2 on the other hand is not able to be activated by any known proteolytic enzyme (Nagase, H. British Connective Tissue Society Meeting 1990).

Control of MMPs by endothelial cell stimulating angiogenic factor

Endothelial cell stimulating angiogenic factor (ESAF) is able to activate all three of the proforms of these enzymes (Tables 2 and 3). This factor has a molecular mass of 600Da, is not a protein or peptide and is a specific mitogen for microvessel cells, having no activity towards large vessel or vein endothelial cells or towards fibroblasts and smooth muscle cells (Keegan et al 1982).

Table 1

The major prometalloproteinases and their activators [a]

ENZYME	SUBSTRATE	NATURAL ACTIVATORS OF THE PROENZYME
MMP-1	Type I, II and II collagens	PLASMIN ESAF
MMP-2	Type IV, V and VI collagens	ESAF
MMP-3	Type III, IV, V collagen Proteoglycans, Laminin Fibronectin	PLASMIN ESAF

(a) MMP-2 is the major type IV (basement membrane) collagen degrading enzyme - previously called a gelatinase (reviewed by Matrisian, 1990.

Table 2

ESAF Activation of Pro MMP-1[a]

Assay	ug degraded/hr
Blank	0.50
Blank + APMA	0.55
Blank + ESAF	0.55
Pro MMP-1	0.05
Pro MMP-1 + APMA (1mM)	5.98
Pro MMP-1 + ESAF	4.89

(a)From Mass spectrometry data a very approximate quantitation of
ESAF can be deduced. We assume that the 75ul of ESAF used in
the assays contains 8 to 10ng of ESAF. In the above table all
appropriate assay blanks have been subtracted.
APMA is an organic mercurial compound - aminophenyl mercuric
acetate which chemically activate the proenzyme.
The method assay is described by Weiss et al (1983)

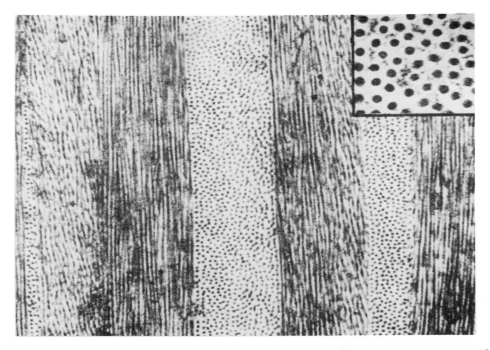

Fig 1. Electron microscope picture of human cornea (courtesy of I Freeman)

Table 3

ESAF Activation of Pro MMP-2[a]

Assay		ug degraded /hr
Blank		0.20
Blank + APMA		0.21
Blank + ESAF (25ul		0.19
Blank + ESAF (50ul)		0.20
Blank + ESAF (75ul)		0.17
Pro MMP-2		0.04
Pro MMP-2	+ APMA	4.03
Pro MMP-2	+ ESAF (25ul)	1.80
Pro MMP-2	+ ESAF (50ul)	2.70
Pro MMP-2	+ ESAF (75ul)	4.13

The relevant assay blanks have been substracted from the cpm
obtained from pro MMP-2 $\pm$ APMA or ESAF. APMA = amino phenyl
mercuric acetate which is an organomercurial compound capable
of activating pro-MMPs. The substrate was gelatin which is
also degraded by this enzyme. The amount of ESAF in 75ul is
of the order of 8-10ng. Method of assay as for Table 2.

ESAF is also a chemoattractant for microvessel endothelial cells. In all
of these processes the amounts of ESAF needed are very small of the order
of nanogrammes. ESAF is also active in vivo on the chick chorioallantoic
and yolk sac membranes and in the rabbit corneal pocket assay (Odedra and
Weiss, 1991).

The synthesis and regulation of the matrix metalloproteinases are
under very strict control. Regulation of synthesis is controlled by
cytokines and gene expression has been shown to occur at the level of
transcriptional activation. Once the mRNA is translated and the proenzyme
(proMMPs) is secreted from the cell activation occurs followed rapidly by
the complexing of the active enzyme with the tissue inhibitors of
metalloproteinases (TIMP) (Matrisian, 1990).

This is of considerable importance since the active enzymes are capable of degrading all the components of the extracellular matrix and it is thought that active enzyme is almost immediately complexed with inhibitor. The complex formed between active enzyme and the inhibitor TIMP is tight and cannot be reversed physiologically with proteinases or chemically by organic mercurial compounds which can activate the proenzyme (Cawston et al 1983). ESAF, again, is able to activate the enzyme inhibitor complex. Table 4 and Table 5 show the reactivation of TIMP inhibited MMP-s -1 and 2 with ESAF.

Table 4.   Reactivation by ESAF of the TIMP/MMP-2 Complex(a)

Assay	ug/collagen degraded /hr
Active MMP-1 .	21.91
Active MMP-1 + APMA	21.98
Active MMP-1 + 1 unit TIMP	2.7
Active MMP-1 + 1 unit TIMP + ESAF	19.94

1 unit of ESAF is approximately 8ng. APMA was used to confirm absence of any proenzyme

Table 5.   Reactivation by ESAF of the TIMP/MMP-2 Complex(a)

Assay	ug collagen degraded /hr
Active MMP-2	5.51
Active MMP-2 + APMA	5.55
Active MMP-2 + 1 unit TIMP	0.2
Active MMP-2+ 1 unit TIMP + 1 unit ESAF	4.35

1 unit of ESAF as approximately 8ng. APMA was used to confirm absence of any proenzyme

That this reactivation is accompanied by splitting of the enzyme inhibitor complex has been demonstrated by McLaughlin et al (1991) using radiolabelled TIMP bound to MMP-1. Free TIMP does not bind to a zinc chelate column although MMP-1 (a zinc metalloenzyme) does as does the MMP-1 complex with TIMP. These workers therefore used zinc chelate chromatography to follow the process occurring after addition of ESAF to the enzyme inhibitor complex. The release of free TIMP from the complex was clearly demonstrated.

Discussion

In all the biochemical activities of ESAF in the activation of proenzymes and reactivation of the enzyme inhibitor complexes the amounts of ESAF needed are very small, never exceeding nanogram quantities. The importance to the understanding of the neovascularisation process of an angiogenic factor able to set in motion a limited degradation of the extracellular matrix is obvious. What is not clear is how such a degradation is indeed limited and does not result in extensive matrix destruction. If the microvessel endothelial cell itself was the source of the proenzyme as has been suggested by Gross et al 1983 and other workers (Presta et al 1985)then a co-operative system between cytokine or growth factor and ESAF would be possible. In this context fibroblast growth factor would fulfill the cytokine role and its own postulated role as an angiogenic factor since it has been shown to stimulate microvessel endothelial cells to synthesise pro MMP-1 together with plasmin (Moscatelli and Rifkin 1988, Moscatelli et al 1985). No information is yet available as to whether FGF or another cytokine can potentiate synthesis of pro MMP2 from the same cells. Also it is not clear how and under what stimulus FGF itself is released from parent cells (D'Amore 1990).

Another reason why the local activation of collagenolytic enzymes is seen to be a key factor in the onset of neovascularisation is the observation that an inhibitor of angiogenesis recently purified from bovine vitreous (Taylor and Weiss 1985) is able to inhibit all the three active matrix metalloproteinases discussed here (MMPs 1,2,3) (Table 6). Vitreous is an avascular tissue and previously early workers had isolated tissue specific inhibitors of collagenase (Kuettner et al 1976) and angiogenesis from these sources (Brem et al 1977, Raymond and Jacobson 1982). The inhibitor described by Taylor and Weiss had molecular mass of 5,700 Da and was a glycoprotein. More recently Thompson and Weiss

Effect of 3.5K vitreous inhibitor on microvessel endothelial cells and fibroblasts

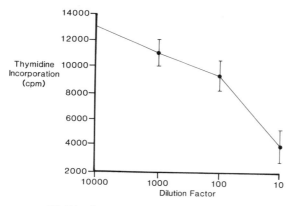

NB. This effect was not observed when cells were grown on gelatin

Fig 2a. Inhibition of microvessel endothelial cells grown on a collagen substratum by 3.5K inhibitor (Thompson 1989).

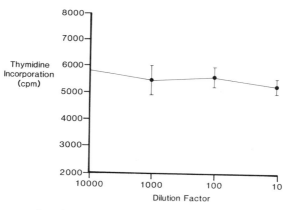

Fig 2b.   As above but using fibroblasts (no inhibition)

confirmed the presence of the 5.7K inhibitor but also detected a second more potent glycoprotein inhibitor with molecular mass of 3,500 (reviewed by Taylor et al 1991). Both factors were able to inhibit microvessel endothelial cell growth on a collagen substratum but were ineffective against fibroblasts under the same conditions (Fig 2a,b). Although ESAF can reverse the vitreal inhibitor/MMP-1 complex the amounts needed to achieve this effect are significantly greater than that required to separate the TIMP/inhibitor complex. TIMP itself is not anti-angiogenic in vivo in the CAM or corneal pocket assay nor does it have any effect in vitro on microvessel endothelial cells in culture. This leads to the supposition that while an involvement with activation or inhibition of matrix metalloproteinases may be a pre-requisite for an angiogenic or anti-angiogenic factor this is insufficient on its own. An involvement with proliferation or migration of the microvessel endothelial cell itself is also of the essence.

In this context ESAF may perhaps justifiably be considered as a true angiogenic factor.

Table 6

Inhibition MMPs by 3.5Kd Vitreous Inhibitor
(Thompson 1989)

Enzyme	Substrate	Inhibition
MMP-1	Type I Collagen	95%
MMP-2	Type I Gelatin	93%
MMP-3	Casein	68%

References

Alexander, C.M. and Werb, Z. (1990) Proteinases and extracellular remodelling. Curr. Opin. Cell Biol. 1: 974-982

Ausprunk, D. (1979) Tumour angiogenesis in "Handbook of inflammation" J.C. Houck (ed) 1: 318-351

Brem, S., Pries, I., Langer, R., Brem, H., Folkman, J. and Patz, A. (1977) Inhibition of neovascularisation by an extract derived from vitreous. Am. J. Ophthalmol. 84: 323-328

Cawston, T.E., Murphy, G., Mercer, E., Galloway, W.A., Hazleman, B.L. and Reynolds, J.J. (1983) The interaction of purified rabbit bone collagenase with purified rabbit bone metalloproteinase inhibitor. Biochem. J. 211: 313-318

D'Amore, P.A. (1990) Models of FGF release in vivo and in vitro. Cancer, Metast. Rev. 9: 227-238

Gross, J.L., Moscatelli, D. and Rifkin, D. (1983) Increased capillary endothelial cell protease activity in response to angiogenic stimuli in vitro. Proc. Nat. Acad. Sci. U.S.A. 80: 2623-2627

Keegan, A., Hill, C., Kumar, S., Phillips, P., Schor, A. and Weiss, J.B. (1982) Purified tumour angiogenesis factor enhances proliferation of capillary but not large vessel endothelial cells in vitro. J. Cell Sci. 55: 261-276

Kuettner, K.E., Hibi, J., Eisenstein, R. and Harper, E. (1976) Collagenase inhibition by catatonic proteins derived from cartilage and aorta. Biochem. Biophys. Res. Commun. 72: 42-46

Matrisian, L.M. (1990) Metalloproteinases and their inhibitors in matrix remodelling. Trends in Genetics 6(4): 121-125

Matrisian, L.M. and Hogan, B.L. (1990) Growth factor regulated proteases and extracellular mnatrix remodelling during mammalian development. Curr. Top. Dev. Biol. 24: 219-259

McLaughlin, B., Cawston, T. and Weiss, J.B. (1991) Activation of the matrix metalloproteinase complex by a low molecular weight angiogenic factor. Biochem. Biophys. Acta. 1073: 295-298

Moscatelli, D. and Rifkin, D.B. (1988) Membrane and matrix localisation of proteinases: a common theme in tumour cell invasion and angiogenesis. Biochem. Biophys. Acta. 948(1): 67-85

Odedra, R. and Weiss, J.B. (1991) Low molecular weight angiogenesis factors. Pharmac. Ther. 49: 111-124

Presta, M., Mignath, P., Mullins, D.E. and Moscatelli, D.A. (1985) Human placental tissue stimulates bovine capillary endothelial cell growth, migration and protease production. Biosci. Rep. 5(9): 783-790

Raymond, L. and Jacobson, B. (1982) Isolation and identification of stimulatory and inhibitory activity cell growth factors in bovine vitreous. Exp. Eye Res. 34: 267-280

Taylor, C.M. and Weiss, J.B. (1985) Partial purification of a 5.7K glycoprotein from bovine vitreous which inhibits both angiogenesis and collagenase activity. Biochem. Biophys. Res. Commun. 133(9): 911-916

Taylor, C.M., Thompson, J.H. and Weiss, J.B. (1991). Matrix integrity and the control of angiogenesis. Int. J. Radiat. Biol. (In press)

Thompson, J.M. (1989) Latency of neutral matrix metalloproteinases and control of neovascularisation. Ph.D. thesis, University of Manchester, Manchester, U.K.

Weiss, J.B., Hill, C.R., Davis, R.J., McLaughlin, B., Sedowofia, K.A. and Brown, R.A. (1983) Activation of a procollagenase by low molecular weight angiogenesis factor. Biosci. Rep. 3(2): 171-177

HYALURONIC ACID PROMOTES AND INHIBITS ANGIOGENESIS

S Kumar, P Kumar, JM Ponting, A Sattar, P Rooney,
D Pye and RD Hunter

Christie Hospital, Manchester, M20 9BX, England

Hyaluronic acid (HA) was first isolated nearly 50 years ago and is a ubiquitous component of the extracellular matrix (ECM) (Meyer and Palmer, 1934). From extrapolation of animal studies it has been estimated that an adult human contains approximately 15g HA, most of which is in skin. Remarkably, the total daily turnover of HA within the human body is in the region of several grams, of which only 0.3 mg is excreted via the kidneys (Laurent and Frazer, 1989). HA is a normal component of body fluids, its mean serum concentration is 50μg/litre and 2.5% of it is eliminated every minute. Interestingly the concentration of HA in lymph is 200-600 fold greater than in serum.

Chemically it is a simple unbranched linear polysaccharide consisting of repeating disaccharide units of N-acetyl-glucosamine and D-glucuronic acid. At physiological pH its carboxyl groups are completely dissociated; hence it is referred to as hyaluronate or more commonly, hyaluronan. Unlike other glycosaminoglycans (GAGs) it is not covalently linked to proteins. Classical studies on the regulation of HA synthesis by Prehm and his colleagues have established that the site of HA synthesis is the plasma membrane and that the elongation of the chain takes place by the addition of sugars to the reducing end, which is in contrast to the synthesis of proteoglycans. Again, it is unique among the GAGs, in being shed from the cells while being synthesized i.e. there is no post-polymerisation modification. Prehm (1989) has identified HA synthase as a 50KDa protein. Rous sarcoma virus induced transformation of fibroblasts activates the synthase and this is accompanied by enhanced transcription and phosphorylation. The synthase is the natural target of pp60 v-src kinase. The catabolism of HA is effected by 3 enzymes viz hyaluronidase and two endoglycosidases: β-D glucuronidase and β-N-acetyl-D-hexosaminidase (Roden et al, 1989).

The list of functional properties of HA is continually growing (Table 1). HA undoubtedly is a molecule of paradoxes and contrasts (Scott, 1989). In tissues like rooster comb, it may be just a space filler, whereas in other situations it can activate early response genes (our unpublished data). The interest of our department in the role of HA in angiogenesis was sheer serendipity. We were puzzled by the fact that wound fluids did not induce angiogenesis in the standard *in vivo* chicken chorioallantoic membrane (CAM) assay. We were aware of a publication by Feinberg and Beebe (1983), wherein HA was noted to inhibit vascularization of chick limb buds. Therefore, we argued that the lack of demonstrable angiogenic activity in wound fluids might be the result of the presence of HA. Indeed, treatment of wound fluid and subsequently of HA with hyaluronidase (HAase) resulted in the angiogenic response in the CAM assay. Fortunately, Ian Hampson, a postgraduate working in our laboratory (together with Dr J Gallagher) had just developed a novel method for sequencing and determining the molecular weight of glycosaminoglycans. Thus, within a matter of weeks, we were able to show that *while both native HA and completely degraded HA were not angiogenic, partially degraded HA was angiogenic* (West et al, 1985). We have highlighted some of our work and reviewed the relevant literature related to the effects of HA on angiogenesis.

Angiogenesis in Health and Disease, Edited by M.E. Maragoudakis *et al.*
Plenum Press, New York, 1992

TABLE 1. A list of putative functions of hyaluronic acid

Space filler and lubricant.

Protection against micro-organisms and immune cells.

Protection against radiation.[*]

Haemostasis; inflammation and wound healing.

Angiogenesis.[*]

Tumour cell invasion?

Embryonic development.

Differentiation of chondrocytes/inhibition of ^{35}S proteoglycan synthesis by chondrocytes.

Induction of myoblast fusion and subsequent muscle differentiation.

Brain development/GABAergic transmission.

Cell aggregation, attachment, detachment, migration and cytoskeletal organization.

Storage of growth factors in the extracellular matrix.

Enhancement of collagen synthesis.[*]

Activation of early response genes.[*]

Maintenance of epithelial polarity.

Maintenance of channels between cells for nutrient flow.

[*]   Our work

Preparation of HA oligosaccharides

Angiogenic fragments of hyaluronate were prepared as described previously (West and Kumar, 1989). 500mg of HA was dissolved in 50ml of 0.1M acetate buffer (pH 5.4), containing 0.15M NaCl and treated with 2000 units of testicular hyaluronidase at 37°C. At 2, 4, 6, 8 and 24 hour intervals, 10ml aliquots were removed and 1ml of 100% TCA was added to each. The mixtures were centrifuged and the supernatants pooled, dialysed, re-centrifuged and lyophilized. The powder was dissolved in 20ml of 0.1M acetic acid and loaded onto a G50 Sephadex column (2.6 x 180cm). 10ml fractions were collected, assayed for uronic acid and individual fractions combined to give 3 pools viz F1, F2 and F3. The size range of oligosaccharides in each pool was determined by incorporating [^{3}H]-glucosamine labelled HA and analysing each fraction by SDS-polyacrylamide gel electrophoresis (SDS-PAGE). F1, F2 and F3 fractions consisted of oligosaccharides of >16, 10-16 and 2-10 disaccharide units respectively. The F3 fraction induced angiogenesis in the standard chicken chorioallantoic membrane assay. Furthermore, the addition of F3 to cultures of bovine aortic endothelial cells (BAEC) and brain derived microvessel endothelial cells stimulated both their proliferation and migration (Fig 1) (West and Kumar, 1989; Sattar et al, 1991). This effect is cell type specific, as no stimulation of fibroblasts, smooth muscle cells or keratinocytes was observed.

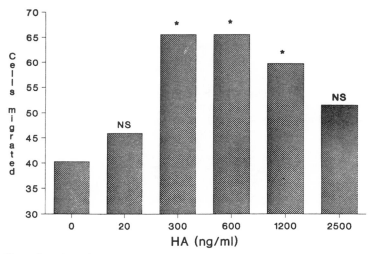

Fig 1 The effect of angiogenic HA oligosaccharides on cell migration of bovine aortic endothelial cells using a polycarbonate membrane migration assy (for further details see Sattar et al., 1991) (*) Statistically significant P < 0.05; (NS) not significant.

HA and skin vascularisation

Angiogenic oligosaccharides of HA (F3 fraction) were incorporated in methyl cellulose and dried into discs of approximately 2mm diameter and 20μm thickness. The discs containing 0, 10 or 100 μg HA were surgically inserted underneath the skin on both sides of the back and in both ears of rabbits. The incisions were sutured and 5 days later skin biopsies from the experimental sites were carefully excised, fixed in 10% formalin and processed for routine histology. The ear skin was invariably inflamed wherein it was not possible to be certain of the exact site of HA application. Therefore, none of these specimens were used for enumeration of blood vessels. The tissue sections from the back skin biopsies were examined (x 400 magnification) and the number of blood vessels in 5 different fields was counted. Implantation of HA resulted in a dose dependent increase in the number of blood vessels (Table 2).

TABLE 2. The effect of application of HA oligosaccharides on vascularisation of rabbit skin

DOSE OF HYALURONIC ACID FRAGMENTS	NO OF BLOOD VESSELS x 10^{-5}/μm (AVERAGE OF 5 FIELD COUNTS)
0 (Control)	4.914
10 μg	5.652
100 μg	7.165 *

*   P < 0.03

In a related collaborative unpublished study with Dr T Fan, the effect of HA on blood flow in subcutaneously located sponge implants in rats was also examined. Five daily injections of 5 or 50 μg HA oligosaccharides resulted in a marked acceleration of sponge vascularisation which was shown by an increased clearance of locally-injected ^{133}Xe. This was accompanied by an increase in protein and DNA content and an increase in the number of blood vessels.

HA in experimentally induced myocardial infarcts

This work was undertaken in collaboration with Professor W Schaper (Bad Nauheim, Germany). He has developed a method of inducing the slow and progressive occlusion of a coronary artery in pig heart, which results in angiogenesis (Fig 2). Semi-purified extracts of tissue specimens representing infarcted, adjacent normal looking and normal heart were examined for their angiogenic activity in the CAM assay (Kumar et al, 1983). Both their HA content and the size of HA was estimated by methods described elsewhere (Kumar et al, 1989). Infarcted (3/3) and normal-looking heart (2/3) induced angiogenesis in the CAM assay whereas only one of 3 specimens of the normal heart gave a response which was very weakly positive. High molecular mass HA (> 10^5 Da) was found in all 3 tissues. It may be important that the proportion of low molecular weight HA (< 10^5 daltons) was significantly higher in the infarcted and normal looking tissues compared with normal heart. In a separate study with Dr D Smith we have observed that the intracardiac infusion of HA oligosaccharides resulted in an increase in the actual number of blood vessels per unit area in treated compared with control heart tissues.

HA and collagen synthesis

Over the years we have employed the CAM assay to examine the angiogenic activity of normal and tumour tissues, myocardial infarcts and synovial fluids (Kumar et al, 1987). The treated and control CAMs were examined using a stereoscan microscope and processed for light and electron microscopy. A comparison of electron micrographs of HA oligosaccharides and CAMs treated with other angiogenic stimuli revealed a striking difference. HA treatment induced a very marked increase in the collagen fibrils (Fig 3). We have investigated the synthesis of collagens by bovine aortic endothelial cells exposed to HA oligosaccharides. The presence of HA resulted in the production of a collagenous polypeptide of 61 KDa, which was reduced in size to 50 KDa by pepsinisation. The size and chemical characteristics of this polypeptide were consistent with the type VIII collagen. A polyclonal antibody to type X collagen (which also reacts with type VIII collagen) produced intense staining in HA activated endothelial cells. The role played by type VIII collagen in angiogenesis is unclear in much the same way as the role of the type X collagen during endochondral ossification. We have proposed that type VIII collagen and type X collagen may play a similar role in non-skeletal and skeletal tissues respectively (Rooney and Kumar, manuscript submitted for publication).

Effect of HA on irradiated endothelial cells

The role of normal tissue damage imposes a severe limitation on radiotherapy. It is recognised that late tissue injury results largely from progressive vascular damage. Mr R Brew in our laboratory examined the effect of angiogenic HA oligosaccharides on irradiated endothelial cells. BAEC were grown in 24 well tissue culture plates and a single dose of X-rays was given so that test plates received a total of 1, 2.5, 5 or 10 Gy. Within 30 minutes after the plates had been irradiated,

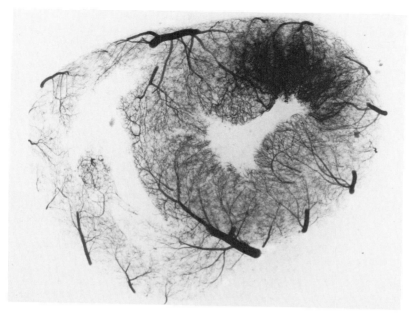

Fig 2 Area of new vessel growth ('dark blush') following slow and progressive occlusion of one coronary artery in a pig heart. Five millimetre slice through the right and left ventricle (courtesy Professor W. Schaper).

Fig 3 HA treated CAM: An electron micrograph showing abundant collagen fibrils both within and around a cell. Bundles of collagen can be seen leaving the cell (↑).

test wells received 1 ml of medium containing 0.1, 1.0 or 10 µg HA/well. Alternatively, HA was added prior to irradiation or 24 hours post-irradiation. Control wells received the same amount of medium but without HA. On the third day the plates were harvested and their cellular DNA measured. From figure 4 it can be seen that the addition of HA into an irradiated plate resulted in a statistically significant increase in DNA (P < 0.001) compared to an irradiated plate that received no HA. The addition of HA had a significant effect in protecting BAEC from post-irradiation damage (P < 0.0001; Fig 5). The effect was more marked if the HA was present before the cells were irradiated. The mechanism of the protective effect is not known. Since HA was able to protect endothelial cells, even if administered several hours post-irradiation, the neutralisation of very short-lived free radicals by HA as the sole underlying mechanism can be ruled out.

HA and tumours

Increased levels of HA have been associated with human and animal tumours. Particularly noteworthy are mesothelioma and Wilms' tumour. It has been suggested that HA production is important in tumour invasion and metastasis. For instance, when rabbit V2 carcinoma invades rabbit muscle there is a nearly 10-fold increase in total GAGs, the greatest (71%) being in HA (Toole et al, 1979). In vitro tumour cells co-cultured with normal fibroblasts induce HA synthesis by the latter. The HA stimulatory activity of tumour cells is a membrane bound, heat sensitive, lipophilic cell surface glycoprotein (Knudson et al, 1989). Knudson's group observed that some tumour cells possess cell surface binding sites for HA, which might permit tumour cells to interact directly with the HA rich ECM. Thus, HA could function to support and guide tumour cell invasion.

Among the childhood renal malignancies, Wilms' tumour is the most frequent and has an excellent prognosis. Another rare type of kidney tumour (bone metastasizing renal tumour of childhood - BMRTC) has a poor prognosis that over recent years has improved quite significantly. In agreement with other reports we found raised levels of HA in the sera of children with Wilms' tumours. However, when we analysed the molecular mass of HA, it was found that while the sera from normal children and Wilms' tumour contained high molecular mass HA, the sera of BMRTC contained almost entirely low-molecular mass HA, similar in size to angiogenic oligosaccharides of HA (Fig 6; Kumar et al 1989). Recently we have demonstrated that low molecular mass HA was synthesised and released into the medium by tissue cultured BMRTC cells (Kumar et al, manuscript in preparation). It is tempting to suggest that low molecular mass HA may facilitate the bone-metastasizing ability of BMRTC.

HA and vascular growth in wound healing models

Lebel and Gerdin (1991) have used the rabbit ear chamber model to examine the effect of HA (KDa 3.9×10^6) on vascular ingrowth in healing granulation tissue. Their results showed that daily injections of HA inhibited, whereas intermittent injections enhanced blood vessel growth. HA was taken up by macrophages, which led Lebel and Gerdin to propose that intermittent injections of HA may have induced the vascularisation via elaboration of angiogenic factors by activated macrophages. The authors postulated that the antiangiogenic effect of daily administration of HA was probably due to its physical hindrance. It is relevant to mention here that Dvorak et al (1987) observed that high (1-4 mg/ml) but not low (20µg/ml) concentrations of HA inhibited fibrin-induced angiogenesis. King et al (1991) have shown in a hamster cheek pouch wound model that the application of HA soaked sponges followed by 4 topical applications over 7 days, decreased the wound size about twice as fast as the control wounds. In our hands, treatment of tissue cultured endothelial cell monolayer with high concentrations of native i.e. high molecular weight HA, had a profound deleterious effect on the integrity of the monolayer. Furthermore, HA oligosaccharides induced activation of early response genes which was abolished by native HA (Deed, Norton and Kumar, unpublished data). In view of these somewhat conflicting reports, it is difficult to draw definite conclusions. It would seem that the effect of HA on angiogenesis is not only dependent on its size, but is also influenced by its concentration and frequency of application.

Internalisation of HA and HA receptor

Dr D West's work from this laboratory, has demonstrated that bovine aortic endothelial cells endocytosed fluorescently labelled macromolecular (10^6 Da) and smaller sized ($10^4 - 10^5$ Da) HA. In comparison, the uptake of labelled HA by fibroblasts and smooth muscle cells was negligible. He also undertook binding studies using ^{3}H-labelled HA. A HA-specific receptor was present on BAEC

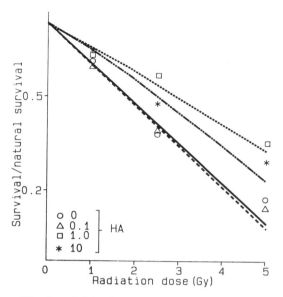

Fig 5 Survival Curves: The beneficial effect of angiogenic HA oligosaccharides on irradiated bovine aortic endothelial cells.

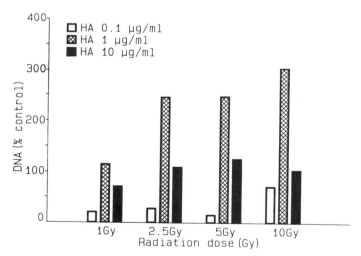

Fig 4 The addition of angiogenic HA oligosaccharides to cultures of irradiated bovine aortic endothelial cells resulted in significant increase in cellular DNA.

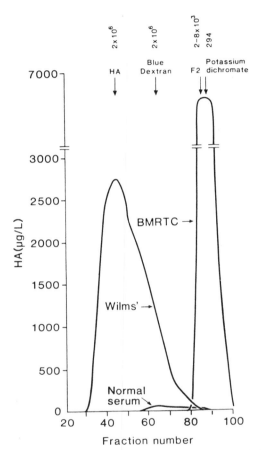

Fig 6 Hyaluronic acid in renal tumours: A calibrated S-1000 Sephacryl Column (1.5 x 113cm) was equilibrated with phosphate-bufferred saline (0.1M pH 7.0) and eluted at 10 ml/hr. The size distribution of HA was estimated using a radiometric assay. The molecular mass of HA in the sera of BMRTC patients was much lower compared with those from normal children and patients with Wilm's tumor (reproduced from Kumar et al., 1989. Int. J. Cancer 44, 445).

(Kd10^{-10}M; with 2000 receptor sites per cell) (West and Kumar, 1989) . Attempts were made to identify the HA receptor. From the preliminary data it seems that there are 5 proteins between 90-125 KDa which may represent a receptor-complex. In other published studies, the number of HA receptors has been noted to be either similar (Madsen et al, 1989) or greater (10^5 per cell; Raja et al, 1988) than seen by us (Sattar et al, 1991). These differences most probably stem from the source of endothelial cells, passage number and cell density. In the past two years, a number of reports have alluded to the CD44 antigen being involved in lymphocyte homing and HA recognition. In more recent publications it indeed has become apparent that CD44 is a polymorphic integral membrane protein which recognises HA and whose proposed roles are lymphocyte activation, matrix adhesion and attachment of lymphocytes to high endothelial venules (Stamenkovic et al, 1991).

CONCLUSIONS

Hyaluronic acid is a molecule of antiquity and is found throughout the animal kingdom from prokaryotes to mammals. Like other ECM components, HA was considered to be an inert space-filler. It may well play that role in certain tissues and situations but certainly it is not an inactive mortar in many other locations in the body. Something with such a large and rapid turnover cannot be a substance without an important role. Nature is rarely wasteful! At the beginning of this chapter we gave a selective list of some of the functions that HA has been alleged to play. Tovard Laurent and his colleagues, who have pioneered many studies on the structural and functional aspects of HA, concluded that HA must be vital for the existence of an organism as no individual or disease with a total lack of HA has been recorded.

Since its discovery, a number of major advances have been made which were summarised in CIBA Symposium No 143 (1989). More recent interest in HA has been in its molecular biology, determination of its role in health and disease, identification and cloning of the HA receptor (particularly CD44 antigen) and cloning of HA binding proteins. Before too long, we will have more definite information as to whether HA or one of its synthetic analogues has beneficial effects in amelioration of radiation induced tissue damage, wound healing, skin vascularisation or hair growth. Undoubtedly this is an exciting time for both established workers and newcomers to the HA field that holds promise for important discoveries.

Angiogenic effects of HA depend on its molecular mass. High concentrations of native i.e. macromolecular HA inhibit angiogenesis whereas its oligosaccharides (2-10 disaccharides) promote angiogenesis *in vivo* and *in vitro* stimulate migration and proliferation of normal endothelial cells from large and microvessels. The HA oligosaccharides have no effect on tissue-cultured human brain tumour derived endothelial cells. The reason for this difference is not known. The endothelial cells are heterogeneous. Could it be that there are fine differences in the amount and molecular mass of HA that the endothelial cells from various sources produce? Elegant studies of Judah Folkman and his colleagues (Ausprunk et al, 1981) have shown that the growing tip of a newly formed blood vessel is rich in hyaluronic acid. Unfortunately, no one else has followed this promising lead.

The availability of reliable and sensitive quantitation assays for HA has provided an opening for determining its relevance in body fluids and tissue extracts. We have shown that although serum levels of HA in patients with two different histological types of renal tumour were similar; the sera of patients with Wilms' tumour, which does not metastasise to bones and has a very good prognosis, contained macromolecular HA. In contrast, low molecular mass HA oligosaccharides were present in the sera of children with BMRTC, which as the name implies has a marked predilection for metastasising to bones. Whether this feature is shared by other primary bone tumours or is a hallmark of tumours that metastasise to bone is not known. What role HA degrading enzymes, HA binding proteins and HA receptors play in tumourigenesis are some of the areas that require further studies.

REFERENCES

Ausprunk DH, Boudreau CL, and Nelson DA. Proteoglycans in the microvasculature II. Histochemical localisation in proliferating capillaries of the rabbit cornea. Am J Pathol, 103, 267-275 (1981)).

Dvorak HF, Harvey VS, Estrella P, Brown LF, McDonagh, J and Dvorak AM. Fibrin containing gels induce angiogenesis: implications for tumour stroma generation and wound healing. Lab Invest, 57, 673-686 (1987).

Feinberg RN and Beebe DC Hyaluronate in vasculogenesis. Science, 220, 1177-1179 (1983).

Fraser JRE and Laurent TC. Turnover and metabolism of hyaluronan. Ciba Foundation Symposium 143 Wiley, Chichester pp 41-59 (1989).

King SR, Hickerson WL and Proctor KG. Beneficial actions of exogenous hyaluronic acid on wound healing. Surgery, 109, 76-84 (1991).

Knudson W, Biswas C, Li X-Q, Nemec RE, Toole BP. The role and regulation of tumour associated hyaluronan. CIBA Foundation Symposium 143, Wiley, Chichester pp 150-169 (1989).

Kumar S, West DC, Shahabuddin S, Arnold F, Haboubi N, Reid H and Carr T. Angiogenesis factor from human myocardial infarcts. Lancet, 2, 364-367 (1983).

Kumar S, West DC and Ager A. Heterogeneity in endothelial cells from large vessels and microvessels. Differentiation, 36, 57-70 (1987).

Kumar S, West DC, Ponting J and Gattamaneni HR. Sera of children with renal tumours contain low molecular-mass hyaluronic acid. Int J Cancer, 44, 445-448 (1989).

Kumar S and West DC. Hyaluronic acid metabolism in childhood renal tumours. J Natl Cancer Inst, 82, 973 (1990).

Lebel L and Gerdin B. Sodium hyaluronate increases vascular ingrowth in the rabbit ear chamber. Int J Exp Pathol, 72, 111-118 (1991).

Madsen K, Schenholm M and Jahke G. Hyaluronate binding to intact corneas and cultured endothelial cells. Invest Opthal Vis Science 30, 2132-2137 (1989).

Meyer K and Palmer JW. The polysaccharide of the vitreous humor. J Biol Chem 107, 629-634 (1934).

Prehm P. Identification and regulation of the eukaryotic hyaluronate synthase. Ciba Foundation Symposium 143 Wiley, Chichester pp 21-40 (1989).

Raja RH, McGary CT, Weigel PH. Affinity and distribution of surface and intracellular hyaluronic acid receptors in isolated rat liver endothelial cells. J Biol Chem, 263, 16661-16668 (1988).

Roden L, Campbell P, Fraser JRE, Laurent TC, Pertoft H and Thompson N. Enzymic pathways of hyaluronan catabolism. Ciba Foundation Symposium 143 Wiley, Chichester pp 60-86 (1989).

Scott DM, Kumar S and Barnes JM. The effect of a native collagen gel substratum on the synthesis of collagen by bovine brain capillary endothelial cells. Cell Biochem and Funct, 6, 209-215 (1988).

Scott JE. Secondary structures in hyaluronan solutions: chemical and biological implications. Ciba Foundation Symposium 143 Wiley, Chichester pp 6-20 (1989).

Sattar A, Kumar S and West DC. Does hyaluronate have a role in endothelial cell proliferation in synovium? Seminars in Arthritis and Rheumatology (1991) (in press).

Stamenkovic I, Aruffo A, Amiot M and Seed B. The haematopoietic and epithelial forms of CD44 are distinct polypeptides with different adhesion potentials for hyaluronate bearing cells. EMBO J 10, 343-348 (1991).

Toole BP, Biswas C, Gross J. Hyaluronate and invasiveness of the rabbit V2 carcinoma. Proc Natl Acad Sci USA, 76, 6299-6303 (1979).

West DC, Hampson IN, Arnold F and Kumar S. Angiogenesis induced by degradation products of hyaluronic acid. Science 228, 1324-1326 (1985).

West DC and Kumar S. The effect of hyaluronate and its oligosaccharides on endothelial cell proliferation and monolayer integrity. Exptl Cell Res, <u>183</u>, 179-196 (1989).

West DC and Kumar S. Tumour-associated hyaluronan: a potential regulator of tumour angiogenesis. Int J Radiat Biol, <u>60</u>, 55-60 (1991).

BRAIN TUMOR ANGIOGENESIS: DRUG DELIVERY AND NEW INHIBITORS

Christopher Guerin, Rafael J. Tamargo, Alessandro Olivi, and Henry Brem

Departments of Neurosurgery, Oncology, and Opthalmology, The Johns Hopkins University School of Medicine, Baltimore, Maryland, 21205, USA

INTRODUCTION

Despite enormous advances in clinical medicine, the prognosis of primary malignant brain neoplasms remains dismal. Currently, the most effective therapies result in a median survival of less than one year, with the majority of patients dying within two years (Black, 1991). In order to develop more effective methods for combatting this disease, we took advantage of two biological properties that are critical to glioma growth: first, the observation that malignant gliomas are predominantly a local disease usually recurring within two centimeters of the original focus (Hochberg and Pruitt, 1980); and second, the increasing evidence that brain tumors, like most solid tumors, are angiogenesis-dependent (Brem, et. al., 1988).

BRAIN TUMOR THERAPY: DRUG DELIVERY

Standard chemotherapy is hindered by difficulties encountered in delivering adequate doses to the tumor. Tumor vessels, especially at the periphery, may retain some of the barrier properties seen in normal brain vessels, thus restricting parenchymal entry of many drugs (Neuwelt, et. al., 1986). The use of high drug doses in an attempt to improve delivery often results in considerable systemic toxicity. Since primary malignant brain tumors are predominantly focal lesions, we developed methods for delivering drugs directly to the tumor bed. Biodegradable polymers impregnated with chemotherapeutic agents can be implanted at the tumor site during surgery and predictably release high local concentrations of drug. This mode of therapy bypasses the blood-brain barrier and minimizes systemic exposure and toxicity (Brem, 1990 and 1990a; Langer, Brem and Langer, 1991; Langer, 1990).

We initially showed that the biodegradeable polyanhydride poly[bis(p-carboxyphenoxy)propane-sebacic acid] coploymer

Angiogenesis in Health and Disease, Edited by M.E. Maragoudakis *et al.*
Plenum Press, New York, 1992

(PCPP-SA) was biocompatible in the brains of rats, rabbits, and monkeys (Brem, et. al., 1989 and 1989a; Tamargo, et. al., 1989). We then showed that polymers impregnated with the standard chemotherapeutic agent used in malignant glioma therapy, BCNU (carmustine), did release intact drug in a predictable manner when implanted intraperitoneally or into the brains of rats (Yang, Tamargo, and Brem, 1989). Autoradiographic studies and HPLC analyses in the rabbit brain demonstrated the localized release of BCNU at very high levels while plasma and contralateral brain levels were negligible (Grossman, et. al., 1988). Finally, we demonstrated that BCNU-impregnated polymers were effective in treating malignant gliomas using the rat 9L glioma model (Tamargo, et. al., 1988). In fact, polymeric delivery of BCNU at the tumor site was more effective than systemic therapy in inhibiting 9L glioma growth in the flank and intracranially.

Based on these results, we began a clinical trial of BCNU-polymer therapy in humans with recurrent malignant gliomas. The initial dose escalation study revealed no evidence of toxicity attributable to polymers (Brem, et. al., 1991). Though not an efficacy study, the post-operative survival of these patients (87 weeks) was considerably higher than that predicted for recurrent glioma patients. Thus, polymer therapy appears to be safe in humans with preliminary evidence of efficacy. To formally assess efficacy, a randomized, double-blind, placebo-controlled trial is underway at multiple centers in North America and Europe.

In summary, we have developed a safe and effective method for delivering drugs to brain tumors. We have used BCNU since it is an established agent. Polymer delivery may prove beneficial for many other drugs as well. We have been interested in developing new drugs which exploit another biological property of brain tumors: the dependence of tumors on angiogenesis for sustained growth.

BRAIN TUMORS: AN ANGIOGENESIS-DEPENDENT DISEASE

Evidence is increasing that the growth of brain neoplasms is angiogenesis-dependent. Solid tumors grow in two phases: avascular and vascular (Folkman, 1972). Several experiments have shown that solid tumors without a vascular supply do not grow to more than 3-4 millimeters in diameter because of the limited distance over which oxygen and nutrients can diffuse (Gimbrone, et. al., 1982). At this size, most tumors are asymptomatic. When such tumors become vascularized, exponential growth ensues and symptoms occur.

Angiogenic activity is a marker of the malignancy of brain tumors. Brem, Cotran, and Folkman (1972) developed a grading system for the angiogenesis response to human neoplasms and found that malignant gliomas were the most "endothelial-rich" of tumors. Among malignant gliomas, increasing degrees of anaplasia were associated with an increased angiogenesis score (Brem, 1976). Post-operative survival also correlated well with the angiogenesis score. We showed that human brain tumor biopsies transplanted to the rabbit cornea induced angiogenesis in proportion to the malignancy of the tumor

(Brem, et. al. 1980). In addition, the ability of CSF to stimulate endothelial cell migration was ten-fold higher in brain tumor patients than in others with nonneoplastic diseases (Brem, Patz, and Tapper, 1983). Recently, several investigators have demonstrated production of the angiogenic growth factors acidic and basic FGF by human brain tumors (Liberman, et. al., 1987; Paulus, et. al., 1990; Takahashi, et. al., 1990; Maxwell, et. al., 1991). Expression of basic FGF has been correlated with the malignancy of brain tumors (Zagzag, et. al., 1990), and acidic FGF has been detected in tumor cyst fluid (Moringlane, Spinas, and Bohlen, 1990). These findings are supported by work on other tumors. The aqueous humor of patients with malignant eye disease and the urine of patients with genitourinary malignancies have angiogenic properties (Tapper, et. al., 1979; Chodak, et. al., 1980). The malignancy and metastatic potential of breast tumors has also been correlated with angiogenic activity (Brem, Jensen, and Gullino, 1978; Weidner, et. al., 1991). Thus, angiogenesis activity appears to reflect the biological aggressiveness of many solid tumors and may be a predictor of malignant behavior. Of more importance, compounds which inhibit the ability of brain tumors to induce neovascularization may have major therapeutic applications in the control of tumor growth (Folkman and Klagsbrun, 1987).

BRAIN TUMOR THERAPY: ANGIOGENESIS INHIBITION

In recent years, many inhibitors of angiogenesis have been discovered, but the development of a clinically useful agent has been hindered by potential toxicity or limited availability. We first demonstrated that angiogenesis induced by the Walker 256 carcinoma, V2 carcinoma, and human brain tumors could be inhibited by cartilage, a tissue which is normally avascular (Brem and Folkman, 1975; Brem, et. al., 1980). Subsequently, Langer, et. al. (1976), isolated a factor from cartilage which inhibited angiogenesis and also inhibited mammalian collagenase (Moses, Sudhalter, and Langer, 1990). Vitreous, another avascular tissue, also contains inhibitors of angiogenesis (Brem, et. al., 1977). Recently, one of these has been isolated and also inhibits collagenase (Taylor and Weiss, 1985). Unfortunately, significant quantities of these tissue-derived inhibitors are not readily available.

In 1981, Gross, et. al., showed that corticosteroids could inhibit angiogenesis, tumor growth, and collagenolysis when delivered by polymers in the rabbit cornea. Subsequently, the combination of heparin and cortisone was shown to powerfully inhibit angiogenesis and tumor growth (Folkman, et. al., 1983). The mechanism of their effect is thought to be related to basement membrane dissolution (Ingber, Madri, and Folkman, 1986). Lee, et. al. (1990), demonstrated that heparin and hydrocortisone could inhibit the vascularization and growth of a human neurofibrosarcoma. Since heparin and cortisone are readily available and in clinical use, we tested their ability to inhibit angiogenesis and tumor growth in a malignant glioma model (Tamargo, Leong, and Brem, 1990). Heparin and cortisone (weight ratio 1:8) were both impregnated into one PCPP-SA polymer from which they were released slowly in high local concentrations. 9L gliomas

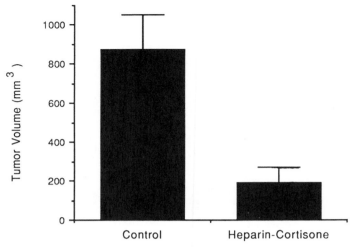

Figure 1. Inhibition of 9L glioma growth by local polymeric delivery of heparin and cortisone acetate. (means ± SEM)

(from Tamargo, et al, J Neurooncol 9:131, 1990, with permission)

were transplanted to the rat flank, and blank or heparin-cortisone loaded polymers were placed adjacent to the tumor. At two weeks heparin-cortisone treated tumors were 78% smaller than controls (p<0.05; Figure 1). This study demonstrated that agents which inhibit angiogenesis can inhibit the growth of malignant gliomas.

BRAIN TUMOR ANGIOGENESIS: NEW INHIBITORS

Despite extensive clinical use, steroid and heparin therapy are associated with significant side effects. In particular, intracranial hemorrhage in brain tumor patients treated with heparin is potentially fatal. Therefore, we began to search for new inhibitors which were more ideally suited for use against intracranial neoplasms. Many known angiogenesis inhibitors disrupt collagen metabolism. In particular, collagenase inhibition is frequently associated with the ability to prevent neovascularization (Moses, Sudhalter, and Langer, 1990). Tetracyclines are antibiotics commonly used in humans which possess several other properties (Table 1), including inhibition of extracellular matrix degrading metalloproteinases (Golub, et. al., 1984 and 1991). One of the inhibitable enzymes is type IV collagenase. Among tetracyclines, the semisynthetic derivative minocycline is a particularly potent collagenase inhibitor. Its low molecular weight and relatively high lipid solubility may also enhance tissue penetration. We have investigated whether minocycline could inhibit angiogenesis (Tamargo, Bok, and Brem, 1991).

VX2 carcinomas were transplanted to the rabbit cornea and blank polymers or polymers containing minocycline (10%-20% by weight) were placed adjacent to the tumor. Vessel length and vascular density in the corneas were assessed. We found that neovascularization induced by the VX2 carcinoma was significantly inhibited by minocycline (p<0.05; Figure 2).

Table 1.   PROPERTIES OF TETRACYCLINE DERIVATIVES[1]

Antibiotic: Inhibits bacterial ribosomal function
                 Also inhibits mammalian protein synthesis
                     in cell free systems
Inhibit matrix metalloproteinases
     collagenases: type IV, interstitial, bacterial
     elastase (macrophage)
Chelate cations
Inhibit other enzymes
     PGE synthetase
     phospholipase A_2
     ornithine decarboxylase
Scavenge free radicals
Inhibit leukocyte phagocytosis and alter migratory
     activity

[1] see Golub, et. al., 1991, for review

This effect was seen in the first week and persisted
throughout the three weeks of the experiment. Inhibition
wasas potent as that observed with heparin-cortisone polymers
used as a positive control. Histological examination of
minocycline treated corneas confirmed that the tumor was
still viable, surviving in an avascular state as a thin sheet
of cells.

    We then assessed minocycline's ability to inhibit the
growth of a malignant glioma (Tamargo, et. al., 1991).   9L
gliomas were transplanted to the rat flank, and blank or
minocycline-loaded (50%) polymers were implanted adjacent to
the tumors.   At eight days, the volume of minocycline treated

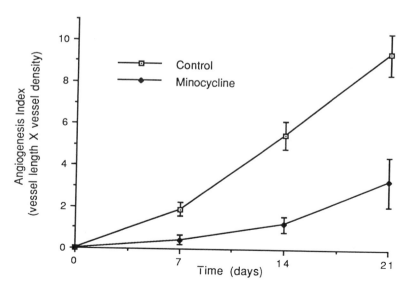

Figure 2. Inhibition of VX2 carcinoma-induced angiogenesis
            in the rabbit cornea by minocycline. (means ± SEM)

(from Tamargo, et al, Ca Res 51:673, 1991, with permission)

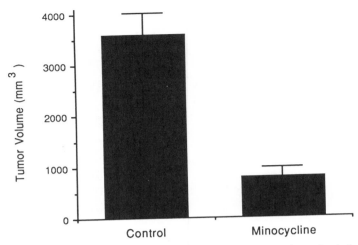

Figure 3.   Inhibition of 9L glioma growth by local polymeric delivery of minocycline. (means ±SEM)

tumors was only one-fourth that of tumors treated with blankpolymer (Figure 3). These experiments showed that: (i) minocycline is an angiogenesis inhibitor, and (ii) minocycline can inhibit malignant glioma growth.

We are currently investigating the effectiveness of minocycline against intracranial glioma models, the optimum method of drug delivery, and the mechanism of minocycline's anti-angiogenic and anti-tumor effects. Tetracyclines have several properties which may contribute to their anti-angiogenic activity. In addition to collagenase inhibition, they also inhibit prostaglandin production and affect several functions of inflammatory cells. Preliminary results suggest that anti-angiogeneic effects are associated with collagenase inhibition and are independent of antibiotic activity. In addition, minocycline does not appear to be directly toxic to glial tumor cell lines. Thus, inhibition of tumor growth is likely due to inhibition of angiogenesis.

CONCLUSION

Malignant glioma remains a devastating disease. In order to develop new treatments, we have taken advantage of two biological properties of malignant gliomas. Sustained release polymers safely deliver drugs to the tumor site at high concentrations both experimentally and clinically. Angiogenesis inhibitors disrupt the neovascularization critical for brain tumor growth. Tetracycline derivatives are clinically safe, readily available, and we have shown them to be potent angiogenesis inhibitors and inhibitors of malignant glioma growth in the laboratory. Using these multiple approaches, we hope to obtain better understanding of the growth and control of brain neoplasms.

ACKNOWLEDGEMENTS

The authors wish to thank Lorne M. Golub, Gary W. Goldstein, John Laterra, Robert Bok, and Michael Pinn for their insights and assistance. This work was supported by a grant from the National Cancer Institute of the National Institutes of Health (CA-52857, H. Brem), the Association for Brain Tumor Research Fellowship in Memory of Steven Lowe (R.J. Tamargo), the Dana Foundation (A. Olivi), and the Poole and Kent Foundation. C. Guerin is a National Research Service Awardee (CA-09574)

REFERENCES

Black PM, 1991, Brain Tumors. N Engl J Med 324:1555-1564

Brem H, 1990, Polymers to treat brain tumors. Biomaterials 11:699-701

Brem H, 1990a, Controlled delivery to the brain. in: "Targetting of Drugs 2: Optimization Strategies", G. Gregoriadis, ed., Plenum Press, New York, pp. 155-163

Brem H, Folkman J, 1975, Inhibition of tumor angiogenesis mediated by cartilage. J Exp Med 141:427-439

Brem H, Kader A, Epstein JI, Tamargo RJ, Domb A, Langer R, Leong K, 1989, Biocompatibility of bioerodable controlled release polymers in the rabbit brain. Selective Cancer Therapeutics 5:55

Brem H, Mahaley S, Vick NA, Black KL, Schold C, Burger PC, Freidman AH, Ciric IS, Eller TW, Cozzens JW, Kenealy JN, 1991, Interstitial chemotherapy with drug polymer implants for the treatment of recurrent gliomas. J Neurosurgery 74:441-446

Brem H, Patz J, Tapper D, 1983, Detection of human central nervous system tumors: use of migration-stimulating activity of the cerebrospinal fluid. Surg Forum 34:532-534

Brem H, Tamargo RJ, Guerin C, Brem SS, Brem H, 1988, Brain tumor angiogenesis. in: "Advances in Neuro-Oncology", PL Kornblith and MD Walker, eds., Futura, Mount Kisco, NY. pp. 89-102

Brem H, Tamargo RJ, Pinn M, Chasin M, 1989a, Biocompatibility of a BCNU-loaded biodegradable polymer: a toxicity study in primates. American Association of Neurological Surgeons, Toronto, p. 381

Brem H, Thopmson D, Long DM, Patz A, 1980, Human brain tumors: differences in ability to stimulate angiogenesis. Surg Forum 31:471-473

Brem S, 1976, The role of vascular proliferation in the growth of brain tumors. Clin Neurosurg 23:440-453

Brem S, Cotran R, Folkman J, 1972, Tumor angiogenesis: a quantitative method for histologic grading. J Natl Ca Inst 48:347-356

Brem S, Jensen HM, Gullino PM, 1978, Angiogenesis as a marker of preneoplastic lesions of the human breast. Cancer 41:239-244

Brem S, Preis I, Langer R, Brem H, Folkman J, 1977, Inhibition of neovascularization by an extract derived from vitreous. Am J Ophthalmol 84:323-328

Chodak GW, Haudenschild C, Gittes RF, Folkman J, 1980, Angiogenic activity as a marker of neoplastic and preneoplastic lesions of the human bladder. Ann Surg 192:762-771

Folkman J, 1972, Anti-angiogenesis: new concept for therapy of solid tumors. Ann Surg 175:409-416

Folkman J, Klagsbrun M, 1987, Angiogenic factors. Science 235:442-447

Folkman J, Langer R, Linhardt RJ, Haudenschild C, Taylor S, 1983, Angiogenesis inhibition and tumor regression caused by heparin or a heparin fragment in the presence of cortisone. Science 221:719-725

Gimbrone MA, Leapman SB, Cotran RS, Folkman J, 1972, Tumor dormancy in vivo by prevention of neovascularization. J Exp Med 136:261-276

Golub LM, Greenwald RA, Ramamurthy NS, McNamara TF, Rifkin BR, 1991, Tetracyclines inhibit connective tissue breakdown: new therapeutic implications for an old family of drugs. Crit Rev Oral Biol Med, in press

Golub LM, Ramamurthy N, McNamara TF, Gomes JB, Wolff M, Casino A, Kapoor A, Zambon J, Ciancio S, Schneir M, Perry H, 1984, Tetracyclines inhibit tissue collagenase activity: a new mechanism in the treatment of periodontal disease. J Periodontal Res 19:651

Gross J, Azazkhan RG, Biswas C, Bruns RR, Hsieh DST, Folkman J, 1981, Inhibition of tumor growth, vascularization, and collagenolysis in the rabbit cornea by medroxyprogesterone. Proc Natl Acad Sci USA 78:1176-1180

Grossman SA, Reinhard CS, Brem H, Brundrett R, Chasin M, Tamargo R, Colvin OM, 1988, The intracerebral delivery of BCNU with surgically implanted bioerodable polymers: a quantitative autoradiographic study. Proceedings of the American Society of Clinical Oncology 7:84

Hochberg FH, Pruitt A, 1980, Assumptions in the radiotherapy of glioblastoma. Neurology 30:907

Ingber DE, Madri JA, Folkman J, 1986, A possible mechanism for inhibition of angiogenesis by angiostatic steroids: induction of capillary basement membrane dissolution. Endocrinology 119:1768-1775

Langer R, 1990, New methods of drug delivery, Science 249:1527-1533

272

Langer R, Brem H, Falterman K, Klein M, Folkman J, 1976, Isolation of a cartilage factor that inhibits tumor neovascularization. Science 193:70-72

Langer LF Brem H, Langer R, 1991, New technologies for fighting brain disease. Technology Review Feb/Mar:63-71

Lee JK, Choi B, Sobel RA, Chiocca EA, Martuza RL, 1990, Inhibition of growth and angiogenesis of human neurofibrosarcoma by heparin and hydrocortisone, J Neurosurg 73:429-435

Libermann TA, Friesel R, Jaye M, Lyall RM, Westermark B, Drohan W, Schmidt A, Maciag T, Schlessinger J, 1987, An angiogenic growth factor is expressed in human glioma cells. EMBO J 6:1627-1632

Maxwell M, Naber SP, Wolfe HJ, Hedley-Whyte ET, Galanopoulos T, Neville-Golden J, Antoniades HN, 1991, Expression of angiogenic growth factor genes in primary human astrocytomas may contribute to their growth and progression. Ca Res 51:1345-1351

Moringlane JR, Spinas R, Bohlen P, 1990, Acidic fibroblast growth factor (aFGF) is present in the fluid of brin tumor pseodocysts. Acat Neurochir (Wein) 107:88-92

Moses MA, Sudhalter J, Langer R, 1990, Indentification of an inhibitor of neovascularization from cartilage. Science 248:1408-1410

Neuwelt EA, Howieson J, Frenkel EP, Specht HD, Weigel R, Buchan CG, Hill SA, 1986, Therapeutic efficacy of multiagent chemotherapy with drug delivery enhancement by blood-brain barrier modification in glioblastoma. Neurosurg 19:573-582

Paulus W, Grothe C, Sensenbrenner M, Janet T, Baur I, Graf M, Roggendorf W, 1990, Localization of basic fibroblast growth factor, a mitogen and angiogenic factor, in human brain tumors. Acta Neuropathol (Berl) 79:418-423

Takahashi JA, Mori H, Fukumoto M, Igarashi K, Jaye M, Oda Y, Kikuchi H, Hatanaka M, 1990, Gene expression of fibroblast growth factors in human gliomas and meningiomas: demonstration of a cellular source of basic fibroblast growth factor mRNA and peptide in tumor tissues. Proc Natl Acad Sci USA 87:5710-5714

Tamargo RJ, Bok RA, Brem H, 1991, Angiogenesis inhibition by minocycline. Ca Res 51:672-675

Tamargo RJ, Epstein JI, Reinhard CS, Chasin M, Brem H, 1989, Brain biocompatibility of a biodegradable controlled release polymer in rats, J Biomed Mat Res 23:253

Tamargo RJ, Leong KW, Brem H, 1990, Growth inhibition of the 9L glioma using polymers to release heparin and cortisone acetate. J Neurooncol 9:131-138

Tamargo RJ, Myseros JS, Brem H, 1988, Growth inhibition of

the 9L gliosarcoma by the local sustained release of BCNU: a comparison of systemic versus regional chemotherapy. American Association of Neurological Surgeons, Toronto, p. 399

Tamargo RJ, Olivi A, Bok RA, Brem H, 1991, Angiogenesis inhibition by minocycline, American Association of Neurological Surgeons, New Orleans, p. 525

Tapper D, Langer R, BellowsAR, Folkman J, 1979, Angiogenesis capacity as a diagnostic marker for human eye tumors. Surgery 86:36-40

Taylor CM, Weiss JB, 1985, Partial purification of a 5.7K glycoprotein from bovine vitreous which inhibits both angiogenesis and collagenase activity. Biochem Biophys Res Commun 133:911-916

Weidner N, Semple JP, Welch WR, Folkman J, 1991, Tumor angiogenesis and metastasis- correlation in invasive breast carcinoma. N Engl J Med 324:1-8

Yang MB, Tamargo RJ, Brem H, 1989, Controlled delivery of 1,3-bis(2-chloroethyl)-1-nitrosourea from ethylene-vinyl acetate copolymer, Ca Res, 49:5103

Zagzag D, Miller DC, Sate Y, Rifkin DB, Burstein DE, 1990, Immunohistochemical localization of basic fibroblast growth factor in astrocytomas. Ca Res 50:7393-7398

EVALUATION OF PROMOTERS AND INHIBITORS OF ANGIOGENESIS USING BASEMENT MEMBRANE BIOSYNTHESIS AS AN INDEX

Michael E. Maragoudakis, Nikos E. Tsopanoglou, Maria Bastaki, and George Haralabopoulos

Department of Pharmacology, University of Patras Medical School
Patras 26500, Greece

INTRODUCTION

Angiogenesis describes the formation of new blood vessels, which arise as outgrowths from existing vessels. Under normal physiological conditions angiogenesis is an extremely slow and highly regulated process. It is involved for a relatively short time in situations such as wound healing, fracture repair and in females during follicle development, ovulation and pregnancy. By contrast, uncotrolled angiogenesis occurs in a variety disease states such as arthritis, psoriasis, diabetic retinopathy, chronic inflammation, hemangiomas, retrolental fibroplasia, solid tumor growth and metastasis etc. (Folkman, 1985). In the aforementioned pathological states angiogenesis proceeds for months or years and in many cases contributes to the pathology of the disease. For example, in diabetic retinopathy the proliferating capillaries bleed and cause blindness. Regarding solid tumors it has been established that their growth and metastasis is angiogenesis-dependent and preventing angiogenesis may be a new approach to cancer therapy (Folkman, 1972).

The need, therefore, is evident for developing substances that inhibit angiogenesis, which are specific, non-toxic, active in vivo and have a known mechanism of action. Such agents may have exciting therapeutic potential in disease states where uncontrolled angiogenesis is involved.

In other conditions such as wound healing, fracture repair, peptic ulcers etc. promoters of angiogenesis could be therapeutically useful.

It is now well recognized that angiogenic capacity is a feature of malignancy, which is absent in normal tissues. In fact angiogenesis is an early event of neoplastic transformation and as such has diagnostic potential for early detection of cancer (Brem et al., 1977; Gullino, 1981).

For all the reasons mentioned a reliable and convenient assay for angiogenesis could be of great value both for diagnostic application and also in the search for specific promoters or inhibitors of angiogenesis.

A major difficulty in attaining these goals and also in the studies of angiogenesis is the available methodology both in vitro and in vivo. All the in vivo assays such as the chick chorioallantoic membrane (CAM), the hamster cheek pouch, the rabbit corneal pocket assay etc, are morpho-

Angiogenesis in Health and Disease, Edited by M.E. Maragoudakis *et al.*
Plenum Press, New York, 1992

logical and as such they are cumbersome, expensive and provide only semi-quantitative results even in expert hands (Gullino, 1981). For this reason these assays are not suitable for testing the large number of samples required for identification and development of inhibitors or promoters of angiogenesis or diagnostic tests. Several in vitro experimental models exist in which isolated endothelial cells can be activated by soluble factors and extracellular matrix proteins to form tube-like structures in tissue culture (Grant et al., 1991). However, this in vitro angiogenesis may not appropriately represent the complex cascade of events leading to neovascularization in vivo.

In this report we discuss the pivotal role of basement membrane (BM) biosynthesis in angiogenesis and the potential of using the rate of BM synthesis during angiogenesis as an index of neovascularization. Furthermore, it will be pointed out that the possibility exists for using the enzymatic steps of BM biosynthesis and assembly as a target for developing specific inhibitors of angiogenesis with a known mechanism of action.

Role of BM in Angiogenesis

BM are highly organized extracellular matrices that separate all epithelial tissues, except liver from stroma. They are found on the basal surfaces of epithelial cells lining of the digestive, respiratory, urinary and reproductive tracts. BM also encircle nerve, muscle, fat and smooth muscle cells. One of the most abundant components of BM is type IV collagen. It is thought to provide the structural three dimensional frame work by joining collagen chains at their ends in an open network that is coated with the other components of BM to form the cords (Terranova et al., 1986). The second most important component of BM is laminin, a cross-shaped glycoprotein, which binds to certain sites on the type IV collagen and through separate domains to heparan sulfate proteoglycan. These components are biologically active and bind to each other to create a charged barrier to the passage of proteins.

In the blood vessels, BM form a continuous layer in the lumen of the vessel and serve as an in vivo foundation for the anchorage of endothelial cells. At certain stage of their formation blood capillaries are endothelial cell tubes lined with BM, which is produced by the activated endothelial cells involved in the process of angiogenesis. All the supporting structures of this blood vessel are formed later by the surrounding mesenchyme (D'Amore and Thomson, 1987). Adhesion of endothelial cells is the most readily observable in vitro result of interaction between endothelial cells and BM. Endothelial cell receptors have been characterized, which recognize specific domains on this matrix proteins (Dejana et al., 1990).

One of the initial steps of angiogenesis is the local dissolution of BM. This is followed by migration and proliferation of endothelial cells (D'Amore and Thomson, 1987). It is likely that this initial dissolution of BM by proteolytic enzymes liberates the endothelial cells from their anchorage, enables them to migrate through the tissue and provides the stimulus for both growth and migration. It has been shown that following angiogenic stimulus in vitro endothelial cells produce proteases and collagenases (Gross et al., 1983). This probably liberates growth factors, which are sequestered in the extracellular matrix (Vlodavsky et al., 1987). Weiss et al. (1986) have shown that degradation products of BM themselves stimulate angiogenesis. In the chick chorioallantoic membrane (CAM) system it was shown that the stage of maximum angiogenesis between days 9-12 of chick embryo development is preceeded by a peak of collagenolytic activity between days 7 and 9 (Missirlis et al., 1990).

In addition to providing a substrate for attachment BM and its components have profound influence on endothelial cell growth and differentiation (Madri and Williams, 1983, Furcht, 1986). The signals for the modulation of endothelial cells by BM may be direct or indirect and could be mediated by effects on the cytoskeleton or via sequestered growth and/or angiogenic factors. It has been shown by Form et al. (1986) that laminin stimulates the proliferation of capillary endothelial cells and this can be modulated by type IV collagen. During the formation of new vessels laminin preceeds the appearance of type IV collagen and appears associated with the migrating and proliferating endothelial cells in the distal area of neovascularization. Type IV collagen appears near the parent vessel. This temporal appearance of BM components has undoubtedly physiological significance in angiogenesis and it may be one of the mechanisms by which the different components of BM regulate the capillary endothelial cells during angiogenesis. The mechanisms by which the endothelial cells perceive the information residing in the extracellular matrix remains unknown.

Because of the modulation of endothelial cell behaviour and differentiation by BM, several experimental model systems have been developed where endothelial cells can be activated by soluble factors and extracellular matrix components to form three-dimentional tube-like structures in tissue cultures (Grant et al., 1991).

Basement Membrane Biosynthesis as an Index of Angiogenesis

In spite of the impressive progress in the field of angiogenesis the past 15 years, we lack detailed knowledge on key molecular and cellular mechanisms that initiate the activation of vascular endothelium and promote or inibit the cascade of events leading to the construction of capillary networks.

From the above discussions, however, one thing seems certain that BM plays a pivotal role in angiogenesis and is synthesized by the proliferating endothelial cells. At the early stages of angiogenesis blood capillaries are bare endothelial cell tubes with their underlying BM. It is logical, therefore, to expect that a quatitative relationship must exist between the rates of BM synthesis and angiogenesis. One would expect a very low background activity of BM synthesis in the absence of angiogenesis, because both BM synthesis and angiogenesis are extremely slow processes under normal physiological conditions. BM biosynthesis is measurable only with tracer methodology in embryonic tissues or in tissues undergoing active angiogenesis (Maragoudakis et al., 1988). Similarly the turnover of vascular endothelium is extremely slow. For example in the normal brain it is estimated as one cell division every 25 years, while in brain tumors it is less than 10 days (Hobson and Denekamp, 1984).

We thought, therefore, that monitoring the rate of BM synthesis by any means would provide a sensitive and quantitative index of angiogenesis.

This was born out experimentally in the CAM system and is currently under investigation in other angiogenesis models. The methodology is detailed in previous reports (Maragoudakis et al., 1988). Briefly stated, in the CAM system (Folkman, 1985) the test materials with the radiolabeled proline ($0.5\mu C$/disc) were placed on sterile round plastic discs 10 mm in diameter and were allowed to dry under sterile conditions. The loaded discs were inverted and placed on the surface of CAM on day 9. Control discs containing equal amounts of radiolabeled proline were placed on the CAM about 1 cm away from the discs containing the test material. A sterile solution of cortisone corresponding to 100 μg/egg was added to all the discs to prevent inflammatory responses. The windows of the eggs were covered with sterile cellophane tape and returned to the incubator. For morphological evaluation of angiogenesis eggs were treated as above with the exception that radiolabelled proline was not added. At day 12 the eggs were flooded with 10% buffered formalin for fixing the CAM in situ. A large area of CAM around the discs was removed, placed in a glass slide and stretched gently to its natural size. Immediately thereafter, the vascular density index was measured by the method of Harris-Hooker et al. (1983) by counting the number of vessels intercepting each circle separately, and then adding the vessels intercepting all three concentric circles which were 4, 5 and 6 mm in diameter.

The extent of collagenous protein biosynthesis in the CAM under the plastic discs was measured as follows: The eggs were placed at 4°C and the tape removed. The CAM under each plastic disc was well defined and was easy to cut off. These round pieces of CAM (10mm in diameter) were transferred into 3 ml plastic centrifuge tubes containing PBS buffer pH 7.3. Protein synthesis and proline and lysine hydroxylations were stopped by the addition of cycloheximide and dipyridyl, 0.11 and 0.17 mg/tube respectively. The tubes were boiled for 10 min in a water bath and cooled at room temperature. Non-protein bound radioactivity was removed by washing with 3 ml of 15% trichloroacetic acid (TCA) and centrifuging at 10,000xg for 10 min. This procedure was repeated four times with 3.0 ml 5% TCA until the radioactivity in the supernatant was equal to background. Pellets containing the protein-bound radioactivity from [U-^{14}C]-proline were suspended in 0.9 ml of 0.1N NaOH and 1.1 ml HEPES buffer pH 7.4 with phenol red indicator and the pH was adjusted to neutral with 0.8 N HCl. For collagenase digestion of the washed CAM, 30λ (7.5 units) of collagenase VII and 10λ of 50 mM CaCl$_2$ was added to each tube and the mixture was incubated at 37°C for 4 hours, when 1.0 ml of 20% TCA solution containing 5 mg tannic acid per tube was added and the mixture was centrifuged at 10,000xg for 10 min. An aliquot of the supernatant was counted in a liquid scintillation spectrometer (Beckman LS1801) to obtain a measure of the radiolabeled tripeptides corresponding to BM collagen and other collagenous materials synthesized by the CAM from [U-^{14}C]-proline. The precipitate was dissolved in 1.0 N NaOH by boiling in a water bath for 5 min. An aliquot of the dissolved CAM was used for total [^{14}C] protein synthesis and another for protein determination. Results are expressed as CPM of [^{14}C]-collagenous or total protein per mg of CAM protein. The collagenous protein synthesized by CAM under these conditions is mostly BM collagen type IV as shown by chromatography and the use of type IV collagenase purified from Walker 256 tumor (Maragoudakis et al., 1990).

The CAM obtained from different days of chick embryo development was used to evaluate the rate of BM collagen biosynthesis in vitro. The same procedure was followed for testing agents in vitro for their effect on BM synthesis. The CAM from 9-60 eggs were dissected out into a medium containing Krebs glucose buffer pH 7.4. Each CAM was cut into pieces of about 10 mg each. From the pooled pieces of CAM suspended randomly, 8-10 pieces (approximately 100 mg wet weight) were transferred to a flask containing 5 ml Krebs glucose buffer pH 7.4, 0.35 mg penicillin and 0.6 mg streptomycin. The membranes were pre-incubated for 60 min at 37°C before adding 0.25 µC [U-^{14}C]-proline per flask and other materials to be tested at the concentrations indicated. Five to eight flasks were used for each treatment group and the flasks were incubated at 37°C for 20 hours. Reactions were stopped by the addition of 0.5 ml buffer containing 0.55 mg cycloheximide and 0.85 mg dipyridyl to each flask and boiling at 100°C for 10 min. The CAM pieces were washed as described above for the in vivo assay and treated with collagenase to determine collagenous and non-collagenous proteins.

Maximum rate of angiogenesis in the CAM system occurs between days 8 and 12 of chick embryo development. After day 12 the vascular density i.e. the number of vessels per given area reaches a plateau. Between days 8-12 the vascular density increases about three-fold (Maragoudakis et al., 1988). At that period the rate of collagenous protein synthesis both in vivo and in vitro was maximum. For example, at day 10 rate of collagenous protein synthesis in vitro was three-fold higher than that observed on day 7 and 11-fold higher than that observed on day 15 when vascular density reached a plateau (Maragoudakis et al., 1985). Thus, we have demonstrated that maximum rate of collagenous protein biosynthesis coincides with the stage of maximum angiogenesis as measured by morphological evaluation of the vascular density of CAM. Monitoring BM collagen biosynthesis may prove to be a convenient quantitative method for assessing angiogenesis in other systems.

Evaluation of Promoters and Inhibitors of Angiogenesis in the CAM system

Using the above methodology for monitoring BM collagen biosynthesis in relation to angiogenesis in the CAM system we evaluated certain promoters and inhibitors of angiogenesis in order to validate this system.

We have shown that hyaluronic acid digests, but not hyaluronic acid itself, causes an increase in collagenous protein synthesis almost two-fold. As pointed out by West et al. (1985) hyaluronic acid digests are strongly angiogenic and this may have physiological significance.

Morris et al. (1988) have shown that the tumor-promoting phorbol ester 12-O-tetradecanoyl phorbol 13-acetate (PMA) stimulates angiogenesis in a dose-dependent manner in the CAM system. As shown in Figure 1 we have confirmed their results and have shown that at doses that PMA promotes angiogenesis there is also a dose-dependent increase in collagenous protein synthesis.

Kanayasu et al. (1989), in order to investigate the mechanisms of angiogenesis involved in the inflammatory process studied the effects of leukotrienes and prostaglandin E_2 on tube formation by cultured vascular endothelial cells in vitro. They have found that leukotriene C_4 stimulated tube formation and endothelial cell migration, while leukotrienes B_4 and D_4 were not effective. As shown in Figure 2 (bottom) we have found that both leukotrienes D_4 and C_4 increase vascular density in the CAM system in vivo, when applied at a dosage of 100 pg/disc. In agreement with Kanayasu et al. (1989) leukotriene B_4 was not effective. When the rates of collagenous protein synthesis were measured under the same conditions (Fig. 2 top) there is a marked increase in the areas under the discs containing leukotrienes C_4 and D_4 (LTC_4 and LTD_4) but not with leukotrienes B_4 (LTB_4) as compared to controls (C) in the same CAM.

One of the most extensively studied inhibitors of angiogenesis is heparin in combination with hydrocortisone (Folkman and Ingber, 1987). We have evaluated the antiangiogenic action of these agents in the CAM and found that under the conditions where heparin plus hydrocortisone suppresses angiogenesis there is a marked decrease in the rate of collagenous protein synthesis in the area of the disc containing the antiangiogenic substances (heparin 50 μg plus hydrocortisone 50 μg). When the amount of hydrocortisone was increased to 150 μg or decreased to 30 μg the antiangiogenic effect was not evident (Folkman and Ingber, 1987), neither was an effect on collagenous protein synthesis (Maragoudakis et al., 1989). The relative concentration of heparin plus hydrocortisone was shown to be critical for demonstrating antiangiogenic activity, which was attributed to a dissolution of BM (Folkman and Ingber, 1987). This mechanism, however, can not explain why only the BM of the growing capillaries is affected, whereas the BM of non-growing capillaries or the BM of the neighboring endothelium remains intact. We investigated the possibility of direct inhibitory effect of heparin plus cortisone on BM synthesis in vitro. With a wide range of concentrations used in the in vitro CAM system (Maragoudakis et al., 1988) we have been unable to demonstrate any inhibitory effect of heparin plus

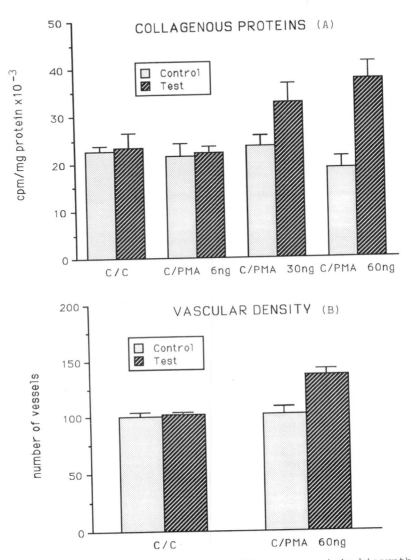

Fig. 1. The effect of PMA on collagenous protein biosynthesis
and the vascular density of the CAM, The specified
amount of PMA were placed on sterile round discs side
by side with control (C) discs as detailed in the text.
In A n=7 and B n=6

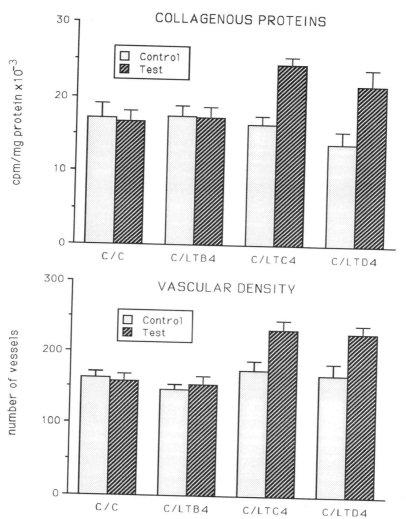

Fig. 2.   The effect of leukotrienes B_4, C_4, D_4 on collagenous
protein synthesis and vascular density in the CAM system.
n=7.   Experimental details as described in the text.

cortisone on BM synthesis. We conclude from these experiments that neither direct inhibition nor dissolution of BM as suggested by Folkman and Ingber is likely to be the mechanism of antiangiogenic action of heparin plus cortisone. These agents prevent angiogenesis by an as yet unknown mechanism and as a result new BM synthesis is suppressed. This provides additional support to our contention that the rate of BM synthesis can serve as an index for neovascularization.

Inhibitors of BM Synthesis as Antiangiogenic Agents

As discussed above, BM has profound effects on the growth and differentiation of endothelial cells and it is one of the basic and first formed structural elements of blood vessels. We thought, therefore, that interfering with biosynthesis of BM components or their assembly into organized extracellular matrix should have inhibitory effects on the formation of new capillary networks.

We tested this hypothesis using GPA 1734 (8,9-dihydroxy-7-methyl-benzo (b)quinolizinium bromide), a previously studied inhibitor of proline and lysine hydroxylation in the BM collagen biosynthesis (Maragoudakis et al., 1978). It was shown that inhibition of the postribosomal hydroxylation reactions of proline and lysine by this agent renders the unhydroxylated product of BM type IV collagen susceptible to proteolytic degradation by endogenous proteases and unsuitable for deposition to preexisting BM of the rat yolk sac. A closely related analog GPA 1967 (9,10-dihydroxyl-7-methyl benzo(b) quinolizinium bromide) had no similar effects on BM synthesis and deposition (Maragoudakis et al., 1978).

We tested the effects of these compounds on the CAM system both in vivo and in vitro. As previously noted in the parietal yolk sac system, GPA 1734 inhibited collagenous protein synthesized by the CAM, while GPA 1967 was without effect at comparable concentrations. At the minute concentrations that GPA 1734 inhibited BM collagen biosynthesis of the CAM in vivo, it also inhibited angiogenesis without any apparent effect on the preexisting blood vessels. On the contrary in the presence of the inactive analog GPA 1967, both BM synthesis and angiogenesis were unaffected (Maragoudakis et al., 1988).

Inhibiting proline and lysine hydroxylation in BM collagen biosynthesis may not be the only way one could interfere with angiogenesis. There are many enzymatic steps in the biosynthesis and assembly of the various components of BM that can be potential targets for inhibiting the overall process.

We have initiated a search for such inhibitors using the methodology in the CAM system as described previously (Maragoudakis et al., 1988). One of the inhibitors that was identified was D609 (tricyclodecan-9-yl-xanthate). This compound inhibits BM biosynthesis by an as yet unknown mechanism different from that of GPA 1734. When 30-50 μg/disc of D609 was placed on the CAM under the in vivo conditions described there was a marked inhibitory effect on collagenous protein synthesis and angiogenesis (Maragoudakis et al., 1990). Preliminary experiments indicate that D609 abolishes the stimulatory effect seen with PMA indicating that protein kinase may be involved in the mechanism of action, see Figure 3.

Antitumor Effect of Angiogenesis Inhibitors

Angiogenesis inhibitors such as protamine (Heuser et al., 1984) or heparin in combination with hydrocortisone (Folkman et al., 1983), which act by an unknown mechanism to suppress angiogenesis, have been shown to inhibit tumor growth and metastasis in experimental models.

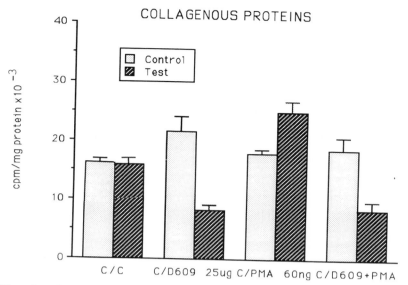

Fig. 3. The effect of D609 (25 µg/disc), PMA (60 ng/disc) and the combination of the two on collagenous protein synthesis in the CAM system. n=7. Experimental details as described in the text.

We have also found that BM synthesis inhibitors, which are antiangiogenic, have antitumor effects in rats bearing Walker 256 carcinosarcoma. Both inhibitors, GPA 1734 and D609, caused dose-dependent antitumor effects (Missirlis et al., 1990 and Maragoudakis et al., 1990). The antitumor effect is unrelated to direct cytotoxic effects on tumor cells. At concentrations where D609 inhibits BM synthesis and angiogenesis it had no effect on tumor cell growth in tissue culture (Maragoudakis et al., 1990).

SUMMARY

A quantitative relationship exists between the rate of BM biosynthesis and angiogenesis. BM collagen type IV, the major component of BM, is easily measured using collagenases. Under conditions of active angiogenesis radiolabeled proline is incorporated into collagenase-solubilized peptides with very low background activity in the absence of angiogenesis. Methodology has been developed in the CAM system for the evaluation of agents that promote or inhibit angiogenesis. Similar methodology can probably be adapted to other models of angiogenesis. Specific inhibitors of angiogenesis can be developed using enzymes involved in BM synthesis as targets. Two such agents have been described GPA 1734 and D609, which suppress angiogenesis in the CAM and tumor growth and metastasis in the rat Walker 256 carcinosarcoma model.

These results suggest that a systematic search for agents that are specific inhibitors of BM synthesis may provide novel angiosuppressors with potential uses in tumor chemotherapy and other angiogenic diseases. Such agents are not expected to have serious side effects for short term treatment, because both BM synthesis and angiogenesis are extremely slow processes under physiological conditions.

ACKNOWLEDGEMENTS

This work was supported by grants from the Greek Ministry of Industry, Energy and Technology. We thank Mrs. Anna Marmara for her secretarial help.

REFERENCES

Brem, S., Gullino, P. M., and Medina, D., 1977, Angiogenesis: A marker of neoplastic transformation of mammary hyperplasia, Science, 195:880-882.

D'Amore, P. A., and Thomson, R. W., 1987, Mechanism of angiogenesis, Ann. Rev. Physiolog., 49:453-464.

Dejana, E., Conforti, G., Zanetti, A., Lampugnani, M. G., and Marchision, P. C., 1990, in: "Vascular Endothelium: Receptors and Transduction Mechanisms", Catravas, J., Gillis, C. N., and Ryan, U. S., ed., Plenum Press, p.p. 141-147.

Folkman, J., 1972, Anti-angiogenesis: New concept for therapy of solid tumors, Ann. Surg., 409-416.

Folkman, J., Langer, R., Linhardt, R. J., Haudenschild, C., and Taylor, S., 1983, Angiogenesis inhibition and tumor regression caused by heparin or heparin fragment in the presence of cortisone, Science (Wash. DC), 221:719-725.

Folkman, J., 1985, Towards an understanding of angiogenesis: Search and discovery, Persp. in Biol. and Med., 29:10-35.

Folkman, J., and Ingber, D. E., 1987, Angiostatic steroids: Methods of discovery and mechanism of action, Ann. Surg., 206:374-383.

Form, M. D., Pratt, B. M., and Madri, J. A., 1986, Endothelial cell proliferation during angiogenesis: In vitro modulation by basement membrane components, Lab. Invest., 55:521-530.

Furcht, L. T., 1986, Critical factors controlling angiogenesis: Cell products, cell matrix and growth factors, Lab. Invest., 55:505-509.

Grant, D. S., Lelkes, P., Fukuda, K., and Kleinman, H. K., 1991, Intracellular mechanisms involved in basement membrane induced blood vessel differentiation in vitro, In Vitro Cell Dev. Biol., 27A:327-336.

Gross, J. L., Moscatelli, D., Rifkin, D. B., 1983, Increased capillary endothelial cell protease activity in response to angiogenic stimuli in vitro. Proc. Natl. Acad. Sci, 80:2623-2627.

Gullino, P. M., 1981, Angiogenesis and neoplasia, New Engl. J. Med., 305:884-885.

Harris-Hooker, S. A., Gajdusek, C. M., Wight, T. N. and Schwartz, S. M., 1983, Neovascular responses induced by cultural aortic endothelial cells, J. Cell Physiol., 114:302-310.

Heuser, L. S., Taylor, S. H., and Folkman, J., 1984, Prevention of carcinomatosis and bloody malignant ascites in the rat by an inhibitor of angiogenesis, J. Surg. Res., 36:244-250.

Hobson, B., Denekamp, J., 1984, Endothelial proliferation in tumors and normal tissues: continuous labelling studies, Br. J. Cancer, 49:405-413.

Kanayasu, T., Nakao-Hayashi, J., Asuwa, N., Morita, I., Ishii, T., Ho, H. and Murota, S., 1989, Leukotriene C_4 stimulates angiogenesis in bovine carotid artery endothelial cells in vitro, Biochem. Biophys. Res. Comm., 159:572-655.

Karakiulakis, G., Missirlis, E., Aletras, A., and Maragoudakis, M. E., 1988, Degradation of intact basement membrane by human and marine tumor enzyme, Biochim., Biophys. Acta, 967:163-175.

Madri, J. A., William, S. C., 1983, Capillary endothelial cell cultures: Phenotypic modulation by matrix components, J. Cell Biol., 97:153-159.

Maragoudakis, M. E., Kalinsky, H. J., Wasvary, J., 1978, Basement membrane biosynthesis: Secretion without deposition of underhydroxylated basement membrane collagen by perietal yolk sac, <u>Biochim. Biophys. Acta</u>, 538:134-154.

Maragoudakis, M. E., Sarmonika, M., and Panoutsacopoulou, M., 1988, Inhibition of basement membrane biosynthesis prevents angiogenesis, <u>J. Pharm. Exp. Therap.</u>, 244:729-733.

Maragoudakis, M. E., Panoutsacopoulou, M. and Sarmonika, M., 1988, Rate of basement membrane biosynthesis as an index of angiogenesis, <u>Tissue and Cell</u>, 20: 531-539.

Maragoudakis, M. E., Sarmonika, M., and Panoutsacopoulou M., 1989, Antiangiogenic action of heparin plus cortisone is associated with decreased collagenous protein synthesis in the chick chorioallantoic membrane, <u>J. Pharma. and Exp. Therap.</u>, 251:679-682.

Maragoudakis, M. E., Missirlis, E., Sarmonika, M., Panoutsacopoulou, M. and Karakiulakis, G., 1990, Basement membrane biosynthesis as a target to tumor therapy, <u>J. Pharm. and Exp. Therap.</u>, 252:753-757.

Missirlis, E., Karakiulakis, G., and Maragoudakis, M. E., 1990, Angiogenesis is associated with collagenous protein synthesis and degradation in the chick chorioallantoic membrane, <u>Tissue and Cell</u>, 22:419-426.

Missirlis, E., Karakiulakis, G., and Maragoudakis, M. E., 1990, Antitumor effect of GPA 1734 in rat Walker 256 carcinoma, <u>Invest. New Drugs</u>, 8:145-147.

Morris, P. M., Hida, T., Blackshear, P. J., Klintworth, G. K., and Swain, J. L., 1988, Tumor-promoting phorbol esters induce angiogenesis in vitro, <u>Am. J. Physiol.</u>, 254 (Cell Physiol. 23):C318-C322.

Terranova, V. P., Hujanen, E., and Martin, G., 1986, Basement membrane and the invasive activity of metastatic tumor cells, <u>J. Nat. Canc. Inst.</u> 77:311-316.

Vlodavsky, I., Folkman, J., Sullivan, R., Fridman, R., Ishai-Michaeli, R., Sasse, J. and Klagsburn, M., 1987, Endothelial cell-derived basic fibroblast growth factor: synthesis and deposition into the subendothelial matrix, <u>Proc. Natl. Acad. Sci. (USA)</u>, 84:2292-2298.

West, D. C., Hampson, I. N., Arnold, F., and Kumar, S., 1985, Angiogenesis induced by degradation products of hyaluronic acid, <u>Science</u>, 228: 1324-1326.

Weiss, J., Orr, F. W., and Honn, K., 1988, Interaction of cancer cells with the microvasculature during metastasis, <u>FASEB J.</u>, 2:12-21.

ON PROMOTERS OF ANGIOGENESIS AND THERAPEUTIC POTENTIAL

Pietro M. Gullino

Dept. Biomedical Sciences
Via Santena, 7
University of Torino, ITALY

The list of molecules able to induce angiogenesis is heterogeneous in chemical characteristics and biological properties[1,2,3,4]. Despite this heterogeneity, it is possible that all have in common the capacity to influence some of the events indispensable for the induction of neovascularization. Stimulation or repression of the angiogenic response may depend on changes in total concentration and relative ratios of modulating molecules normally present in the organism. As an approach to identify them we analyzed changes in tissue composition that occur in the rabbit cornea undergoing angiogenic stimulation.

The rabbit cornea was selected as a model for the following reasons: (1) it is avascular, i.e. the background to assess an angiogenic response is zero and spurious changes of tissue composition due to hypo- or hyperemia during sampling are avoided, (2) it is transparent i.e. the time sequence of vessel promotion can be followed in vivo with a stereo microscope and any inflammatory response occuring during angiogenesis stimulation produces opacity that can easily be detected, (3) it is bilateral i.e. a treated and a control specimen can be obtained from the same animal, and (4) it is sufficiently large for surgical manipulations and analytical sampling.

Prostaglandins, copper and angiogenesis

The fluid surrounding the neoplastic cells of a growing tumor was removed in vivo with the micropore chamber technique[5], and the lyophilized material from 1.0 ml of this fluid was incorporated into a slow-release polymer pellet[6]. The complete fluid as well as the ethanol extract were angiogenic[7]. One peculiarity of this fluid is the elevated content of type E prostaglandins (PGE) which is about double that of the interstitial fluid of the subcutaneous tissue where the tumor was transplanted (1.0 versus 0.4 ng/ml on average). Indeed, both natural and chemically induced growing tumors are known to produce PGE[8].

The possibility that PGE was involved in the angiogenic response of neoplastic cells was tested in 2 separate experiments. Doses from 0.1 to 10 ng of PGE1, PGE2,[6] PGI2 and PGF2α were incorporated into a slow release polymer pellet[6] and tested with the rabbit cornea assay[9] Only PGE were found to be angiogenic and PGE1 at the lowest doses[7]. Very heavy treatment of the rabbit with indometacin given by OS and topically, succeeded in preventing corneal angiogenesis by neoplastic Balb/c fibroblasts suggesting that production of PGE could indeed be a determinant event in the angiogenic response of the cornea[7].

These findings gave us the opportunity of inducing corneal vascularization with a reasonable predictability utilizing a well defined molecule and a precise dosage. Consequently, it was plausible to ask whether the composition of the cornea was changed by the PGE1 angiogenic stimulation. With this objective in mind, the content of Cu, Ca, Fe, Mg, P, Zn, Co and K was determined in two groups of corneal fragments removed at the end of day 3 after pellet implantation when capillary buds started to appear on the limbal vessels i.e. the angiogenic response was initiated but the cornea was not yet vascularized. One group of fragments was constituted by the tissue underlying the pellet containing PGE1 and the second group consisted of equal fragments underlying a pellet without PGE1 or containing a non-angiogenic molecule like PGI2. Each rabbit had one cornea treated with PGE1 and the contralateral as control[7,10]. It was found that copper ions were the only ones showing a consistent increment of about 1/3 over the control value.

The relevance of this finding for the angiogenic process was evaluated with 3 experimental approaches. Rabbits were kept on a copper deficient diet for 4 - 5 weeks until serum copper was reduced to about one half the physiological level. These animals became unable to vascularize the cornea when treated with angiogenic doses of PGE1. Return to normal diet and to physiological levels of serum copper sustained the expected high frequency of angiogenic response to appropriate doses of PGE1[7].

Ceruloplasmin, the copper carrier of serum, or fragments of the ceruloplasmin molecule still bound to copper were found to be angiogenic for the rabbit cornea. If copper was removed from the ceruloplasmin molecule or from the fragments thereof, the angiogenic capacity was lost. Heparin or glycyl-L-Histidyl-Lysine tripeptide are not angiogenic but when bound to copper became angiogenic[10,11].

Evidence of altered copper metabolism is well documented in the literature for human subjects[12,13] as well as for experimental animals[14]. Whether the leading cause of it is the elevated angiogenesis of the growing tumor is still unclear, although the removal of the tumor is rapidly followed by return to normal ceruloplasmin and copper levels in serum. At this time it seems reasonable to state that copper compounds are substantially involved in the angiogenesis process but their role in the global biological picture of neovascularization in the adult organism is still unclear. The possibility of interfering with copper metabolism using chelating agents or depriving the tumor bearing organism of copper with an appropriate diet is a theoretical possibility. To my knowledge results of a controlled clinical trial in this sense are not available.

Gangliosides and angiogenic response

The ganglioside content of the corneal tissue under angiogenic stimulation was found to be about twice that of the controlateral cornea taken as control[15]. The tissue samples were obtained at the end of day 3 after application of the angiogenesis trigger, therefore they derived from corneas ready to be colonized by capillaries but were not yet vascularized. The change in ganglioside composition could indicate that the tissue to be penetrated by the newly-formed microvessels was "modulating" the angiogenic response. As an approach to corroborate this hypothesis the effect of gangliosides added to cultures of capillary endothelium was evaluated.

Addition of GM1 or GT1b to clones of bovine capillary endothelium in serum-free medium increased migration as measured with the Boyden chamber procedure[16]. The increment in motility was optimal when fibronectin was added to the medium and preincubation of endothelium with gangliosides* was as effective as the addition of the gangliosides to the medium of the Boyden chamber[16,17]. This suggested that cell motility and enrichment of gangliosides, probably at the cell surface, were two interrelated events. The best "mobilizer" was found to be GT1b, the richest in sialic acid among the gangliosides available to us. Indeed, the removal of sialic acid from the molecule nullified the effect on motility and sialic acid by itself was ineffective[17].

Survival of capillary endothelium was also improved by addition of gangliosides to the culture medium: best results were obtained with GT1b. The combination of gangliosides with basic fibroblastic growth factor (bFGF) was particularly efficient in the sense that doses of gangliosides or bFGF, ineffective by themselves, preserved cell viability and occasionally increased cell number when in combination. Again, removal of sialic acid from the ganglioside molecule nullified the improvement on viability and sialic acid by itself was ineffective[17].

The results on motility and survival of capillary endothelial cells in culture indicated that the increment of ganglioside concentration observed in vivo for corneas under angiogenic stimulation, could represent a condition favouring colonization by capillaries. Thus, the angiogenic molecule was able to modify tissue composition to the effect of positively influencing the morphogenesis of a capillary-like network.

In order to ascertain whether this interpretation was also valid in vivo the following experiment was performed. A copolymer pellet with 50 ng PGE1 was implanted in the cornea and a second pellet was implanted close by. This second pellet contained 50 ng GM1 in one cornea and nothing in the contralateral cornea of the same rabbit. When 50 ng PGE1 acted alone there was no angiogenic response. When the PGE1 bearing pellet was flanked by the GM1 pellet, an angiogenic response was obtained

---

*Abbreviations according to L. Svennerholm, J. Lipid Res. 5; 145:155, 1964.

in 4 out of 5 implants. Local enrichment of gangliosides obtained by the presence of the second pellet promoted an angiogenic response from a PGE1 dose insufficient to do so when acting alone.

Ganglioside enrichment of culture media improved growth and motility of capillary endothelium, ganglioside enrichment of corneas in vivo promoted an angiogenic response when the dose of the effector was insufficient to elicit neovascularization. Gangliosides by themselves were not angiogenic.

In order to corroborate this interpretation, the behaviour of the vessels induced after GM1 promotion was tested with the following experimental approach. At day 5 after PGE1 + GM1 pellets were implanted and a brush of capillaries started to penetrate the cornea, the GM1-bearing pellet was removed. Within 5 days the vessels disappeared. At this time the pellet bearing 50 ng GM1 was reintroduced in the initial corneal pocket. Newly-formed vessels rapidly reappeared and within 6 days surrounded both pellets.

If the procedure was repeated but the PGE1 rather than the GM1 pellet was removed, the results were the same i.e. the vessels regressed because GM1 alone was unable to preserve their integrity, but reintroduction of the PGE1 pellet renewed neovascularity. The local enrichment of gangliosides was necessary and sufficient to elicit neovascularization when the angiogenesis effector was present but unable to do so by itself because of the low dose.

When the experiment was repeated using bFGF (5.0 ng/pellet) as the angiogenesis trigger, and GM1 as promoter, the same sequence of events occurred. The modulating effect of GM1 on the angiogenic response appears to potentiate the action of angiogenic molecules[14].

The increment of the ganglioside content observed in the angiogenesis stimulated corneas involved all 3 major gangliosides of the tissue, (GM3, GM2, GD3) but to a different degree: GM3 increased almost one third above the level of the untreated cornea, GM2 almost doubled and GD3 almost tripled. The end result was an overall doubling of the total ganglioside content of the cornea but an even larger difference in the relative proportions among the components. As an approach to understanding the significance of the difference in ratios, three experimental procedures were followed.

First, cultures of capillary endothelium were challenged with the 3 corneal gangliosides added singularly in different doses. We observed that GM3 impaired cell survival at doses about 8-fold smaller than GM2 and 30-fold smaller than GD3. In the second experiment the endothelial cell cultures were challenged with GM3 + GM2 + GD3 mixtures. A clear difference in growth and motility was observed according to the relative proportion of the 3 gangliosides in the mixture. When the proportion among GM3, GM2 and GD3 mimicked that of the normal cornea, growth and motility remained at control level. As the GM3/GD3 ratios decreased, both growth and motility increased to levels twice the control value. In vitro

GM3 appears to have a cytostatic effect on capillary endothelium (manuscript in preparation).

In the last experiment we planned to see whether the results in vitro could be reproduced in vivo i.e. whether the cytostatic effect of GM3 in vitro could be translated into an angiostatic effect in vivo. In the same rabbit one cornea received one copolymer pellet bearing 50 ng PGE1 and close by a second pellet bearing 50 ng GM1 while the contralateral cornea received one PGE1 pellet (50 ng) and a second pellet bearing GM3 (50 ng). Out of 14 rabbits, 9 corneas bearing PGE1 + GM1 gave a positive response, but only 1 cornea bearing PGE1 + GM3 showed a limited angiogenic response. More experiments are ongoing comparing the effect of GM3 versus GD3, the actual ganglioside present in the cornea. At this writing it appears that GM3 has an angiostatic effect in vivo that mimicks the cytostatic effect observed in vitro on microvascular endothelium.

Concluding remarks:

Our working hypothesis is that an angiogenesis factor modifies an equilibrium among a set of molecules that are normally present in the adult tissues. Changes in this equilibrium i.e. modification of total concentration and/or relative proportions among these molecules stimulates or prevents angiogenesis. The experimental approach we are following in trying to corroborate this hypothesis consists of measuring changes in composition of the corneal tissue under angiogenic stimulation and in assessing their pertinence to the process of neovascularization.

Copper ions and gangliosides are two tissue components so far found to undergo concentration changes, and for the reasons given above, are believed to be determinant events in the angiogenesis process. Obviously other tissue components could be involved. Assuming that they can be identified, what next?

One possibility is that the identification of a sufficient number of tissue molecules changing concentration and/or relative proportions will permit a rational mapping of the events that may control angiogenesis. Even if this mapping is impossible, any manipulation that will interfere with the changes in concentration of any molecule involved in the angiogenic response of the cornea may alter the process. These manipulations may theoretically be of two kinds, addition of one molecule to the system with consequent possibility of speeding up the neovascularization, or addition of a molecule with a sufficiently close structure as to act like a fraudolent intruder and stop the process.

The improvement of angiogenesis could benefit: (a) wound, ulcer or fracture healing in persons who for old age or diseases like diabetes have difficulty in completing the natural process or (b) "takes" of skin transplants or (c) any other condition requiring rapid neovascularization such as tissue infarcts.

The impairment of angiogenesis may slow down or arrest tumor growth or alter the evolution of diseases like diabetic retinopathy, psoriasis, so called vascular glaucoma and chronic arthropathies of different nature.

In the adult organism new formation of vessels does not occur except under peculiar conditions such as renewal of uterine mucosa or corpus luteum formation. Therefore, an impairment of angiogenesis should not substantially interfere with any vital process but effectively influence target tissues where angiogenesis is a very active process such as in the tumor or in the pathologic processes listed above.

References

1. P. A. D'Amore and S. J. Braunhut, Stimulatory and inhibitory factors in vascular growth control in: "Endothelial cells", U. S. Ryan, ed., CRC Press, Vol.2, pp. 13-37, (1987).
2. B. R. Zetter, Angiogenesis, state of the art., Chest, 93:159S-166S (1988).
3. J. Folkman and M. Klagsbrun, Angiogenic factors, Science, 235:442-447 (1987).
4. P. M. Gullino, Angiogenesis factors, in: "Handbook Exp. Pharmacology", R. Baserga, ed., Springer Verlag, New York, Vol.57, Ch.15, pp 427-449 (1981).
5. P. M. Gullino, S. H. Clark and F. H. Grantham, The interstitial fluid of solid tumors, Cancer Res., 24:780-798 (1964).
6. J. S. Goodwin, G. Husby and R. C. Williams jr., Prostaglandin E and cancer growth, Cancer Immunol. Immunother., 8:3-7 (1980).
7. R. Langer and J. Folkman, Polymers for the sustained release of proteins and other macromolecules, Nature, 263:797-800 (1976).
8. M. Ziche, J. Jones and P. M. Gullino, Role of prostaglandin E1 and copper in angiogenesis, J. Natl. Cancer Inst., 69:475-482 (1982).
9. M. A. Gimbrone jr., R. S. Cotran, S. B. Leapman and J. Folkman, Tumor growth and neovascularization: an experimental model using the rabbit cornea, J. Natl. Cancer Inst., 52:413-427 (1974).
10. K. S. Raju, G. Alessandri, M. Ziche and P. M. Gullino, Ceruloplasmin, copper ions and angiogenesis, J. Natl. Cancer Inst., 69:1183-1188 (1982).
11. G. Alessandri, K. S. Raju and P. M. Gullino, Angiogenesis in vivo and selective mobilization of capillary endothelium in vitro by heparin-copper complex, Microcirculation, Endothelium and Lymphatics, 1:329-346 (1984).
12. M. Shifrine and G. L. Fisher, Ceruloplasmin levels in sera from human patients with osteosarcoma, Cancer, 38:244-248 (1976).
13. M. C. Linder, J. R. Moor and K. Wright, Ceruloplasmin assays in diagnosis and treatment of human lung, breast and gastro-intestinal cancers, J. Natl. Cancer Inst., 67:263-275 (1981).
14. H. Ungar-Waron, A. Gluckman, E. Spira, M. Waron and Z. Trainin, Ceruloplasmin as a marker of neoplastic activity in rabbits bearing the VX2 carcinoma, Cancer Res., 38:1296-1299 (1978).
15. M. Ziche, G. Alessandri and P. M. Gullino, Gangliosides promote the angiogenic response, Laboratory Invest., 61:629-634 (1989).
16. G. Alessandri, K. S. Raju and P. M. Gullino, Interaction of

gangliosides with fibronectin in the mobilization of capillary endothelium, *Invasion Metastasis*, 6:145–165 (1986).

17. G. De Cristan, L. Morbidelli, G. Alessandri, M. Ziche, A. P. M. Cappa and P. M. Gullino, Synergism between gangliosides and basic fibroblastic growth factor in favouring survival, growth and motility of capillary endothelium, *J. Cell. Physiology*, 144:505–510 (1990).

THE DEVELOPMENT OF THERAPEUTIC ANGIOSUPPRESSION:

PROBLEMS AND PROGRESS

Steven Brem

Neurosurgual Oncology, North Western Medical Faculty Found.
Inc.
Streetville Center, 233 East Evie, Suite 500, Chicago
Ill. 60611, U.S.A.

Research in tumor angiogenesis has progressed rapidly in the past 25 years, marked by four investigative phases: 1) first, the study of the pathophysiology of tumor-host interactions to reveal the link between tumor blood flow and its internal mileu;[1] 2) second, the paradigm of neoplastic and neovascular interdependence, regulated by a chemical substance, a "tumor angiogenesis factor",[2] 3) third, the explosion in the New Biology led to the purification, cloning, and widespread availability of several classes of angiogenic polypeptide growth factors, receptors, monoclonal antibodies, oncogenes, and suppressive oncogenes;[3-8] 4) fourth the successful clinical treatment of an angiogenic disease using a biological response modifier,[9] and the call for clinical trials for cancer patients using inhibitors of angiogenesis, with the promise of less toxicity than standard chemotherapy.[10-11] With a growing list of inhibitors from which to choose,[8,11-20] what generic problems and possibilities do these agents pose?

An insight comes from recent experiments where copper depletion and penicillamine treatment (CDPT) effectively inhibited tumor growth by a selective reduction of endothelial cell (EC) turnover of approximately 50%.[18] Because a small change in a mitotic regulatory circuit, with time leads to a major change in the overall growth of a cell population, a subtotal reduction in endothelial cell mitotic rate may be sufficient to suppress neovascularization.[21] The intracerebral nodules of VX2 carcinoma showed a remarkable histologic resemblance to dormant VX2 carcinomas maintained artificially in the avascular state by anatomic separation from an available blood supply. Despite proximity to preformed vessels, the VX2 carcinoma was consigned pharmacologically to the angiosuppressed state, demonstrating the feasibility of therapeutic suppression. Furthermore, it is becoming clear that angiogenesis is not simply controlled by the presence of a single factor, e.g. bFGF, but rather by balance of several angiogenic inducers and inhibitors. In physiological angiogenesis, such as that of embryonic development, the down-regulation of EC proliferation is concomitant with neovascularization, lest the embryo become a hemangioma.[22] For these reasons, the term "angiosuppression" was suggested to describe the physiological or pharmacologic down-regulation of angiogenesis.[18]

Angiosuppression, because it arrests turnover growth at the tiny prevascular stage, will reduce tumor burden and therefore be effective with other forms of adjuvant therapy (chemotherapy, immunotherapy, and

Angiogenesis in Health and Disease, Edited by M.E. Maragoudakis *et al.*
Plenum Press, New York, 1992

radiation therapy) that work best when the tumor population is small.[23,24] Most drugs act by first-order kinetics, but the pharmacokinetics for angiosuppressive drugs remains to be determined. Such factors as the route of administration, tissue absorption and distribution, drug metabolism, biliary and kidney clearance, equilibrium constants for protein binding in the plasma and for non-specific binding in tissue,[25] all need to be evaluated.

The side effects of angiosuppression may be less daunting than conventional chemotherapy. The turnover time of normal tissue endothelium is estimated to be 20-2000 times longer than parenchymal tissue cells, and does not vary between slow turnover tissues, e.g. lung and brain, from that where the parenchyma is rapidly turning over, e.g. jejunum.[26] Therefore, angiosuppression might not have the gastrointestinal toxicity of general, antimotic chemotherapy. Because of the importance of angiogenesis to both wound-healing and tumor growth, there is concern that wound healing might be impaired. Empirically, however, we observed that scalp[18] and flank[27] wounds healed in angiosuppressed animals,[18] possibly because of the involvement of distinct, nonmitogenic, angiogenic factors.[26] A relatively minor problem, nevertheless, might be alopecia since angiogenesis plays an important role in the cyclical growth of the hair follicle;[28] furthermore, we observed shedding of hair in a few angiosuppressed rabbits. Angiogenesis is essential for normal ovulatory function;[29] therefore, chronic angiosuppression would render female patients infertile and oligomenorrheic.

Given the wide range of structurally different inhibitors of angiogenesis, each compound would have, in addition, its own peculiar side effects. Inhibitors of collagen biosynthesis, for example, might predispose to osteoporosis. Given the "redundancy in angiogenic regulation", i.e. the multiplicity of angiogenic stimulators and inhibitors, Zetter suggested that tissues that require an angiogenic response will elaborate multiple mechanisms to provoke such a response.[30] By the same token, it follows that an angiosuppressive agent, to be truly effective, may need to inhibit more than one angiogenic molecule or interfere with more than one regulatory mechanism. This "therapeutic plurality" may account for the efficacy of penicillamine as an angiosuppressive agent, because it works at several postulated points in the angiogenesis cascade: 1) removal of copper, a trace metal bound to at least 12 angiogenic molecules; 2) inhibition of collagen biosynthesis; 3) inhibition of proteolytic enzymes; plasminogen activator and collagenase. Furthermore, an angiosuppressive molecule may potentiate the effects of another agent. For example, the effect of angiostatic steroids can be potentiated by an anticollagen agent,[12] an inhibitor of arylsulfatase,[31] or suramin.[32] Therefore, combinations of angiosuppressive drugs may prove more useful than a single agent alone.

There are many variables in human tumor biology that in vitro and in vivo models, such as the rabbit cornea, cannot address. Not all vessels are created equally, and angiosuppression, like angiogenesis,[33] may be organ-specific with a differential response influenced by the anatomic site. As with other forms of therapy, human tumors can be expected to differ in their angiogenesis-dependency. Brain tumors, for example, because of their extraordinary vascularity might be the most angiogenesis-dependent.[34] Angiogenesis is a dynamic process and varies according to structural features such as tumor size,[35] stage of development,[35] and proximity to microvasculature.[36,37] The known heterogeneity of tumors in terms of cytokinetics, antigenecity, chemosensitivity, karyotype, and drug resistance might conspire to limit angiosuppression.

Angiosuppression, since it is not cytotoxic directly to tumor cells, will not destroy tiny, dormant colonies, and necessarily will become a long-term treatment. If the levels of the drug should become subtherapeutic, and if remaining tumor is not eradicated by other means, then a clinical recurrence might ensue. The ideal schedule appears to first surgically excise the tumor, decreasing the tumor burden, and removing a large depot of angiogenic substances that could overwhelm an inhibitor. Surgical removal of the tumor results in disappearance of the tumor-induced vasculature. Thus, angiosuppression could be particularly useful in the central nervous system (CNS), to prevent recurrence of tumor once the original lesion is excised. In our copper-depletion experiments,[18] it was necessary to pretreat the animals; acute therapy was ineffective. This could be due to the time required to remove an angiogenic stimulus, such as copper salts in the tissues, or alternatively suggest that angiosuppression may be more effective against freshly formed vessels in contrast to non-proliferative, well-differentiated microvessels, already within the scaffold of the tumor.

Even in an established tumor, most of the endothelium is noncycling and statin-positive.[38] The vascular tree exists as a quiescent monolayer of endothelial cells with a nonterminal, differentiated phenotype that can be "switched" into the mitotic cycle by the local release of angiogenic mitogens.[5] Although malignant human brain tumors are well-vascularized, only a small fraction of vascular endothelial cells are actually replicating at a given moment.[39] By contrast, continuous labeling of small 3.5-5.0 mm, experimental tumors shows that between 25 and 100% of the endothelial cells turnover during a period of one week,[40] supporting the view that angiosuppression would be more effective for these smaller, younger, proliferative tumors about to acquire new vessels, in contrast to the older, larger, quiescent tumors.

It is unknown what will happen to the vascular infrastructure of a large, established tumor once angiosuppression begins. There have been only rare reports of regression of established tumors by angiosuppression. Acute vascular injury, or even infarction, might be undesirable for a large, established tumor, and could lead to necrosis and swelling, with unwanted clinical consequences. The recently discovered angiogenic protein, VPF (vEFG),[7] also acts as a permeability factor, and the effects of angiosuppressive agents on vascular permeability are unknown. For brain tumors, where tumor-associated vasogenic edema is a major cause of morbidity, any treatment that blocked the hyperpermeability of brain tumor vessels would be welcome. On the other hand, we and others noted that angiosuppression increased cerebral edema.[18,40] Tumor vessels are bidirectional conduits that not only permit the efflux of metabolites and serum proteins, but also clear interstitial proteins back into the systemic circulation.[42] Clinically, angiogenesis ofter appears in areas previously edematous. We suggested that the increased permeability and vasogenic edema represent a forme fruste of angiogenesis in the brain.[18] Edema resulting from angiosuppression has not been reported in organs except the brain, but these tissues contain lymphatics that could clear the intersitial proteins.

Drug delivery to the endothelial cell will be as important to the success or failure of angiosuppression as it is to current chemotherapy. The microcirculation is known to be impaired in the center of large tumors because of elevated interstitial pressure,[43] even though the center of the tumor still contains focii of endothelial cell proliferation,[44] another reason that angiosuppression might be more effective for smaller tumors. Because angiogenesis is a local process, polymer implants in patients may prove valuable to release angiosuppressive agents in high concentration directly at the tumor site, analogous to the polymeric implants in rabbits that have been used successfully in the cornea to inhibit angiogenesis.[12,16,20,31,45-47] Slow-release, biodegradable polymers containing standard chemotherapy (BCNU) are now being used in patients with malignant gliomas and these could be further developed to release angiosuppressive drugs. Systemic angiosuppression[11,15,18,27,48,50] has also proved successful in inhibition of tumor growth; it has the advantage of reaching nearly all areas of the tumor and the host organ through the microcirculation, whereas an implant could thwart disease locally, but allow remaining cells to escape by forming new colonies at the edge of the therapeutic zone. Because it is known that the ultimate effects of an angiogenic polypeptide, eg. FGF or TNF, depend greatly on its route of administration,[51] it follows that an angiosuppressive agent could have differential effects related to whether it is given intravascular (systemic) or extravascular (local). The variability of the endothelium to angiogenic polypeptides is also determined by: cell density, the isoform of the growth factor, the concentration of the mitogen, the 3-dimensional growth pattern, the size and site of the parent vessel, the composition of the extracellular matrix, and the availability and interactions with other growth factors.[52]

In view of the highly variable, even paradoxical, effects predicated on the complex interplay of numerous intracellular and microenvironmental factors, there will be a need to interpret disparate, conflicting, or negative data with caution. Because a rigorous test of an angiogenesis inhibitor is to directly implant the tumor in vascularized tissue,[48] there is a need for a quantitative method to reproducibly and rapidly screen potential inhibitors in situ. Because endothelial cell mitosis is a key step in angiogenesis, we have developed a rapid immunocytochemical method to measure angiogenesis in the rabbit brain using a monoclonal antibody to bromodeoxyuridine.[44] For clinical trials, it would be of immense benefit to have a serum marker or protein to measure angiogenic activity in the circulation, but no such test is currently available.

One putative angiosuppressive agent, suramin, is already trials for cancer therapy. Suramin is a 1400 dalton, polyanionic compound with a known structural formula[53,54] that is the prototype of anticancer drugs that works by binding of growth factors,[55,56] including angiogenic molecules, vasculotropin (vEGF, VPF);[6,67] EGF, TFG-β; PDFG;[55] and FGF.[58] The autocrine transformation by chimeric signal peptide-basic fibroblast growth factor is reversed by suramin.[58] Suramin reverts the malignant phenotype of the simian sarcoma virus,[59] an acutely transforming retrovirus with a capacity to induce experimental sarcomas and gliomas. Suramin can replace heparin in angiosuppression experiments,[32] suppress tumor growth rates,[32,55] and increase the life span of mice inoculated with B16 melanoma.[32] Basic fibroblast growth factor (bFGF)-induced neovascularization in vivo and the bFGF-induced growth of human prostatic cells in vitro is specifically inhibited by suramin.[60]

In addition to the preclinical evidence of growth factor antagonism and angiosuppressive activity, suramin also binds to a wide variety of intracellular and cell membrane enzymes[61,62] as well as histones, globulins, fibrinogen and albumin.[61] Furthermore, it is an anti-parasitic, immunosuppressive, demyelinating, and adrenocorticolytic agent;[61] thus, its actions are hardly confined to angiosuppression. Not surprisingly, its toxicity-serious infections (37%), peripheral neuropathy (30%), coagulopathy (73%), a fatigue-malaise syndrome (34%), or skin rash (42%) is formidable.[61] Tested in 35 patients with metastatic prostatic cancer refractory to hormonal manipulation, its objective response was modest, with only three patients showing complete disappearance of soft tissue tumors, and a few others showing partial (>50%), temporary, or variable site-related responses.[61] In a second trial, suramin was found to be inactive when administered to 12 patients with metastatic renal cell carcinoma by continous parenteral infusion to a peak plasma level greater than 200 μg/ml. No objective radiographic responses were observed, although greater than 40% necrosis of multiple tumor sites and hemorrhages were noted in one patient, suspected by the clinical investigators to be related to angiogenesis inhibition.[62] Unfortunately, there was no measure of angiogenesis activity either clinically or in the tissues examined at autopsy, but the need for optimization of the drug administration is emphasized.[62]

The precarious initial experience with suramin illustrates the problems ahead in the next decade in bringing angiosuppression from the laboratory to the clinic. It should be remembered that the first trials for the mainstay of adjuvant chemotherapy were also frustrating, as noted by Burchemal, "Historically, first trials are perilous and can set back an entire field, especially if the first drugs are too weak, given in too small a dose, and for far too short a period."[24] The field of angiogenesis research and endothelial biology, once the province of a few laboratories, has enjoyed spectacular growth since the early work of its pioneers. A few years ago, the National Academy of Sciences identified the study of angiogenesis as one of a few national scientific priorities.[63] In addition to N.I.H. funding, currently more than 15 biotechnology and pharmaceutical corporations have committed considerable material and human resources to develop inhibitors of angiogenesis. The meeting here, sponsored by the Advanced Scientific Institute of NATO, reflects the worldwide interest and high expectations demanded of its present leaders. It is no longer a question "if" angiosuppression will work, but rather "when", "what indications", "which compound", "how much", "how fast", "what route", "what risk", and "how long". Much work remains, but the final chapter of the angiogenesis story should be its most challenging and rewarding.

REFERENCES

1. P. M. Gullino and F. H. Grantham, Studies on the exchange of fluids between host and tumor. II. the blood flow of hepatomas and other tumors in rats and mice. JNCI 27:1465-1491 (1961).
2. J. Folkman, Anti-angiogenesis: New concept for the therapy of solid tumors. Ann. Surg. 175:409-416 (1972).
3. J. Folkman and M. Klagsbrun, Angiogenic factors. Science 235:442-447 (1987).
4. D. Gospodarowicz, N. Ferrara, L. Schweigerer and G. Neufeld, Structural characterization and biological functions of fibroblast growth factor. Endocrine Rev. 8:95-114 (1987).
5. T. Magiag, Molecular and cellular mechanisms of angiogenesis, in "Important Advances in Oncology", Jr. DeVita, S. Hellman, S. A. Rosenberg, eds. Philadelphia: JB Lippincott, 85-89 (1990).
6. N. Vaisman, D. Gospodarowicz, G. Neufeld, Characterization of the receptors for vascular endothelial growth factor. J. Biol. Chem. 265:19461-19466 (1990).
7. K. Schulze-Osthoff, W. Risau, E. Vollmer and C. Sorg, In situ detection of basic fibroblast growth factor by highly specific antibodies. Am. J. Pathol. 137:85-92 (1990).
8. N. Bouck, Tumor angiogenesis: The role of oncogenes and tumor suppressor genes. Cancer Cells 2:179-185 (1990).
9. J. Folkman, Successful treatment of an angiogenic disease. New Engl. J. Med. 320:1211-1212 (1989).
10. P. Denekamp, Vasculature as a target for tumor therapy. Prog. Appl. Microcirc. 4:28-38 (1984).
11. D. Ingber, T. Fujita and S. Kishimoto, Synthetic analogues of fumagillin that inhibit angiogenesis and tumor growth. Nature 348:555-560 (1990).

12. J. Folkman and D. E. Ingber, Angiostatic steroids: Method of discoveryand mechanism of action. Ann. Surg. 206:374-383 (1987).

13. M. E. Maragoudakis, M. Sarmonika and M. Panoutsacopoulou, Inhibition of basement membrane biosynthesis prevents angiogenesis. J. Pharmacol. Exper. Therap. 244:729-733 (1988).

14. H. Brem, R. J. Tamargo, C. Guerin, S. S. Brem, and H. Brem, Brain tumor angiogenesis, in: "Advances in Neuro-Oncology", Mount Kisko, NY, Futura Publishing Company, P.L. Kornblith and Walter M.D. (eds), pp. 89-102.(1988).

15. P. Madarnas, O. Benrezzak, V. N. Nigam, Prophylactic antiangiogenic tumor treatment. Anticancer Res. 897-902 (1989).

16. T. E. Maione, G. S. Gray, and J. Petro, Inhibition of angiogenesis by recombinant human platelet factor-4 and related peptides. Science 247:77-79 (1990).

17. Y. Shing, J. Folkman and J. Weisz, Affinity of fibroblast growth factors for b-cyclodextrin tetradecasulfate. Anal. Biochem. 185: 108-111 (1990).

18. S. Brem, D. Zagzag, A. M. Tsanaclis, Inhibition of angiogenesis and tumor growth in the brain: Suppression of endothelial cell turnover by penicillamine and the depletion of copper, an angiogenic cofactor. Am. J. Pathol. 137:1121-1142 (1990).

19. J. Murata, I. Saiki, and T. Makabe, Inhibition of tumor-induced angiogenesis by sulfated chitin derivatives. Cancer Res. 51:22-26 (1991).

20. R. Tamargo, R. A. Bok and H. Brem, Angiogenesis inhibition by minocycline. Cancer Res. 51:672-675 (1976).

21. D. M. Noden, Embryonic origins and assembly of blood vessels. Am. Rev. Resp. Dis. 140:1097-1103 (1989).

22. R. M. Shymko and L. Glass, Cellular and geometric control of tissue growth and mitotic instability. J. Theor. Biol. 63:355-374 (1976)

24. J. H. Burchenal, Adjuvant therapy - theory, practice and potential. Cancer 37:46-56 (1976).

25. C. A. Nichol, Pharmacokinetics: Selectivity of action related to physiochemical properties and kinetic patterns of anticancer drugs. Cancer 40:519-528 (1977).

26. M. J. Banda, D. R. Knighton, T. K. Hunt and Z. Werb, Isolation of non-mitogenic angiogenesis factors from wound fluid. Proc. Natl. Acad. Sci., USA 79:7773-7777 (1982).

27. J. L. Gross, D. Hertel, W. F. Herblin, M. Neville and S. S. Brem, Inhibition of basic fibroblast growth factor-induced angiogenesis and glioma tumor growth in vivo. Proc. Am. Assoc. Cancer Res. 32:57 (Abstr. #338) (1991).

28. K. S. Stenn, L. A. Fernandez and S. J. Tirrell, The angiogenic properties of the rat vibrissa hair follicle associate with the bulb. J. Invest. Dermatol. 90:409-411 (1988).

29. D. Gospodarowicz, and K. Thakral, Production of a corpus luteum angiogenic factor responsible for proliferation of capillaries and neovascularization of the corpus luteum. Proc. Natl. Acad. Sci. USA 75:847-851 (1978).

30. B. R. Zetter, Angiogenesis, State of the art. Chest 93 (3):159S-1665 (1988).

31. N. T. Chen, E. J. Corey and J. Folkman, Potentiation of angiostatic steroids by a synthetic inhibitor of arylsulfatase. Lab. Invest. 59:453-455 (1988).

32. J. W. Wilks, T. F. Dekoning, J. M. Cocuzza, Suramin plus angiostatic steroids suppress tumor growth (Abstr.) Proc. Am. Assoc. Cancer Res. 31:60 (1990).

33. R. Auerbach, L. Alby and L. W. Morrisey, Expression of organ-specific antigens on capillary endothelial cells. Microvasc. Res. 29:401-411 (1985).

34. S. Brem, R. S. Cotran, and J. Folkman, Tumor angiogenesis: a quantitative method for histologic grading. J. Natl. Cancer Inst. 48: 347-356 (1972).

35. D. Zagzag, S. Brem, and F. Robert, Neovascularization and tumor growth in the brain. A model for experimental studies of angiogenesis and the blood-brain barrier. Am. J. Pathol. 131:361-372 (1988).

36. S. Brem, H. Brem and J. Folkman, Prolonged tumor dormancy by prevention of neovascularization in the vitreous. Cancer Res. 36:2807-2812 (1976).

37. G. H. Hepner and B. E. Miller, Tumor heterogeneity: biological implications and therapeutic consequences. Cancer Metast. Rev. 2:5-23 (1983).

38. A. M. Tsanaclis, S. S. Brem, S. Gately, H. M. Schipper and E. Wang, Statin immunolocalization in human brain tumors: Detection of noncycling cells using a novel marker of cell quiescence. Cancer in press (1991).

39. T. Nagashima, T. Hoshino and K. G. Cho, Proliferative potential of vascular components in human glioblastoma multiforme. Acta Neuropathol. (Berl) 73:301-305 (1987).

40. B. Hobson and J. Denekamp, Endothelial proliferation in tumours and normal tissues. Continous labelling studies. Br. J. Cancer 49: 405-413 (1984).

41. J. L. Olson, D. W. Beck, D. S. Warner and H. Coester, The role of new vessels and macrophages in the development and resolution of ederma following a cortical freeze lesion in the mouse. J. Neuropath. Exper. Neurol. 46:682-694 (1987).

42. B. R. Deane, J. Greenwood, P. L. Lantos, and O. E. Pratt, The vasculature of experimental brain tumours: Part 4. The quantification of vascular permeability. J. Neurol. Sci. 65:59-68 (1984).

43. T. Jain. Determinants of tumour blood flow: A review. Cancer Res. 48:2641-2658 (1988).

44. S. E. Brien, D. Zagzag, S. Brem, Rapid in situ cellular kinetics of intracerebral tumor angiogenesis using a monoclonal antibody to bromodeoxyuridine. Neurosurgery 25:715-719 (1989).

45. H. Brem and J. Folkman, Inhibition of tumor angiogenesis mediated by cartilage. J. Exper. Med. 141:427-438 (1975).

46. S. Brem, I. Preis, R. Langer, Inhibition of neovascularization by an extract derived from vitreous. Am. J. Ophthalmol. 84:323-328 (1977).

47. J. Folkman, R. Langer, R. J. Linhardt, Angiogenesis inhibition and tumor regression caused by heparin or a heparin fragment in the presence of cortisone. Science 221:719-725 (1983).

48. S. Taylor and J. Folkman, Protamine is an inhibitor of angiogenesis. Nature 297:307-312 (1982).

49. H. Brem, M. S. Mahaley Jr., N. A. Vick, Interstitial chemotherapy with drug polymer implants for the treatment of recurrent gliomas. J. Neurosurg. 74:441-446 (1991).

50. H-L. Peterson, Tumor angiogenesis inhibition by prostaglandin synthetase inhibitors. Anticancer Res. 6:251-254 (1986).

51. J. Folkman and M. Klagsbrun, A family of angiogenic peptides. Nature 329:671-672 (1987).

52. J. R. Merwin, W. Newman and L. D. Beall, Vascular cells respond differentially to transforming growth factors Beta$_1$ and Beta$_2$ in vitro. Am. J. Pathol. 138:37-51.(1991).

53. M. Sjölund, and J. Thyberg, Suramin inhibits binding and degradation of platelet-derived growth factor in arterial smooth muscle cells but does not interfere with autocrine stimulation of DNA synthesis Cell Tissue Res. 256:35-43 (1989).

54. K. Ono, H. Nakane and M. Fukushima, Differential inhibition of various deoxyribonucleic and ribonucleic acids by suramin. Eur. J. Biochem. 172:349-353 (1988).

55. C. A. Stein, R. V. LaRocca and R. Thomas, Suramin: An anticancer drug with a unique mechanism of action. J. Clin. Oncol. 7:499-508 (1989).

56. J. Fantini, T.-J. Guo, J. Marvaldi and G. Rougon, Suramin inhibits proliferation of rat glioma cells and alters N-CAM cell surface expression. Int. J. Cancer 45:554-561 (1990).

57. J. Plouët and H. Moukadari, Characterization of the receptor to vasculotropin on bovine adrenal cortex-derived capillary endothelial cells. J. Biol. Chem. 265:22071-22074 (1990).

58. A. Yayon, and M. Klagsbrun, Autocrine transformation by chimeric signal peptide-basic fibroblast growth factor: Reversal by suramin Proc. Natl. Acad. Sci. USA 87:5346-5350 (1990).

59. C. Betzholtz, A. Johnsson, C.-H. Heldin, and B. Westermark, Efficient reversion of simian sarcoma virus-transformation and inhibition of growth factor-induced mitogenesis by suramin. Proc. Natl. Acad. Sci, USA 83:6440-6444 (1986).

60. M. Ciomei, E. Pesenti and F. Sola, Antagonistic effect of suramin on bFGF: In vitro and in vivo results (Abstr.). 16th LH Gray Conference: Vasculature as a target for anticancer therapy, Manchester England, Sept. 17-21 (1990).

61. R. V. LaRocca, M. R. Cooper, M. Uhrich, Use of suramin in treatment of prostatic carcinoma refractory to conventional hormonal manipulation. Urol. Clin. NA 18:123-129 (1991).

62. R. V. LaRocca, C. A. Stein, R. Danesi, A pilot study of suramin in the treatment of metastatic renal carcinoma. Cancer 67:1509-1513 (1991).

63. Committee on Science, Engineering and Public Policy. In:"New Pathways in Science and Technology" National Academy-Random House Press, New York (1985).

ANTI-ANGIOGENIC VERSUS ANTI-VASCULAR APPROACHES
TO CANCER THERAPY

Juliana Denekamp

CRC Gray Laboratory
Northwood
Middlesex, U.K.

This conference has brought together clinicians and scientists from very diverse areas of pathophysiology in which abnormal vascular proliferation is a problem. The adult body is finely adjusted to have an adequate vasculature in all normal tissues, to provide the nutritional requirements both at rest and in response to stress. That stress may be transient, e.g. exercise, when collateral vessels open up and/or the main vessels vasodilate. Alternatively it may require the production of additional vessels by a process of cell proliferation, e.g. in granulation tissue in a wound or in placenta. A precise balance between the tissues' need for nutrients and the development of new vessels is maintained under normal conditions, but a deviation from that balance towards hyper-vascularisation or hypovascularisation is a characteristic of many disease states. Much of the emphasis, e.g. in diabetes and arthritis, is on preventing abnormal neovasculature developing.

In oncology the development of new vessels in response to tumour derived angiogenic factors marks the important transition step for carcinomas from carcinoma in situ to frank stage 1 carcinoma. Likewise for any metastasis seeded from a primary tumour, a new network of vessels is required if the secondary deposit is to grow to more than approximately 1 mm diameter. These facts, and the recognition that many cancers are already disseminated at the time the patient presents with symptoms, have led a number of groups to consider anti-angiogenic strategies as a form of cancer therapy. This work was pioneered by Folkman and his colleagues in Boston (Folkman et al. 1974; 1987). Table 1 summarises some of the approaches to this problem and contrasts them with the concepts of vascular targeting of cancer therapy.

The three main steps of angiogenesis, i.e. basement membrane dissolution, endothelial migration and endothelial proliferation are all being studied to find inhibitors or competitors for essential pathways. This requires an understanding of the factors and co-factors needed for a response to angiogenic stimuli. If the angiogenic stimulus is inactivated or blocked, no new vessels should be formed. The existing vessels might then regress if they are surplus to the metabolic needs of the tumour. However, if the metabolic demand is not reduced, it seems unlikely that they will regress.

The poor neovascular network has also been recognised as a target for treatments designed to **kill** the tumour cells, rather than simply preventing further tumour expansion

Angiogenesis in Health and Disease, Edited by M.E. Maragoudakis *et al.*
Plenum Press, New York, 1992

Table 1

Anti-angiogenic strategies	Anti-vascular strategies
– Prevent dissolution of the basement membrane	– Attack and kill proliferative endothelial cells
– Prevent endothelial cell migration	– Use proliferative endothelium to act as a focus for a coagulative event
– Prevent endothelial cell proliferation	– Increase new vessel permeability so that intravascular : extravascular pressures become equal
– Prevent collagenase activity in tissues	– Increase platelet adherence to immature endothelium
– Prevent invasion of new vessels between tumour cells	– Alter cytokine production at the surface of immature endothelium
– Prevent anastomosis of new vessel buds	– Alter tissue factor and factor VIII production by immature endothelium
	– Take advantage of differential antigen expression on mature and immature endothelial cells
	– Take advantage of differential expression of membrane bound enzymes in mature and immature endothelium
	– Use hypoxic or acidic micro-environment in tumour blood vessels to activate prodrugs
	– Alter regional blood flow because of lack of vessel musculature and innervation
	– Alter systemic blood pressure so that intravascular : interstitial pressure ratio becomes unity in tumours

(Denekamp 1982, 1990). These approaches are all aimed at causing occlusion or collapse of the vessels with consequent tumour cell death. This can be achieved by vessel rupture (haemorrhage), localised thrombosis, or by equalising the intravascular and interstitial pressures within the tumour so that the driving force for flow is lost. The target must clearly represent a difference between tumour vessels and all the other vascular networks in the body. The possible targets that are now recognised are listed in Table 1. They can be summarised as endothelial cell differences (proliferation, differentiation, surface markers), or vessel differences (structure and innervation within the tumour or in the supplying and draining main vessels). The poor quality of the blood in tumour vessels may also allow prodrug activation, and the sluggish flow may encourage local thrombus formation. Experimental studies indicate that complete cessation of flow for 12-24 hours leads to the death of all the cells surrounding a capillary. Shorter periods of stasis may be useful if combined with prodrugs that are activated to toxic intermediates only in the absence of oxygen (bioreductive drugs). This approach of indirectly causing ischaemic or haemorrhagic necrosis could be effective on any solid tumour with a newly formed vascular network. It should, however, be even more effective on large tumour masses in which the nutritional supply is already more compromised. Many agents are now being shown to have anti-vascular activity as part of their mechanism in causing tumour damage (Moore & West 1991; Denekamp 1991).

These two approaches to therapy via the tumour vasculature are conceptually quite different. Agents that are anti-angiogenic can be screened in systems like the rabbit cornea or the chorio allantoic membrane where vessel growth towards a tumour fragment is scored, or even on proliferating endothelial cell cultures. By contrast the vascular targeted agents need to be tested on established vascularised solid tumours *in vivo* and their effect should be detectable as haemorrhagic or ischaemic necrosis or a gross reaction of the tumour, e.g. shrinkage or even local control. It is, of course, possible that agents derived from the anti-angiogenic screens will eventually turn out to have effects on pre-existing vascular networks, or even cause direct cytotoxicity to the tumour cells.

REFERENCES

Denekamp, J. (1981) Endothelial cell proliferation as a new means of targeting tumour therapy. Brit.J.Cancer 45, 136-139.

Denekamp, J. Vascular attack as a therapeutic strategy for cancer. Cancer and Metastasis Reviews 9: 267-282, 1990.

Denekamp, J. The current status of targeting tumour vasculature as a means of cancer therapy: an overview. Int.J.Radiat.Biol. 1991, vol. 60, nos.1/2, 401-408.

Moore, J.V., West, D.C. Vasculature as a target for anti-cancer therapy. Int.J.Radiat.Biol. 60: 1-421. (Proceedings of 16 L.H. Gray Conference). 1991.

Folkman, J. Tumor angiogenesis factor. Cancer Research 34: 2109-2113. 1974.

Folkman, J. and Klagsbrun, M. Angiogenic factors. Science, 235, 442-447. 1987.

Folkman, J. How is blood vessel growth regulated in normal and neoplastic tissue? GHA Clowes Memorial Award Lecture. Cancer Research, 46: 467-473. 1986.

ASSAYS FOR ANGIOGENIC FACTORS

Jacqueline B. Weiss

Wolfson Angiogenesis Unit
Rheumatic Diseases Centre
Clinical Sciences Building
Hope Hospital, Salford M6 8HD, UK

INTRODUCTION

It may be a controversial statement but it must be said that no single assay for angiogenic factors described to date is totally satisfactory. In this context it may seem inappropriate to discuss these assays at all but it is with this proviso in mind that this brief and not comprehensive review is undertaken. Basically any single assay for an angiogenic factor is likely to be insufficient.

IN VITRO ASSAYS

In vitro assays utilise the ability of the factor to stimulate proliferation of endothelial cells, or to act as a chemo-attractant for such cells or to influence cell motility. A major problem often overlooked particularly by newcomers to the field is the choice of endothelial cell type. As angiogenesis is confined to the micro-vasculature - microvessel endothelial cells should and must be used for these assays.

Chemokinesis (cell motility)

Chemokinesis is observed by plating microvessel endothelial cells onto coverslips already coated with gold particles (McAuslan and Reilly 1983, Zetter, 1980). Motility is measured by assessing the tracks left in the coated cover slip by endothelial cell ingestion of the particles

Angiogenesis in Health and Disease, Edited by M.E. Maragoudakis *et al.*
Plenum Press, New York, 1992

during movement. Total track area is measured before and after addition of a test substance. This method has been modified by Obeso and Auerbach (1984) who replaced the gold particles by negatively charged polystyrene beads. Much simpler is the system of "wounding" endothelial cell cultures by scraping a trough in a confluent monolayer culture of cells and monitoring cell movement into the cleared area of the plate (Scholley et al 1977).

Cell Proliferation

Proliferation of microvessel endothelial cells (MVEC) is assessed by culturing cells in the presence or absence of test material and counting the cells at a fixed time in a cell counter. Alternatively the uptake of tritiated thymidine into the DNA of the cell may be used as a measure of cell numbers.

Chemotaxis

Assays for chemotaxis use a system originally devised by Boyden (1966). Cells are plated onto the upper surface of a Nucleopore filter held in place between two plastic plates containing wells. The filter acts as an interface between the two compartments (wells). Culture medium is added to both compartments and is in contact with the filter. Test material is added to the lower compartment. This system in known as a Boyden chamber. The chamber is incubated for an appropriate time eg 7hrs (Banda et al 1982) and cell migration assessed by counting the number of nuclei on the underside of the filter. The original Boyden chamber was unicellular but nowadays multiwell chambers are used. Hoover and Wright (1983) devised a checkerboard system to ensure a true chemotactic effect rather than a chemokinetic or random cell movement effect.

IN VIVO ASSAYS

The in vivo tests fall into two categories - those which are performed in a potentially host angiogenic environment and those which are performed in an environment which is normally avascular. A factor giving a positive result in the first category may do so because it is acting directly on the microvessel cells or indirectly because it is stimulating a host cell to produce its own angiogenic factor(s). In the second case a putative angiogenic factor may summon cells into the avascular tissue which carry their own angiogenic substance(s) (Odedra

308

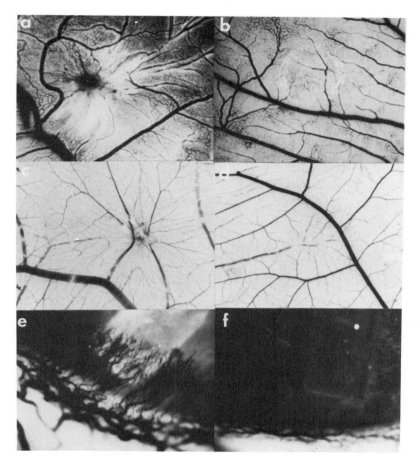

Fig. 1.   In vivo tests for angiogenesis using Endothelial Cell
           Stimulating Angiogenesis Factor (ESAF) as a test substance

           a) Positive result on yolk sac membrane (YSM)
           b) YSM negative control
           c) Positive result on CAM
           d) CAM negative control
           e) Positive result on rabbit cornea
           f) Negative corneal control

and Weiss, 1991). In any event in this tissue natural inhibition of angiogenesis will need to be overcome.

The Transparent Chamber

One of the earliest test systems used in the study of angiogenesis was the transparent chamber. This was originally developed by Sandison (1924) for use on the rabbit ear and modified later by Clark and Clark (1932). Two metal plates are fitted over each other to create a circular healing chamber. The method was later modified for use in the hamster cheek-pouch by Warren and Shubik (1966).

Rat Dorsal Air Sac

This technique was developed in 1953 by Selye in a study of inflammation and was subsequently modified and utilised by Folkman and his colleagues in 1971. A bolus of air is injected subcutaneously between the skin and the underlying fascia of the rat dorsal skin. Test materials in millipore chambers are implanted into the air sacs and a vascular response is observed in the fascia in 48 hours. The vascular response consists of vasodilation and capillary proliferation. Capillary proliferation is verified by the presence of endothelial cell mitosis in histological sections as well as by injection locally of tritiated thymidine which results in labelling of the endothelial cell. When soluble materials are to be tested then a silicone rubber tube is implanted into the air sac which is infused with test material inter-mittently. The angiogenic response is determined at 48 hours by microscopic observation of the fascia.

Corneal Pocket Test System

The system using the rabbit cornea was developed by Gimbroni et al (1974). Test fractions are implanted into a small pocket formed within the cornea a few millimeters from the edge of the cornea and the resident limbal blood vessels (limbal plexus). The material to be tested is incorporated into a stable polymer pellet and the cornea is examined at 5 to 7 days after application of the sample. An angiogenic response is characterised by the directional growth of microvessels from the limbal plexus into the avascular cornea towards the site of implantation

and the extent is determined by microscopic observations of the cornea
in situ or sometimes histologically after excision of the cornea (Fig. 1
e,f). The corneal assay has now been developed for a number of animals
including the rabbit (Kissun and Garner 1977, Folkman 1985), the guinea
pig (Poliverini et al 1977), the mouse (Muthukkarappan and Auerbach
1979) and rat (Fromer and Clintworth 1976). As has been mentioned
before, this is the only test system which takes advantage of a totally
avascular tissue.

Sponge Implant Model

Recently, a new model has been devised (Andrede et al 1987) in
which sponges are implanted subcutaneously in a rat. A small segment of
polythene tubing is fixed to the interior of each sponge disc by means
of silk sutures so that each disc had a central cannula. The discs are
sterilised by boiling before implantation. Supposed angiogenic samples
are administered via this cannula. Blood flow measurements are
determined using a radio-labelled Xenon clearance technique.
Reproducible techniques were obtained by this method. This method is
claimed to be an improvement on that of Edwards et al (1960) who used
polyvinyl alcohol sponge implants to study the effects of tissue
extracts on granulation tissue growth in rabbits. Sponge implants using
the technique of Edwards et al are examined for microvessel invasion
histologically. (Fig. 2).

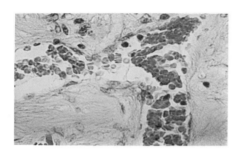

Fig 2.   Sponge implant showing sprouting vessel at day 7.
                (Test substance ESAF - courtesy Prof. D.J. Jackson)

311

Intradermal Injections

Lymphocyte induced angiogenesis was first discovered during studies which involved intradermal injection of allogeneic or semi-allogeneic lymphocytes into mice by Sidky and Auerbach (1975). This system has been used subsequently for an attempt at quantitation of lymphocyte induced angiogenesis (Sidky and Auerbach 1979, Majewski et al 1985, Kaminski et al 1984).

The Chick Chorioallantoic Membrane (CAM)

Probably the most commonly used test system for angioigenesis is the CAM. It is relatively simple and cheap and it permits screening of a large number of samples. A false air-sac is created in the fertilised hen egg which allows the CAM to fall away from the egg shell membrane (Zwilling 1959). This is done by aspirating through a small hole in the end of the egg approximately 2-3ml of albumin. The CAM itself is exposed through a window made in the shell. This is cut into the air-sac created by the removal of the albumin and is approximately 1cm square. On day 10 or 11 a pellet containing the test material is placed on the CAM and the window is sealed with tape. Two to three days later the membrane is examined and the angiogenic response is assessed (Ausprunk 1979, Folkman 1985) (Fig. 1 a,b). An essential precaution for the CAM is to prevent any dust from the cut egg shell falling onto the membrane as this tends to give an inflammatory and therefore inappropriate result. It is best to cut the window using a dental cutter under a vacuum provided by a small, hand held vacuum cleaner.

The Yolk Sac Membrane

The use of the chick yolk-sac membrane has considerable advantages over the CAM (Taylor et al 1986). Chick embryos are cultured ex-ovo after three days of incubation and the samples are applied after a further two days of incubation. An angiogenic response is visible after 24 hours and is characterised by a convergence of small vessels towards the site of the sample application in exactly the same manner as is seen in the CAM (Fig. 1 c,d. A great advantage of this method is that four pellets can be applied to the same egg. The egg is cracked and dropped very gently into cradles of plastic film suspended over short lengths of plastic piping approximately 8cm in diameter. It is essential to crack the eggs no later than 68 hours after incubation or the yolk-sac will stick to the egg shell membrane. The egg in its cradle is covered with a petrie dish lid and incubated in a humidified air incubator. After 48 hours samples can be applied to the membrane. It is useful to examine the eggs after six hours or eight hours for any damage to the membrane which may occur with salty samples and if any damage is seen at this stage the eggs are rejected. At 24 hours the eggs are examined and any angiogenic response noted. The particular advantage of the yolk-sac membrane (YSM) is that the immune system is not functional at this time and therefore immune responses will not be observed. A comparison of results on CAM and YSM showed 26 comparable results in both tests. (Table 1) (Wiseman, Taylor and Weiss in preparation).

Table 1.   Response of Chick Chorioallantoic Membrane (CAM) and Yolk Sac Membrane to Purified Samples of ESAF

No. of samples tested	=	26
No. of identical results from both tests	=	23
No. of positive CAM/Negative yolk sac membrane responses	=	2
No. of positive yolk sac membrane/ Negative CAM responses	=	1

DISCUSSION

As has been said before, any single test for angiogenesis is probably not satisfactory and a positive in vitro as well as in vivo test is needed to determine what is a true angiogenic factor. It is also helpful if in the in vitro test the factor is specific for the microvascular endothelial cells. An interesting example of an angiogenic factor which is highly angiogenic in vivo but is unable to effect any

changes in microvessel endothelial cells _in vitro_ is angiogenin a factor
which is related to ribonuclease (Fett et al 1985). The mechanism by
which it stimulates angiogenesis is not clear. Anomalies of this kind
contribute to the fascination of research in angiogenesis.

References

Andrade, S.P., Fan, T-P.D., and Lewis, G.P. (1987) Quantitative _in vivo_
studies and angiogenesis in a rat sponge model. Br. J. Exp. Path. 68:
755-766

Ausprunk, D.H. (1979) 'Tumour Angiogenesis' in: Chemical messengers of
the inflammatory process, handbook of inflammation. J. Houck (ed)
Elsevier, North Holland, Amsterdam, 1: 317-351

Banda, M.J., Knighton, D.R., Hunt, T.K. & Webb, Z. (1982) Isolation of a
non-mitogenic angiogenesis factor from wound fluid. Proc. Natl. Acadm.
Sci. (USA) 79: 7773-7777

Boyden, S. (1962) J. Exp. Med. 115: 453-466

Clark, E.R., & Clark, E.L. (1932) Observations on living preformed blood
vessels as seen in transparent chambers inserted in the rabbit's ear. Am
J. Anat. 49: 441-447

Edwards, R.H., Sarmenta, S.S. & Hass, G.M. (1960) Stimulation of
granulation tissue growth by tissue extracts. A.M.A. Arch. Pathol. 69:
286-302

Folkman, J. (1971) Tumour angiogenesis: therapeutic implications. New
England J. Med. 285: 1182-1186

Folkman, J. (1985) Tumour angiogenesis. Adv. Cancer Res. 43: 197-203

Fromer, C.H., Klintworth, G.K. (1976) An evaluation of the role of
leukocytes in the pathogenesis of experimentally induced corneal
vascularisation III. Studies related to the vasoproliferative
capability of polymorphonuclear leukocytes and lymphocytes. Am J.
Pathol. 82: 157-170

Gimbrone, M..A, Jr., Cotran, R.S., Leapman, S.B., & Folkman, J. (1974)
Tumour growth and neovascularisation: an experimental model using the
rabbit cornea. J. Natl. Cancer Inst. 52: 413-427

Hoover, R.L. & Wright, T.C. (1983) Chemotaxis of endothelial cells to
formyl-methionyl-leucyl phenylalanine. Microvasc. Res. 26: 122-125

Kaminski, M.J., Majewski, S. Jablonska, S., Pawinska, M. (1984) Lowered
angiogenic capability of peripheral blood lymphocytes in progessive
systemic sclerosis (scleroderma) J. Invest. Dermatol. 82: 239-243

Kissun, R.d. & Garner, A. (1977) Vasoformative properties of normal and
hypoxic retinal tissue. Br. J. Ophthalmol. 61: 394-398

Majewski, S., Skopinski-Rosewska, E., Jablonska, S., Polakowski, I., Pawinska, M., Marczak, M., Szurolo, A. (1985) Modulatory effect of sera from scleroderma patients on lymphocyte-induced angiogenesis. Arthritis, Rheum. 28: 1133-1139

McAuslan, B.R., Reilley, W.G., Hannan, G.N., Gole, G.A. (1983) Angiogenic factors and their assay: activity of formyl leucyl phenylalanine, adenosine diphosphate, heparin, copper, and bovine endothelium stimulating factor. Microvasc. Res. 26: 323-338

Muthukkarappan, V.R., Auerbach, R (1979) Angiogenesis in the mouse cornea. Science 205: 1416-1418

Obeso, J.L., & Auerbach R. (1984) A new micro technique for quantitating cell movement in vitro using polystyrene bead monolayers. J. Immunol. Meth. 70: 141-152

Odedra, R. & Weiss, J.B. (1991) Low molecular weight angiogenesis factors. Pharmac. Ther. 49: 111-124

Polverini, P.J., Cotran, R.S., Gimbrone, M.A., Unanue, E.R., (1977) Activated macrophages induce vascular proliferation. Nature, 269: 804-806

Sandison, J.C. (1924) Anat. Rec. 28: 281-287

Selye., H. (1953) On the mechanism through which hydrocortisone affects the resistance of tissues to injury. An experimental study with the granuloma pouch technique. J. A. Med. Assoc. 152: 1207-1219

Sholley, M.M., Gimbron, M.A., Cotran, R.S. (1977) Cellular migration and replication in endothelial regeneration: a study using irradiated endothelial cultures. Lab. Invest. 36: 18-25

Sidky, Y.A., & Auerbach, R. (1975) Lymphocyte induced angiogenesis: a quantitative and sensitive assay of the graft-us-host-reaction. J. Exp. Med. 141: 1084-1100

Sidky, Y.A., & Auerbach, R. (1979) Response of the host vascular system to immunocompetent lymphocytes: effect of pre-immunisation of donor or host animals.. J. Proc. Soc. Exp. Biol. Med. 161: 174-177

Taylor, C.M., Weiss, J.B., Kissun, R.D. and Garner, A. (1986) Effect of oxygen tension and the quantities of procollagenase activating angiogenic factor present in the developing kitten retina. Br. J. Ophthalmol. 70: 162-165

Warren, B.A., & Shubik, P. (1966) The growth of the blood supply to melanoma transplants in the hamster cheek pouch. Lab. Invest. 15: 464-478

Zetter, B.R. (1980) Migration of capillary endothelial cells is stimulated by tumour-derived factors. Nature 285: 41-43

Zwilling, E. (1959) A modified chorioallantoic procedure. Transplant Bull. 6: 115-116

DEVELOPMENT AND VALIDATION OF A SPONGE MODEL

FOR QUANTITATIVE STUDIES ON ANGIOGENESIS

T.-P.D. Fan, D.-E. Hu and C.R. Hiley

Department of Pharmacology, University of Cambridge
Tennis Court Road, Cambridge CB2 1QJ, UK

INTRODUCTION

The assays for angiogenesis in current use include the corneal micropocket technique (Gimbrone et al., 1974; Leibovich et al., 1987), chick embryo chorioallantoic membrane (Auerbach et al., 1974; Crum et al., 1985), the hamster cheek pouch (Greenblatt & Shubik,1968; Schrieber et al., 1986), and rat dorsal air sac (Folkman et al., 1971; Ingber et al., 1990). The development of these methods in the 1970s was instrumental in the discovery, purification and biochemical characterisation of angiogenic factors and inhibitors. However, these models, with the exception of the corneal assay, are only qualitative or semi-quantitative. Thus Vallee et al. (1985) concluded that "none of the procedures available for studying angiogenesis is ideal and that the design and verification of specific and reproducible methodology remains an imperative of the highest priority". In the last five years, several new models have been developed, e.g. subcutaneous implantation of plastic chambers (Dvorak et al., 1987) or porous polytetrafluoroethylene tubes (Sprugel et al., 1987). In addition, techniques involving the subcutaneous implantation of sterile sponges into experimental animals have become popular (Davidson et al., 1985; Fajardo et al., 1988; Kusaka et al., 1991).

In 1987, we described an objective and reproducible method of measuring angiogenesis in sponge implants in rats (Andrade et al., 1987). The method not only allows histological and biochemical studies to be made but also has the major advantage of using a simple ^{133}Xe wash-out technique for repeated quantitation of *relative* blood flow changes through the sponges over a period of weeks. Since the sponges originally contained no blood vessels, the development of blood flow was considered to represent a neovascularisation. This assumption was supported by histological observations that the sponges were gradually infiltrated by host blood vessels.

In a recent study, we carried out detailed analysis of the ^{133}Xe clearance curves and some morphometric analysis of sponge sections (Fan et al., in press). Here, we report on two additional approaches to validate this technique. First, we measured *absolute* blood flow in the sponges using a ^{113}Sn microsphere wash-in technique. Second, we established that the level of haemoglobin in the implants correlated well with ^{133}Xe measurements. In addition, some data obtained with angiogenic factors and inhibitors are presented in support of the general applicability of this model in angiogenesis research.

Angiogenesis in Health and Disease, Edited by M.E. Maragoudakis *et al.*
Plenum Press, New York, 1992

THE CANNULATED SPONGE MODEL

Circular sponge discs (1.2 cm diameter) were prepared from a sheet of 5 mm thick polyether foam (E R Carpenter & Co, Haverhill, Suffolk, UK). A 1.2 cm segment of polythene tubing (1.4 mm internal diameter; Portex Ltd, UK) was secured to the interior of each sponge disc by means of 5/0 silk sutures so that every sponge disc has a central cannula. Before implantation, sponge discs were soaked in 70% ethanol for 2-3 hours and then rinsed in sterile phosphate buffered saline (PBS, without calcium and magnesium). After squeezing the sponges in a 20ml syringe to remove the excess PBS, they were sterilised by overnight ultraviolet light irradiation.

Implantation of sponge discs was performed with aseptic techniques. Hypnorm (0.5 ml/kg; 0.315 mg/ml fentanyl citrate and 10mg/ml fluanisone, Janssen Pharmaceuticals, UK) was used to induce neuroleptanalgesia in male Wistar rats weighing 180-200 g. After the dorsal side had been shaved and wiped with 70% ethanol, a 1 cm dorsal, midline, vertical skin incision was made approximately 4 cm caudal to the occipital ridge. Using a pair of curved scissors, two subcutaneous air-pockets were fashioned, one anterior and the other posterior to the incision. Two needle punctures (5 cm apart) were made on top of the pockets. A sterile sponge disc was then inserted into each air-pocket, with the free end of its cannula being exteriorised through the needle puncture. To immobilise the sponge implant, the base of each cannula was sutured to the rat skin. Finally, the skin incision was sutured with two interrupted 5-0 silk stitches, and the cannula was plugged with a sterile polythene stopper so as to prevent overt infection and evaporation of ^{133}Xe-saline during blood flow measurements. The stopper was changed every day and if infection of the implants became apparent, the animals were excluded from the experiments. To prevent the rats from tampering with the cannulae, they were housed individually in plastic cages and provided with a normal diet and water.

Using rats prepared in this manner, it was possible to gain access to the sponge implant via its cannula for (i) administration of putative angiogenic factors or inhibitors, (ii) collection of wound fluid samples for biochemical analysis, and (iii) repeated blood flow measurements over two to three weeks using the ^{133}Xe clearance technique described below.

BLOOD FLOW MEASUREMENTS USING ^{133}Xe CLEARANCE

[133]Xe measurements were carried out every two days, starting on Day 4 after sponge implantation. The animals were treated with Hypnorm as before. ^{133}Xe-saline injection (10 mCi in 3 ml; Amersham International, UK) was diluted in sterile physiological saline according to its activity and 10 µl was injected into the sponge implant via the cannula. The cannula was immediately plugged to prevent evaporation of ^{133}Xe and the washout of radioactivity from the implant was monitored every 40 s for 6 min using a collimated gamma scintillation detector coupled to an SR8 scaler-ratemeter (NE Technology Ltd, Reading, UK) and an Epson LX850 printer. The % [133]Xe clearance after 6 min was calculated by the formula:

$$[(\text{initial count} - \text{count at 6 min})/\text{initial count}] \times 100\%.$$

Since the background radioactivity was always less than 0.1% of the initial count, it was ignored in the calculation.

Fig. 1 shows that the ability of the sponge implants to clear ^{133}Xe increased as a function of time. The clearance of between 19-22% that occurred during the first 6 days was due to passive diffusion of the ^{133}Xe from the sponge. By Day 8 the clearance reached 30% indicating a commencement of blood flow in the sponge which continued to increase over the 14-day period. Morphometric

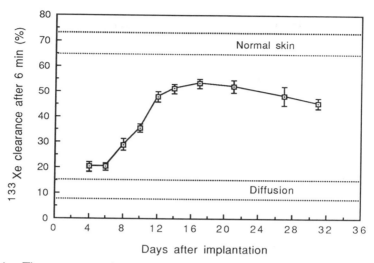

Fig. 1. Time course of sponge neovascularisation in rats as measured by the 6 min ^{133}Xe clearance technique. Data represent mean ± s.e.m. n = 40.

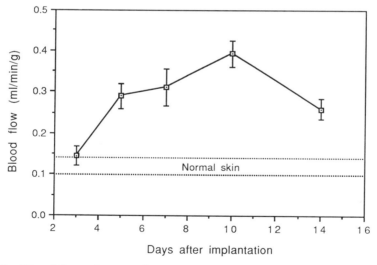

Fig. 2. Blood flow changes in sponge implants as measured by ^{113}Sn microspheres. n = 6.

analysis of the sponge sections showed a good correlation between the degree of tissue infiltration and ^{133}Xe clearance (Fan et al., in press).

BLOOD FLOW MEASUREMENTS USING [113]Sn MICROSPHERES

To assess further the function of the developing vasculature in sponge implants, we adopted a [113]Sn microsphere wash-in technique for measurements of *absolute* blood flow. The rationale of the microsphere method for quantitation of blood flow is that an increase in flow to any particular anatomical region will lead to the trapping of a greater number of microspheres which have been injected into the heart. Since the diameter of capillaries (1-8 μm) is less than that of the microspheres (15 ± 5 μm), the latter become trapped in the tissue and sponge implants and the radioactivity in that particular tissue/sponge represents the amount of cardiac output which it has received.

On Days 3, 5, 7, 10 and 14 after sponge implantation, rats were anaesthetised with sodium pentobarbital (60 mg/kg, i.p.). Two polythene cannulae were inserted into each of the femoral arteries; one was used for recording systemic blood pressure with a Bell & Howell type 4-422-0001 pressure transducer connected to a Grass model 7D polygraph and the other for the withdrawal of blood at a constant rate (0.5 ml/min) during the microsphere injection. After dissecting the neck muscle medially, another cannula was passed down the right carotid artery into the left ventricle for the determination of cardiac output and organ blood flow by the microsphere technique as described by Hiley et al. (1980). When the animal was stable, [113]Sn-labelled microspheres in 300 μl of 0.01% Tween 80/isotonic saline, were injected into the left ventricle over 7-10 s. During, and for 70 s after, the microsphere injection, blood was withdrawn from one of the femoral arteries. The animals were then killed with an overdose of sodium pentobarbital and the sponges, the heart, lungs, both kidneys, pectoral muscles and samples of dorsal and ventral skin were removed, weighed and counted in a Packard Auto-Gamma 500 gamma counter. Cardiac output and sponge and organ blood flows were calculated as described by McDevitt and Nies (1976). The amount of radioactivity in the blood samples withdrawn from the femoral artery was also counted. The radioactivity in lungs was checked to make sure that there was no arteriovenous shunting of blood in the periphery and symmetry of kidney flow served as an indicator of mixing.

As shown in Fig. 2, blood flow in normal rat skin was found to be 0.12 ± 0.02 ml/min/g. In the sponges, the blood flow was 0.15 ± 0.02 ml/min/g on Day 3 and increased to 0.29 ± 0.03 ml/min/g on Day 5. By Day 7, the blood flow was 0.31 ± 0.04 ml/min/g, reaching a maximum of 0.39 ± 0.03 ml/min/g on Day 10 and then it declined to 0.26 ± 0.02 ml/min/g on Day 14. These data are similar to those of Parnham et al. (1979) who measured the distribution of cardiac output in carrageenan-impregnated sponges in rat over a period of 10 days.

Taken with the ^{133}Xe data, the [113]Sn data suggest that the ^{133}Xe clearance technique not only gives an indication of the rate of perfusion of the sponges with blood but also relates to the vascular density, particularly the available exchange surface area, of sponge implants. Hence the measurement of ^{133}Xe clearance in the implant gives a good estimate of its functional vascularity. Histological studies confirmed that the sponges were gradually infiltrated by host blood vessels. Since the sponges originally contained no blood vessels, the development of blood flow must result from the ingrowth of new blood vessels into the sponges.

HAEMOGLOBIN CONTENT OF VASCULARISED SPONGE IMPLANTS

Plunkett & Hailey (1990) described an in vivo quantitative angiogenesis model using tumour cells entrapped in alginate beads. These beads sequestered cells from direct contact with their immediate environment, but allowed diffusible angiogenic factors to pass through to induce neovascularisation in the host. It was shown that measurements of pooled ^{51}Cr-labelled red blood cells and local haemoglobin (Hb) concentration at the alginate-tumour pellet site can provide a quantitative means of assessing angiogenesis. More recent studies by Plunkett and co-workers showed that inhibitors of angiogenesis such as protamine sulphate and the angiostatic steroid tetrahydro S inhibited tumour-induced angiogenesis as determined by these two methods and confirmed by gross morphology and histological analysis of the alginate pellet site (Robertson et al., 1991).

In view of these findings, we decided to establish the time-course of change of sponge Hb concentration and to see if there is any correlation between ^{133}Xe clearance and Hb concentration in the sponge implants. Fig. 3 shows that there is indeed a good correlation between the Hb concentrations in the sponges and ^{133}Xe clearance. In addition, Figs. 4 and 5 show consistently higher levels of ^{133}Xe clearance and Hb concentrations in sponges treated with daily doses of 10pmol bradykinin plus 0.3pmol interleukin 1. The combination of these two peptides has previously been shown to stimulate angiogenesis in this model (Hu & Fan, 1991). Assuming that all the Hb measured is derived from red blood cells present in the neovasculature, the good correlation between Hb contents and ^{133}Xe clearance values further suggests that measurements of relative blood flow changes in sponge implants by the latter method do provide an accurate means to assess new blood vessel formation.

ASSESSMENT OF ANGIOGENIC FACTORS OR INHIBITORS

Using this model, we have been studying the role of basement membrane components, fatty acid derivatives, neuropeptides and other vasoactive peptides in angiogenesis. It was found that the neovascular response can be modified by selective modulation of basement membrane biochemistry. For example, daily administration of 5µg hyaluronic acid fragments (4-25 disaccharides; West et al., 1985) enhanced sponge neovascularisation (data not shown). In contrast, two basement membrane synthesis inhibitors D609 and GPA 1734 (Maragoudakis et al., 1988) suppressed angiogenesis as shown in Fig. 6.

Experiments with receptor antagonists of platelet-activating factor (PAF) inhibited the angiogenic response, suggesting PAF may be involved in some aspects of angiogenesis (Smither & Fan, 1990). However, cyclo-oxygenase and/or lipoxygenase inhibitors did not modify the time-course of neovascularisation. These data indicate that endogenous prostaglandins and leukotrienes do not play a significant role in this model (Smither & Fan, 1991). Table 1 shows a list of agents which are not active in this model.

Figures 7 and 8 show that daily administration of either human interleukin 1-alpha (IL-1α), or tumour necrosis factor-alpha (TNF-α) accelerated the sponge-induced neovascularisation in a dose-dependent manner. These findings confirm the work of Mahadevan et al. (1989) on the angiogenic properties of these two cytokines in a similar model.

In an interesting series of experiments, Ford and his co-workers (1989) systematically examined the characteristics of cells infiltrating polyurethane sponge implants in mice over a period of 13 days. It was found that on Day 3 neutrophils reached a peak of 72% of the total cell infiltrate and declined gradually to 38% on Day 8 and 18% by Day 13. In contrast, macrophage number on Day 3 represented 18% and increased to 47% on Day 8 and 62% by Day 13. The corresponding values for lymphocytes were 10%, 15% and 20%, respectively. More importantly, these workers detected significantly higher levels of IL-1, IL-6, TNF and macrophage colony stimulating factor (M-CSF) in the sponge wound fluid compared with basal serum levels in non-wounded mice. On the other hand, IL-2, IL-3 and IL-4 could not be detected. Time course

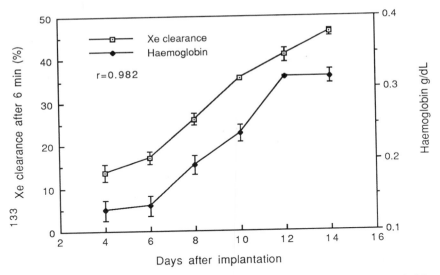

Fig. 3. Correlation between ^{133}Xe clearance and haemoglobin
contents of sponge implants. n = 8.

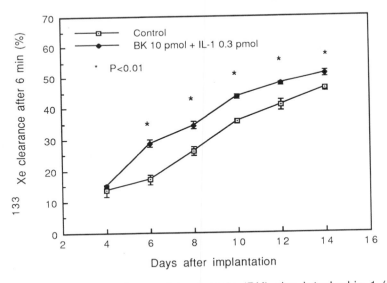

Fig. 4. Effect of daily doses of bradykinin (BK) plus interleukin 1 (IL-1)
on ^{133}Xe clearance from sponge implants. n = 8.

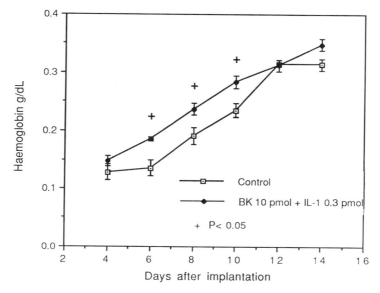

Fig. 5. Effect of daily doses of bradykinin (BK) plus interleukin 1 (IL-1) on the haemoglobin contents of sponge implants. n = 8.

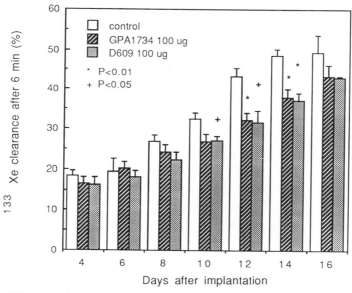

Fig. 6. Effect of basement membrane synthesis inhibitors GPA 1734 and D609 on sponge-induced neovascularisation. n = 6.

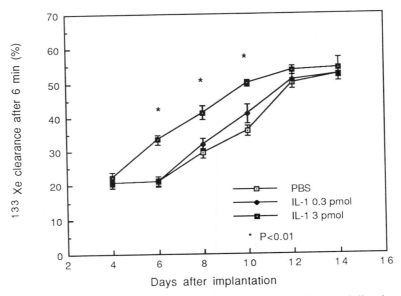

Fig. 7. Enhancement of sponge-induced angiogenesis by daily doses of interleukin 1 (IL-1). n = 8.

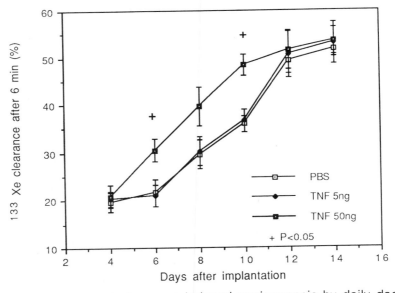

Fig. 8 Enhancement of sponge-induced angiogenesis by daily doses of tumour necrosis factor-alpha (TNF). n = 8.

Table 1. Agents which are inactive in modulating sponge-induced angiogenesis.

Agent	Amount	Reference*
Amines		
Histamine	0.01-1 µmol	Thompson & Brown 1987
Vasoactive peptide		
Endothelin	200 pmol	Takagi et al., 1990
Cytokines		
Interleukin 6	0.3 -3 pmol	Motro et al., 1990
PG/LT synthesis inhibitors		
Indomethacin	0.25-2.5 µg	Haynes et al., 1989
Quercetin	2.5-25 µg	
BW755C	2.5-25 µg	
Angiotensin-converting enzyme inhibitor		
Captopril	1-100 µg	Wang & Prewitt, 1990

The substances were injected into the sponges in 50 µl volume, from Day 1 to Day 14 after sponge implantation. $n \geq 6$ in each experiment. PG/LT = prostaglandins/leukotrienes.
*The reference indicates published work where an effect of the particular agent in angiogenesis has been suggested.

studies revealed that TNF, M-CSF and IL-6 peaked earlier than IL-1 and the levels of these cytokines had fallen by Day 13. It was proposed that these cytokines interact to promote tissue remodelling and that the eventual decrease in their levels may be the result of a regulatory process by the healed wound.

Recent studies have suggested that under certain situations, some neuropeptides also behave as growth factors (Woll and Rozengurt, 1989). Several vasoactive neuropeptides potentiated the ability of IL-1α to stimulate fibroblast proliferation (Kimball and Fisher, 1988). To explore the possibility that vasoactive neuropeptides and cytokines interact to modulate neovascularisation, selected neuropeptides were injected into the sponges, alone or in combination with IL-1α. Daily administration of 1nmol of the neuropeptides substance P (SP), calcitonin gene related peptide (CGRP), or angiotensin II (AII) caused an enhancement of angiogenesis. These data confirmed the angiogenic activity of these peptides reported by others (Ziche et al., 1990; Haegerstrand et al., 1990; Ferandez et al., 1985; Le Noble et al., 1991). Furthermore, 1nmol vasoactive intestinal polypeptide (VIP) or bradykinin (BK) also stimulated sponge-induced angiogenesis. When combined, subthreshold doses of IL-1α (0.3pmol) and the neuropeptides (10pmol) or BK (10pmol), but not AII, were sufficient to promote neovascularisation.

It might be argued that vascular reactivity to these agents might interfere with the assessment of angiogenesis by 133Xe measurement. However, the new vessels will have to be developed before being influenced by vasoactive agents. Furthermore, the test substances are always injected via the cannulae into the sponges 24h before 133Xe measurements so as to exclude their possible acute effects on the microvasculature (dilatation or constriction). Laser Doppler flowmetry was also used to examine the effects of these substances on

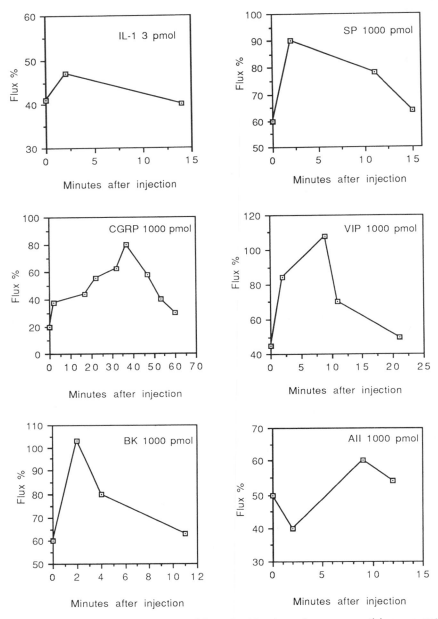

Fig. 9   The transient effects of interleukin 1 and neuropeptides on rat skin microvasculature as determined by laser Doppler flowmetry.

rat skin vasculature. For this series of experiments, the peptides were injected intradermally in 50 μl volume. Flux % was measured immediately after injection for the duration indicated, using an MBF3D Laser Doppler Blood Flow Monitor (Moor Instruments Ltd, Devon, England). The data in Fig. 9 show that at the highest doses studied, these peptides only produced transient (< 60 min) blood flow changes in normal rat skin. Parallel studies using the ^{133}Xe clearance technique to follow the effects of these peptides at 6h, 12h and 24h after injection also failed to show any protracted vascular responses (data not shown). Furthermore, these peptides did not show any vasoactivity in 14-day old sponges.

Preliminary histological studies of the implants have shown increased vascularity in the test sponges as compared to controls. Furthermore, the effects of SP, VIP, BK and AII were abolished by the selective receptor antagonists [D-Pro^2, D-Phe^7, D-Trp^9]-SP, VIP-(10-28), des-Arg^9, [Leu^8]-BK and saralasin, respectively (Fan & Hu, 1991; Hu & Fan, 1991). Thus it is apparent that the angiogenic responses to PAF and the neuropeptides examined here are receptor-mediated. Future strategies to discover selective agonists or antagonists on these receptors will hopefully lead to better drugs for the management of angiogenic diseases such as rheumatoid arthritis, atherosclerosis and cancer.

OTHER IMPLANT MODELS

Because of its low cost and technical simplicity, many groups have been using sponge implants as a matrix to study angiogenesis. Different types of sponge materials have been used: viscous cellulose, polyvinyl alcohol (Ivalon), polyester, polyether, polyether polyurethane, and gelatin (Gelfoam). The differences in sponge material, porosity, shape and size make direct data comparison difficult.

Davidson et al. (1985) and Buckley et al. (1985, 1987) implanted Ivalon sponges subcutaneously in rats to study the angiogenic properties of cartilage-derived growth factor and epidermal growth factor by measuring the DNA, protein and collagen content of the implants. More recently this group of workers showed that sponge implants containing slow-release pellets releasing a neutralising antiserum directed against basic fibroblast growth factor (bFGF) had significantly reduced content of DNA, protein and collagen. These data suggest endogenous bFGF is involved in wound repair (Broadley et al., 1989).

Fajardo et al. (1988) devised a disc angiogenesis system with the same sponge material. It consists of a small disc of the foam covered on both flat sides by Millipore filters, leaving only the edge as the area for cell penetration into the disc. Factors to be tested are incorporated in ethylene-vinyl acetate co-polymer to facilitate their sustained release from the centre of the disc. After a period of growth, the implanted disc is removed, fixed and embedded in paraffin, then stained and angiogenesis measured by simple morphometric technique. Ivalon sponges were also used by Smaje et al. (1988) to examine the changing sensitivity to histamine H_1- and H_2-receptor agonists in the growing vasculature.

Using FGF-coated Gelfoam in rats, Thompson et al. (1988) first demonstrated that neovascularisation could be site-directed. Gelfoam has also been employed by Ciomei et al. (1991) and Presta et al. (1991) in their studies on the antagonistic effect of suramin on bFGF-induced neovascularisation, and on the angiogenic activity of mutant bFGF in mice. Mahadevan et al. (1989) demonstrated that laminin and fibrinogen, but not collagen type IV, can stimulate polyether polyurethane sponge-induced neovascularisation. Richardson et al (1987) have implanted double velour Dacron sponges into mice and removed them at fixed times for examination by light microscopy, autoradiography and electron microscopy.

In 1987, Dvorak et al. devised a new method for quantitative studies of angiogenesis; a special plastic chamber is prepared by gluing two round Thermanox coverslips on either surfaces of a plexiglass ring. Twelve regularly spaced pores are bored in one of the coverslips. Such chambers are then filled with varying contents and implanted subcutaneously in experimental animals. It was shown that chambers filled with cross-linked homologous fibrin or plasma induced an angiogenic response within 4 days. New blood vessels grew into the chambers through the coverslip pores, flaring out radially from each pore, both laterally and vertically. The distinct advantage of this assay is that angiogenesis can be scored under a dissecting microscope and quantitated by standard morphometric techniques.

ADVANTAGES OF THE CANNULATED SPONGE MODEL

In comparison with other assays for angiogenesis, several advantages of the cannulated sponge model are apparent. (i) Because the injected ^{133}Xe is cleared by expiration within 2 h, it is possible to make repeated blood flow measurements in the same animal over a period of two to three weeks. (ii) The sponge wound fluids can be extracted easily for biochemical analysis of the levels of inflammatory mediators (Smither and Fan, 1991), angiogenic factors and cytokines (Ford et al., 1989; 1990). Such information will help elucidate the role of these factors in angiogenesis. (iii) The functional vascularity of the neovasculature can be correlated with histological findings. (vi) The system allows testing of compounds which can either be administered locally into the sponges or systemically. Furthermore, this model has been adapted for mice (see Fig. 10). The miniaturisation of the sponge model in mice would allow tumour-induced angiogenesis to be studied more widely. Similar work has recently been carried out in mice by Mahadevan et al. (1990) and in hamsters by Thorpe et al. (1991).

LIMITATIONS

The limitations of this system are: (i) Animals have to be kept singly to prevent damage to the cannulae. Therefore, the model uses up experimental animal accommodation. (ii) Unlike the corneal implant assay and chorioallantoic membrane assay, gross inspection of sponge neovascularisation is not possible. (iii) The technique is based on the assumption that the neovasculature in the sponge is responsible for alterations in blood flow. However, local haemodynamics can also be influenced by other factors such as the properties of the extravascular compartment within which the ^{133}Xe is injected; the capacity of lymphatic drainage. The cellular composition (Ford et al., 1989; 1990) of the sponges may vary according to the treatment received and that the quantity and quality of the extracellular matrix (Kurkinen et al., 1980) can be different. Thus it is important to carry out supporting histological studies on the sponge sections.

CONCLUSION

It is clear that no totally predictable models of human angiogenic diseases exist. In practice, investigators use a battery of different models (in vivo and/or in vitro) to answer specific questions. In addition to quantitation of blood vessel formation, models involving subcutaneous implants (plastic chambers, steel mesh chambers and sponges) provides an opportunity for measuring the levels of biological response modifiers (e.g. prostaglandins, cytokines) and cellular components. However, such analyses can only be carried out post mortem. Incorporation of the ^{133}Xe clearance technique for rapid measurements of blood flow in sponge implants makes this model a simple and objective method for routine studies of angiogenesis modifiers. Biochemical analysis of sponge fluids, immunohistochemistry, morphometric evaluation of sponge sections and perhaps in situ hybridisation for the detection of growth factors and inhibitors will increase our understanding of the molecular mechanisms of angiogenesis.

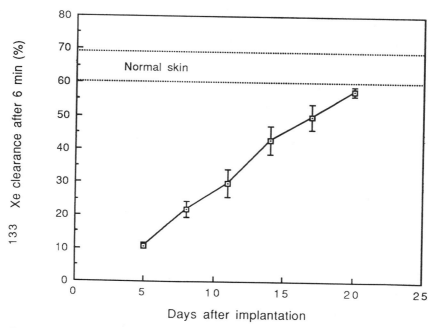

Fig.10   Time course of sponge neovascularisation in mice as measured
by the 6 min ^{133}Xe clearance technique. Data represent mean $\pm$
s.e.m. n = 10.

ACKNOWLEDGEMENTS

This work was supported by the British Heart Foundation and the Wellcome
Trust. We thank Drs J. Saklatvala, D. West, S. Kumar and Professor M.
Maragoudakis for the gifts of IL-1α, TNF-α, hyaluronic acid fragments, D609 and
GPA 1734 respectively.

REFERENCES

Andrade, S.P., Fan, T.-P.D., and Lewis, G.P., 1987, Quantitative in-vivo studies
on angiogenesis in a rat sponge model, Br. J. exp. Path., 68:755.
Auerbach, R., Kubai, L., Knighton, D., and Folkman, J., 1974, A simple
procedure for the long term cultivation of chicken embryos. Dev. Biol.,
41:391-394.
Broadley, K.N., Aquino, A.M., Woodward, S.C., Buckley-Sturrock, A., Sato, Y.,
Rifkin, D.B., and Davidson, J.M., 1989, Monospecific antibodies implicate
basic fibroblast growth factor in normal wound repair, Lab. Invest.,
61:571-575.
Buckley, A., Davidson, J.M., Kamerath, C.D., and Woodword, S.C., 1987,
Epidermal growth factor increases granulation tissue formation dose
dependently, J. Surg. Res., 43:322-328.
Buckley, A., Davidson, J.M., Kamerath, C.D., Wolt, T.B., and Woodword, S.C.,
1985, Sustained release of epidermal growth factor accelerates wound
repair, Proc. Natl. Acad. Sci., USA; 82:7340-7344.
Ciomei, M., Pesenti, E., Sola, F., Pastori, W., Mariani, M., Grandi, M., and
Spreafico, F., 1991, Antagonistic effect of suramin on bFGF: in vitro and
in vivo results, Int. J. Radiat., 60:78.

Crum, R., Szabo, S., and Folkman, J., 1985, A new class of steroids inhibits angiogenesis in the presence of heparin or a heparin fragment, Science, 230:1375-1378.

Davidson, J.M., Klagsbrun, M., Hill, K.E., Buckley, A., Sullivan, R., Brewer, P.A., and Woodward, S.C., 1985, Accelerated wound repair, cell proliferation, and collagen accumulation are produced by a cartilage-derived growth factor, J. Cell Biol., 100:1219.

Dvorak, H.F., Harvey, V.S., Estrella, P., Brown, L.F., McDonagh, J., and Dvorak, A.M., 1987, Fibrin containing gels induce angiogenesis. Implications for tumor stroma generation and wound healing, Lab. Invest., 57:673-686.

Fajardo, L.F., Kowalski, J., Kwan, H.H., Prionas, S.D., and Allison, A.C., 1988, The disc angiogenesis system, Lab. Invest., 58:718-724.

Fan, T.-P.D., and Hu, D.-E., 1991, Modulation of angiogenesis by inflammatory polypeptides, Int. J. Radiat. Biol., 60:71.

Fan, T.-P.D., Hu, D.E., Smither, R.L. and Gresham, G.A., (in press) Further studies on angiogenesis in a rat sponge model, in: "Angiogenesis," P.B. Weisz., R. Langer and R. Steiner, ed., Karger AG, Basel, New York and Tokyo.

Fernandez, L.A., Twickler, J. and Mead, A., 1985, Neovascularization produced by angiotensin II, J. Lab. Clin. Med., 105:141-146

Folkman, J., Merler, E., Abernathy, C., and Williams, G., 1971, Isolation of a tumor factor responsible for angiogenesis, J. Exp. Med., 133:275-288.

Ford, H.R., Hoffman, R.A., Wing, E.J., Magee, D.M., McIntyre, L.A., and Simmons, R.L., 1989, Characterization of wound cytokines in the sponge matrix model, Arch. Surg., 124:1422-1428.

Ford, H.R., Hoffman, R.A., Wing, E.J., Magee, D.M., McIntyre, L.A., and Simmons, R.L., 1990, Tumor necrosis factor, macrophage colony-stimulating factor, and interleukin 1 production within sponge matrix allografts, Transplantation, 50:460-466.

Gimbrone, M.A. Jr., Cotran, R.S., Leapman, S.B., and Folkman, J., 1974, Growth and neovascularisation: an experimental model using the rabbit cornea, J. Natl. Cancer Inst., 52:413-427.

Greenblatt, M., and Shubik, P., 1968, Tumor angiogenesis; transfilter diffusion studies in the hamster by the transparent chamber technique, J. Natl. Cancer Inst., 41:111-124.

Haegerstrand, A., Dalsgaard, C.-J., Jonzon, B., Larsson, O. and Nilsson, J., 1990, Calcitonin gene-related peptide stimulates proliferation of human endothelial cells, Proc. Natl. Acad. Sci. USA, 87:3299-3303.

Haynes, W.L., Proia, A.D. and Klintworth, G.K., 1989, Effect of inhibitors of arachidonic acid metabolism on corneal neovascularization in the rat, Invest. Ophthalmol. Vis.Sci., 30:1588-1593.

Hiley, C.R., Yates, M.S., Roberts, P.J., and Bloom, A.E., 1980, Alterations in liver blood flow during glycerol-induced acute renal failure in the rat, Nephron, 26:244-248.

Hu, D.E., and Fan, T.-P.D., 1991, Synergistic interaction between bradykinin and interleukin 1 in angiogenesis, Br. J. Pharmacol., 104:83P.

Ingber, D., Fujita, T., Kishimoto, S., Sudo, K., Kanamaru, T., Brem, H., and Folkman, J., 1990, Synthetic analogues of fumagillin that inhibit angiogenesis and tumor growth, Nature, 348:555-557.

Kimball, E.S., and Fisher, M.C., 1988, Potentiation of IL-1-induced BALB/3T3 fibroblast proliferation by neuropeptides, J. Immunol., 141:4203-4208

Kurkinen, M., Vaheri, A., Roberts, P.J., and Stenman, S., 1980, Sequential appearance of fibronectin and collagen in experimental granulation tissue, Lab. Invest., 43:47-51.

Kusaka, M., Sudo, K., Fujita, T., Marui, S., Itoh, F., Ingber, D., and Folkman, J., 1991, Potent anti-angiogenic action of AGM-1470: comparison to the fumagillin parent, Biochem. Biophys. Res. Comm., 174:1070-1076.

Le Noble, F.A.C., Hekking, J.W.N., Van Straaten, H.W.M., Slaaf, D.W. and Struyker Boudier, H.A.J., 1991, Angiotensin II stimulates angiogenesis in the chorioallantoic membrane of the chicken embryo, Eur. J. Pharmacol., 195:305-306.

Leibovich, S.J., Polverini, P.J., Shepard, H.M., Wiseman, D.M., Shively, V., and Nuseir, N., 1987, Macrophage-induced angiogenesis is mediated by tumour necrosis factor-alpha, Nature, 329:630-632.

Mahadevan, V., Hart, I.R., and Lewis, G.P., 1989, Factors influencing blood supply in wound granuloma quantitated by a new in vivo technique, Cancer Res., 49:415-419.

Mahadevan, V., Malik, S.T.A., Meager, A., Fiers, W., Lewis, G.P., and Hart, I.R., 1990, Role of tumour necrosis factor in flavone acetic acid induced tumour vasculature shutdown, Cancer Res., 50:5537-5542.

Maragoudakis, M.E., Sarmonika, M., and Panoustacopoulou, M., 1988, Inhibition of basement membrane biosynthesis prevents angiogenesis, J. Pharmacol. Exp. Ther., 244:729-733

McDevitt, D.G., and Nies, A.S., 1976, Simultaneous measurements of cardiac output and its distribution with microspheres in the rat, Cardiovasc. Res., 10:494-498.

Motro, B., Itin, A., Sachs, L., Keshet, E., 1990, Pattern of interleukin 6 gene expression in vivo suggests a role for this cytokine in angiogenesis. Proc. Natl. Acad. Sci., USA, 87:3092-3096.

Parnham, M.J., De Leve, L.D., and Saxena, P.R., 1979, Development of enhanced blood flow responses to prostaglandin E1 in carrageenan-induced granulation tissue, Agents and Actions, 9:510-515.

Plunkett, M.L., and Hailey, J.A., 1990, An in vivo quantitative angiogenesis model using tumour cells entrapped alginate, Lab. Invest., 62:510-517.

Presta, M., Rusnati, M., Statuto, M., Maier, J.A.M., Gualandris, A., Pozzi, A., and Ragnotti, R., 1991, Basic fibroblast growth factor (bFGF) and endothelial cells: receptor interactions, signal transduction, and cellular response. This issue.

Richardson, T.C., Humphreys, J.A.H., and Townsend, K.M.S., 1987, Subcutaneous implantation of double velour Dacron into the mouse: infiltration and angiogenesis, Br. J .exp. Path., 68:359-366.

Robertson, N.E., Discafani, C.M., Downs, E.C., Hailey, J.A., Sarre, O., Runkle, R.L.Jr., Popper, T.L., and Plunkett, M.L., 1991, A quantitative in vivo mouse model used to assay inhibitors of tumor-induced angiogenesis, Cancer Res., 51:1339-1344.

Schreiber, A.B., Winkler, M.E., and Derynck, D., 1986, Transforming growth factor-alpha: a more potent angiogenic mediator than epidermal growth factor, Science 232:1250-1253.

Smaje, L.H., Noor, N.M., and Clough, G.F., 1988, Changing sensitivity to H_1 and H_2 receptor agonists in the growing vasculature, in: "Vascular Endotheium in Health and Disease," Shu Chien, ed., Advances in Experimental Medicine and Biology, Plenum, New York. vol 242, pp 145-150.

Smither, R.L., and Fan, T.-P.D., 1990, PAF antagonists inhibit angiogenesis in a rat sponge model, Br. J. Pharmacol., 99:87P.

Smither, R.L., and Fan, T.-P.D., 1991, Role of arachidonic acid metabolites in a rat sponge model of angiogenesis, Int. J. Radiat. Biol., 60:48.

Sprugel, K.H., McPherson, J.M., Clowes, A.W., and Ross, R., 1987, Effects of growth factors in vivo. I. Cell ingrowth into porous subcutaneous chambers, Am. J. Pathol., 129:601-613.

Takagi, Y., Fukase, M., Takata, S., Yoshimi, H., Tokunaga, O., and Fujita, T., 1990, Autocrine effect of endothelin on DNA synthesis in human vascular endothelial cells, Biochem. Biophys. Res. Comm., 168:537-543.

Thompson, J.A., Anderson, K.D., DiPietro, J.M., Zweibel, J.A., Zametta, M., Anderson, W.F., and Maciag, T., 1988, Site-directed neovessel formation in vivo, Science, 241:1349-1352.

Thompson, W.D., and Brown, F.I., 1987, Quantitation of histamine-induced angiogenesis in the chick chorioallantoic membrane: mode of action of histamine is indirect, Int. J. Microcir. Clin. Exp., 6:343-357.

Thorpe, P.E., Wallace, P.M., Knyba, R.E., Watson, G.J., Mahadevan, V.A., Land, H., Yerganian, G., and Brown, P.J., 1991, Selective killing of proliferating endothelial cells by an anti-fibronectin receptor immunotoxin, Int. J. Radiat. Biol., 60:24.

Vallee, B.L., Riordan, J.F., Lobb, R.R., Higachi, N., Fett, J.W., Crossley, G., Buhler, R., Budzik, G., Breddam, K., Bethune, J.L., and Alderman, E.M., 1985, Tumor-derived angiogenesis factors from rat Walker 256 carcinoma: an experimental investigation and review, Experientia,41:1-15.

Wang, D.H., and Prewitt, R.L., 1990, Captopril reduces aortic and microvascular growth in hypertensive and normotensive rats, Hypertension, 15:68-77

West, D.C., Hampson, I.N., Arnold, F., and Kumar, S., 1985, Angiogenesisinduced by degradation products of hyaluronic acid, Science, 228:1324-1326.

Woll, P.J., and Rozengurt, E., 1989, Neuropeptides as growth regulators. Br. Med. Bull., 45:492-505.

Ziche, M., Morbidelli, L., Pacini, M., Geppetti, P., Alessandri, G. and Maggi, C.A., 1990, Substance P stimulates neovascularisation in vivo and proliferation of cultured endothelial cells, Microvasc. Res., 40:264-278.

NEW APPROACHES TO THE MEASUREMENT

OF PROLIFERATION RATES

Juliana Denekamp

CRC Gray Laboratory
Northwood
Middlesex, U.K.

INTRODUCTION

The field of cell proliferation kinetics has expanded enormously since the original description in 1953 that DNA synthesis in preparation for the next mitosis occupies a discrete portion of the intermitotic life cycle. Howard & Pelc (1953) defined the phases of the cell cycle, i.e. the postmitotic gap (G1), DNA synthesis (S) and the premitotic gap (G2). This nomenclature has become very commonly adopted. Specific precursors of DNA have been developed which allow cells in this phase of DNA synthesis to be labelled. The most extensively used has been radioactive tritiated thymidine, but this is now being replaced by the non-radioactive halogenated pyrimidines BUdR and IUdR. The radioactive precursors were detected in autoradiographs by the grains they induce in a photographic emulsion as the isotope decays. The overall uptake can also be detected in a scintillation counter. By contrast the BUdR or IUdR labelled cells can be detected with specific monoclonal antibodies, which can then be tagged with coloured or fluorescent stains and identified in histological sections or by flow cytometry (Gray et al. 1986).

It would appear at first sight that proliferation rate is an easy thing to define. But in practice there are several parameters that could all be construed or misconstrued as measures of this rate (Figure 1).

The cell cycle time, T_c, is the time taken to go from one mitosis to the next, and the reciprocal of this is the rate of cell division, <u>amongst the proliferating cells</u>. This qualifying phrase is important. In some cell populations all cells may be actively engaged in cell production, i.e. the growth fraction is unity. However, in other populations, some cells are differentiated, doomed to be sterile, or in the quiescent G_o phase. In this case, cell production in the <u>total</u> population is slower than that measured for the cycling cells. This difference leads to the concept of a turnover time, T, (if no nett growth is occurring and 1 cell is lost for each new cell born) or a potential doubling time, T_{pot}. This is the reciprocal of the birth rate. If all cells are dividing $T_{pot} = T_c$.

If no cells are being lost from the population the potential doubling time will equal the volume doubling time (T_d) of the tumour, organ or culture population. However, in practice, cells are often dying, being shed or actively leaving the tissue and thus the volume

Angiogenesis in Health and Disease, Edited by M.E. Maragoudakis *et al.*
Plenum Press, New York, 1992

growth rate may be much slower than the potential growth rate. In normal adult epithelial tissues cell loss and production are exactly balanced. In many carcinomas, which have derived from epithelia, the cell loss characteristic is partially retained and 90-95% of the cell production does not contribute to local growth, but is lost from the tumour volume. By contrast, connective tissues, which give rise to sarcomas, have very little production or loss in the steady state condition. In the tumours arising from these tissues, cell loss is also a less dominant characteristic.

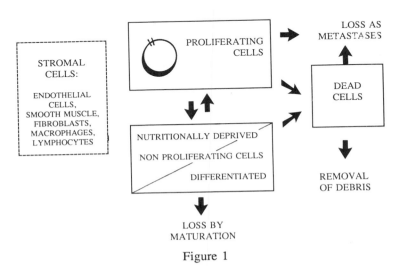

Figure 1

The proliferation kinetic characterisation of a tumour or normal tissue involves measurements of both production and loss of cells. Usually only a fraction of the cell population is actively engaged in proliferation (i.e. the growth fraction). Cell loss can occur from either proliferating or non proliferating pools and in many instances the transition from proliferating to non proliferating is reversible. A separate analysis of the parenchymal cells and of the stromal cells, e.g. endothelium, is needed for a full description of the tissue.

To obtain a full kinetic description of a tissue, it is necessary to measure the duration of the cell cycle, and its constituent phases, the growth fraction and the cell loss rate. This is very labour intensive and time consuming and requires multiple specimens from the tissue over a protracted time (see Steel 1979; Denekamp 1982; Gray + Darzynkiewicz 1986).

For many experimental purposes it is enough to calculate the cell production rate, or its reciprocal, the turnover time or potential doubling time. This can be done from two simple parameters, the fraction of cells in DNA synthesis at any instant, LI, and the duration of the DNA synthetic phase, T_S. The labelling index, LI, can be quite readily obtained with either the radioactive ^{3}HTdR or the immunologically recognisable precursors, BUdR or IUdR, either in autoradiographs, in histochemically stained sections or using fluorescent tags in a flow cytometer.

Measurement of T_S is more difficult. It was originally obtained with serial samples by following a cohort of labelled cells through mitosis, using the percent labelled mitosis technique (Steel 1979; Denekamp 1982). Since T_S is commonly between 8 and 15 hours, frequent samples must be taken, e.g. hourly for at least 15-20 hours. This is tedious and time consuming, especially since the autoradiographs must be exposed to the decaying isotope for several weeks or months.

An alternative approach was to use two different labels to identify how many extra cells take up the second label if a fixed interval, e.g. 3-4 hours is allowed between the application of the two precursors. In the past, this has been done with ^{3}H and ^{14}C labelled thymidine, using a 2-layered emulsion technique to distinguish the short range emissions of ^{3}H (located very close to the source material) from the long range emission from ^{14}C (which give a halo around the nucleus in the thicker more distant emulsion). This is technically difficult and hence has never become widely used. Using BUdR and IUdR, monoclonal antibodies have been developed that are sufficiently specific to distinguish between them. The same double labelling approach can therefore be used histochemically. Again this has not been widely adopted.

A dramatic step forward was taken in the mid 1980s when it was realized that flow cytometry could give both LI and T_S from a single sample (Begg et al. 1985). The technique is quick, easy and does not require the administration of a radioactive isotope. It can now be routinely performed on human material, which can be fixed in alcohol at the time of excision, and then sent to an appropriate flow cytometry laboratory (Wilson et al. 1988; Begg 1989).

The technique depends upon the simultaneous recognition of two parameters as each individual cell passes through the laser beam of a sophisticated flow cytometer. A specific stain for DNA is used (ethidium bromide or propidium iodide) which then fluoresces in proportion to the DNA content of that cell. Cells in G_2 fluoresce exactly twice as brightly as those in G_1 because their DNA content has doubled in preparation for mitosis. BUdR is supplied to the cells before the sample is taken, and the monoclonal antibody recognises only those cells which have taken up the BUdR (or IUdR). These appear as a cloud of high fluorescence for the second colour and lie uniformly between G_1 and G_2 if the sample is taken soon after labelling. If however an interval of several hours elapses between labelling and removing the sample, these labelled cells will all have increased their DNA content and therefore moved towards the level characteristic of G_2 cells. This relative movement is the feature that can be used to estimate the duration of S. The median DNA content of the BUdR labelled cells should be halfway between G_1 and G_2 at time zero, and its actual value can be determined at time = 4 hours (for example). By extrapolation from 0.5 through this value, it is possible to predict when all labelled cells would have reached G_2, i.e. the duration of S. (Begg et al. 1985).

This technique has revolutionised the clinical possibilities for cell kinetics. It is now possible to give a low dose of BUdR or IUdR (200 mg) to a patient, take a single biopsy 4-6 hours later and within 24 hours have the information needed to calculate the potential doubling time of the constituent cells in the biopsy. If tumours are being studied, a further

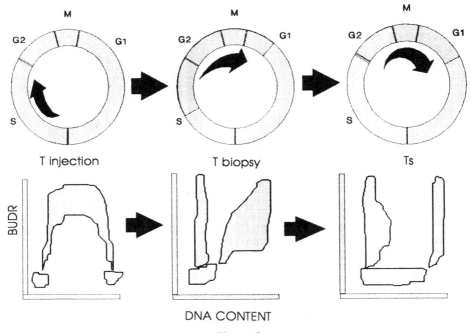

Figure 2

Illustration of the method by which the labelling index and duration of DNA synthesis can be assessed from a single biopsy. Cells labelled with BUdR are uniformly distributed between G_1 and G_2 at the time of injection (left hand panel). After a time equal to T_s, all of the BUdR labelled cells have moved into G_2 and some have even moved through mitosis into G_1 (right hand panel). The extent to which this movement has occurred at the time of the biopsy (centre panel) is used to estimate T_s. (Schematic diagram kindly provided by G.D. Wilson, 1991). For details of this relative movement technique see Begg et al. 1985.

advantage is gained if the tumour is aneuploid, for then it is possible to distinguish between the diploid normal cells and the hyperploid tumour cells. So far it has not been possible to use an additional antibody to recognise a cellular subtype, e.g. endothelial cells within a tumour, but with the more sophisticated flow cytometers it is likely that this will be possible using a three parameter analysis instead of the present dual parameters. For any cell culture or pure preparation of endothelial cells, the existing technique should be easy to apply.

REFERENCES

Begg, A.C., McNally, N.J., Shrieve, D.C., Kärcher, H. A method to measure the duration of DNA synthesis and the potential doubling time from a single sample. Cytometry 6: 620-626. 1985.

Begg, A.C. Derivation of cell kinetic parameters from human tumours after labelling with bromodeoxyuridine or iododeoxyuridine. Brit.J.Radiol. Report 19: 113-119. 1989.

Denekamp, J. Cell kinetics and Cancer Therapy, pp.1-162. Pub. C.C. Thomas, USA. 1982.

Gray, J.W., Darzynkiewicz (editors). Techniques in Cell Cycle Analysis, pp.1-407. Humana Press. Clifton N.J., USA. 1986.

Howard, A. and Pelc, S.R. Synthesis of desoxyribonucleic acid in normal and irradiated cells and its relation to chromosome breakage. Heredity, suppl.6: 261-273. 1953.

Steel, G.G. Growth kinetics of tumours. Oxford University Press. 1977.

Wilson, G.D., McNally, N.J., Dische, S., Bennett, M.H. Cell proliferation in human tumours measured by *in vivo* labelling with bromodeoxyuridine. Brit.J.Radiol. 61: 419-422. 1988.

"FACTITIOUS ANGIOGENESIS": ENDOTHELIALIZATION OF ARTIFICIAL CARDIOVASCULAR PROSTHESES

Peter I. Lelkes, Mark M. Samet, Carl W. Christensen[#] and David L. Amrani[*]

Lab. Cell Biol. and [#] Lab. Cardiovasc., Dept. Medicine,
Univ. Wisc. Med. School, MCC, and
[*]Dept. Allied Health Sci. Univ. Wisc., Milwaukee, WI

Introduction: There is an increasing demand for permanent cardiovascular prostheses, such as vascular grafts, cardiac assist devices or total artificial hearts. Currently, their longterm usage is thwarted by a number of divergent problems, such as thromboembolism, or neointimal hyperplasia. These complications can be frequently traced to inadequate hemocompatibility of the synthetic, blood contacting surfaces. This problem may be ameliorated by *factitious angiogenesis*, namely the generation of nonthrombogenic, artificial blood conduits by precoating their luminal surfaces with a functional monolayer of autologous endothelial cell. In this presentation we will discuss the endothelialization of three cardiovascular prostheses: small caliber vascular grafts, artificial blood pumps, and autologous skeletal muscle ventricles.

Endothelialization of small caliber vascular grafts: The high rate of failure of synthetic, small caliber vascular grafts ($\leq$ 6mm inner diameter) impedes their use as arterial substitutes. Rapid occlusion of many grafts occurs due to either acute macroscopic clot formation and/or massive, neointimal hyperplasia[1]. Traditionally, these complications have been attributed to the inadequate hemocompatibility of the synthetic, blood contacting surfaces, caused by the intrinsic thrombogenicity of synthetic biomaterials [2]. These arguments explain the activation of the blood coagulation system and consequent thrombus formation, but they fail to rationalize the good patency rate of larger grafts (>6 mm inner diameter), when made from the same material, as a failing small caliber graft. The problem of vascular graft patency is complex and related, in part, to the limited adaptivity of man-made vascular walls to local hemodynamics[3,4]. Another cause for graft failure might be the mismatch in vessel wall compliance, in particular at the sites of anastomosis between natural and synthetic blood conduits[5]. The thrombogenic potential of such "disturbances", is much more deleterious in small vessels, than in the large arteries.

In addition to fluid mechanical considerations, the insufficient hemocompatibility of synthetic blood conduits is the result of the lack of a functional endothelial cell(EC) lining[6]. In vivo, the ECs, lining the luminal surface of the entire vasculature (including the heart), play a pivotal role in maintaining the hemostatic balance between procoagulant and anti-coagulant properties of the blood stream. Under non-pathological conditions, ECs are quiescent and express a thromboresistant surface with minimal fibrin deposition and platelet adhesion. In pathological situations, such as trauma, inflammation or infection, ECs become activated and their surface becomes thrombogenic, e.g. via enhanced platelet adhesion and activation. Indeed, the surface of activated ECs plays a pivotal, catalytic role in promoting blood clotting [7]. If the endothelial lining is damaged or missing, for example, in atherosclerosis [8] or after injury during angioplasty [9], the subendothelial basement membrane presents an extremely thrombogenic surface.

The concept of "endothelialization" was introduced, about 15 years ago [10], at the time, when the importance of endothelial cells for attaining and maintaining a hemostatically inert blood vessel wall was recognized. However, at the same time, the art of procuring and culturing endothelial cells was in its infancy. In particular, nothing was known about EC-biomaterial compatibility. Therefore, initial experiments were aimed at understanding the basic requirement for growing ECs on biopolymeric materials, such as polytetrafluoroethylene (PTFE), polyurethane (PU) and Dacron. These studies yielded information on the biological factors and physical forces that govern the interactions of cultured endothelial cells with polymeric surfaces under static conditions, such as short term attachment and long term growth[2,11,12].

Since a (static) cell culture dish is an inadequate representation of the diverse hemodynamics within the vascular tree, attention has turned to examining the effects of hemodynamic parameters, such as pulsatile blood-flow and cyclic deformation as effectors of EC biology[13-15]. Some of the pertinent findings can be summarized as follows:

The extent and kinetics of EC attachment to various biomaterials depends on the wettability of the surface and the presence of adhesive proteins, such as fibronectin, laminin, and various types of collagens[2,11,12]. The initial adhesion, besides being partially electrostatic in nature, is mediated in a complex fashion by receptors of the integrin family[16-18], and shows predilection for certain substrates (e.g. fibronectin > > laminin). For long-term attachment and growth, however, there seems to be no particular substrate preference[19], presumably since the cells synthesize and deposit their own adhesive, basement membrane proteins[20].

These in vitro studies did not yield a consensus as to the optimal approach for endothelializing vascular grafts[11,12]. During the past decade, numerous preclinical and clinical studies reported a certain "semiprogress" in maintaining graft patency and/or reduced thrombogenicity. Progress was also hampered by the limited availability of autologous endothelial cells, which until recently were harvested from larger vessels and seeded at low densities onto the grafts[11,12].

A breakthrough was achieved upon developing techniques suitable for an operating room setting and has lead to the clinical realization of endothelialization of vascular grafts[21,22]. There are two essential components that contributed to this breakthrough: 1.) Identification of a suitable source for harvesting large numbers of autologous ECs, namely subcutaneous adipose tissue, which is easily accessible and harbors large numbers of microvascular ECs[23], and 2.) Initial "Sodding" of the cells at high densities, so that a confluent monolayer is achieved almost instantaneously prior to implantation[24,25].

Albeit this procedure is still experimental, several centers around the world have successfully implanted endothelialized grafts, using similar techniques[22,26]

Based on these initial, promising, results, the problem of endothelializing small caliber grafts seems to have been solved. However, it is unclear whether these techniques are adequate to provide long term solutions. This cautionary note is certainly appropriate, given the fact that autologous grafts are prone to biodegradation, late stenosis, atherosclerosis and, aneurysm [27].

If the initial promise is fulfilled, endothelialization of vascular grafts may solve several major problems in cardiovascular surgery. Furthermore, endothelialized vascular grafts could be used to implant genetically engineered endothelial cells[28]. For example, genes for missing enzymes, or deficient blood clotting (co)factors (such as Factor VIII or tPA), could be introduced into ECs by stable transfection. Thus, vascular grafts, endothelialized with genetically altered ECs, could become useful vehicles for gene therapy[29].

Endothelialization of blood contacting surfaces in devices used for cardiac assist and replacement. Thromboembolic complications, such as strokes and multiple organ failures continue to limit the permanent or even temporary use (as a bridge-to-transplantation) of currently available artificial cardiac prostheses[30,31]. Thromboembolism is believed to be caused by the intrinsic thrombogenicity of the blood-contacting prostate surfaces, which in most blood pumps are made out of polyurethane (PU)-based elastomers, such as Biomer®. These polymers combine excellent mechanical stability (the devices are supposed to last several years, with over 30 million "beats"/year) and relative benign in vitro thrombogenic potential (e.g in terms of platelet adhesion/activation). However, once these surfaces are in contact with circulating blood in vivo, they cause fibrin deposition and platelet activation[32]. Inadequate hemocompatibility leads to either massive local thrombus formation inside the blood chambers, or to the generation of thromboembolic showers. Circulating thrombi can cause stroke and multiple organ failure[30,31].

The improvement of the blood compatibility of synthetic biomaterials is a major research focus[32,33]. Chemical modifications of the base materials resulted in changes in bulk properties or physico-chemical surface characteristics and reduced the thrombogenicity of the blood contacting surfaces[33]. An alternate approach, successfully applied for improving graft patency, is to endothelialize the luminal surface of artificial blood pumps. While the use of endothelialized grafts is on the brink of clinical feasibility[22], little progress has been reported in passivating the luminal surface of a complex cardiovascular prosthesis, such as a total artificial heart, or a ventricular assist device using autologous EC. Early attempts at endothelialization were hampered by the lack of an adequate source for ECs and/or inadequate in vitro EC culture techniques[10-13].

Recent advances in both areas have encouraged us and others, to tackle the problem of device-related thrombogenicity by endothelialization, using autologous microvascular ECs[15,34]. The feasibility of our approach is epitomized by the pioneering experiments of W.F. Bernard and colleagues, who successfully lined the blood sac of a ventricular assist device in vitro with fetal fibroblasts and, upon implantation, could demonstrate retention of the cells and formation of a thrombus-free neointima[35]. However, the use of autologous EC might be preferable, since allogeneic cells might evoke immune responses.

In practical terms, endothelialization of an artificial heart requires lining the luminal surface of a blood pump with a functional endothelial cell monolayer, capable of withstanding pulsatile flow-induced shear stress, cyclic deformations of the polymeric substrate and oscillating blood pressures. A prerequisite for successful endothelialization of an artificial heart is to re-create in vitro these dynamic conditions, and to study the cellular and molecular mechanisms of EC adaptation to a "hemodynamically correct" environment[15].Indeed, in vitro studies have revealed distinct, dynamic EC responses to flow-induced shear stress or mechanical deformation[36-39]. Our approach to evaluating EC adaptation to a complex dynamic environment is exposure of the cells to the combined effects of hydrostatic pressure, shear stress and mechanical strain. Cellular responses are compared to cells exposed to each of the variables alone. EC properties evaluated include morphology, mechanisms of signal transduction, and gene/protein expression. We believe that only by exposing the ECs to a combination of experimental variables can we fully account for the dynamic environment inside the luminal sac of a beating, artificial heart. Some of our concepts, model systems and results have been presented elsewhere[15,40]. In the following, we will briefly describe the techniques we use and discuss newer results.

Since our goal is to endothelialize an artificial blood pump for human use, most of our experiments were performed with human microvascular endothelial cells (HAMVECs) obtained from subcutaneous fat, according to the techniques introduced by Jarrell, Williams and their colleagues[23-25]. Subcutaneous fat, obtained in large amounts by liposuction, renders this technique suitable for procuring autologous ECs from the prospective recipients of cardiovascular prostheses. For some experiments we have also used a recently isolated line of sheep cardiac endothelial cells (SHECs)[40].

Endothelial cell attachment and long-term growth were used to assess the suitability of a given polymer as the base material for blood sacs [15]. In long term studies (up to 11 days) we found that rapid and effective initial adhesion did not necessarily assure subsequent cellular growth. These data suggest that the surface coating of a given biomaterial might be important for establishing the EC monolayer, but that the physico-chemical nature of the biomaterials, underneath the adhesive protein-coating determines the long-term maintenance of the EC lining ("Shine-through effect"). Furthermore, we found that the substratum which provides for optimal growth of the ECs, is also the one which optimally stimulates the synthesis and deposition a well-defined, complex extracellular matrix (not shown).

Corona treatment increases the wettability of biomaterials by conferring surface charges and results in enhanced EC growth[41]. We have compared the growth potential of SHECs seeded onto Corona treated vs. fibronectin coated suspension culture dishes, which, if untreated, provide a very poor substrate for culturing anchorage dependent cells. Following Corona treatment, the cells formed colonies of typical cobble stone morphology (**Figure 1a**). An identical morphology was attained when the cells were seeded onto dishes precoated with fibronectin (not shown). SHECs grow reluctantly on untreated suspension culture dishes. After 7 days in culture, only a few cell adhered and were mainly elongated in shape. (**Figure 1b**)

To study EC growth under flow conditions, we have built a series of flow-chambers which mimic, with increasing complexity and realism, the dynamic flow conditions inside an artificial ventricle [15,40]. We also have set up a complex flow visualization system to quantitatively analyze the flow kinetics in these chambers. We can directly correlate (macroscopic) flow kinetics and (microscopic) cellular responses in selected areas of the chambers. As we have previously described, endothelial cells growing in our axisymmetrical cell culture chambers, will become resistant to flow-induced shear stress, provided the cells are initially "sodded" at high enough densities and left, under static conditions, for ca. 6 days, to develop a complex basement membrane[15,40]. In addition, after approx. 24 hours, the cells were found to be oriented according to the local flow patterns.

The durability of the EC monolayer was reduced, when the cells were exposed to pulsatile

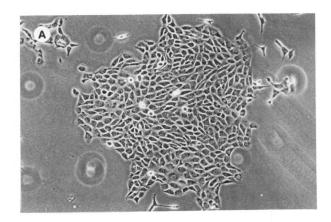

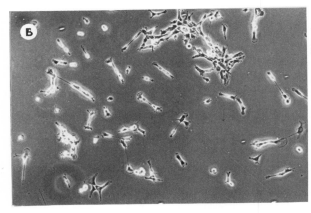

Figure 1: Effect of Corona treatment on endothelial cell growth on polystyrene suspension culture dishes. 1000 Sheep heart endothelial cells were seeded in 35 mm into either corona treated (A) or untreated suspension culture dishes. The phase contrast micrographs were taken after 7 days in culture. Original magnification 100 x.

flow prior to 6 days in static culture. The localization of lesions to the EC monolayer was related to the local flow field. Shown in **Figure 2a** are monolayers of SHECs that were cultured under static conditions for 48 hours prior to exposing them for another 20 hours to pulsatile flow at 100 ml/min. The endothelial cell monolayer in the center of the chamber albeit randomly oriented, in line with the complex flow patterns observed [15,40]. Similar endothelial cell disorientation by turbulent flow has been reported by others[42]. By contrast, in the area of the recirculating eddy [40], the integrity of the endothelial cell monolayer is compromised (**Figure 2b**). Obviously these areas of recirculation and stagnation are "danger zones", in which endothelial denudation might lead to platelet deposition and activation.

In designing appropriate cell culture chambers, attention has to be paid to the slightest detail, which may invalidate the chosen model system. For example, we experienced our own "Space-Shuttle-Syndrome". Our flow-through culture systems were initially fitted with "medical grade" O-rings. These O-rings, made out of silicone, might be classified as biocompatible, however, we observed that they were cytolytic for endothelial cells. We established that these O-rings, in spite of extended rinsing prior to use, were leaching out some as yet unidentified toxic substance, which rapidly (within 2-3 hours) caused the endothelial cells to detach and eventually to die (**Figure 3**).

Similarly, both the material used for the model chambers, and their size and conduit configuration is important. For example, we have described in detail the flow patterns in a particular growth chamber[40]. However, detailed analysis of the flow patterns in a scaled-down version of this chamber yielded significant differences in the trend of the major, parameters that characterized the nature of the flow field [15,40]. These findings (not shown) are instructive in that they emphasize the importance of using appropriate model systems for studying the actual problem.

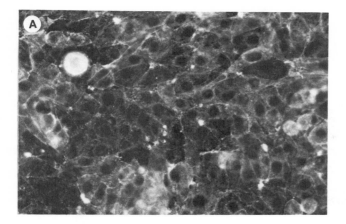

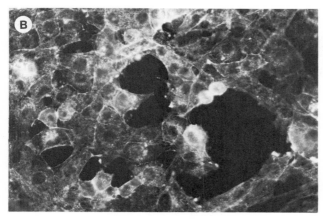

Figure 2. Effect of the flow field on the integrity of Sheep heart endothelial cell (SHEC) monolayers. SHEC were pre-cultured for 48 hours and the exposed to pulsatile flow (details see text). After 20 hours, the cells were fixed and their cytoskeletal arrangement visualized by staining with rhodamine phalloidine. A:Intact SHEC monolayer, at the center of the chamber B: Perturbed monolayer, at an area of the recirculation and stagnation. Original magnification: 630 x.

We have been studying the response of ECs to cyclic tensional deformations ("cell exercising") using a commercial product (Flexercell®), in which the cells are grown on flexible silastic membranes forming the bottom of 6 well plates[43,44]. By applying a computer-controlled cyclic regimen of negative pressure, the silastic membranes are deflected and relaxed in a cyclic fashion, generating a maximal stretching of 25% at the edges of the plates.

We established conditions for a long-term maintenance of confluent monolayers of HAMVEcs. If cyclic deformations commenced before the cells grew on the substratum under static conditions for at least 48 hours, most of the EC dislodged within the first hour of the experiment. Optimal retention of the EC monolayer was obtained if the cells were left to deposit their extracellular matrix for at least 6 days. To date, we have extended our durability tests for up to 30 days (> 2.5 million cycles).

Visual inspection of the plates indicated that during 10 days of stretching, HAMVECs in the areas of minimal flexion formed a dense confluent monolayer, exhibiting cobblestone morphology. By contrast, after only 24 hours of flexing, the cells in the periphery of the wells (areas of maximal flexing) had aligned perpendicular to the vector of deflection (not shown). Similar results have been presented for different ECs by other investigators[43]

Long-term effects of cyclic mechanical deformations on the thrombogenicity of ECs, were assessed by measuring the expression of several indicators of the procoagulant/anticoagulant properties of the EC surface. Quiescent EC, presenting an anticoagulant cell surface, express high levels of prostacycline (PGI_2)[39]. By contrast, "activation" of endothelial cells increases the procoagulant competence of the cell surface as assessed by an increase in the plasminogen activator

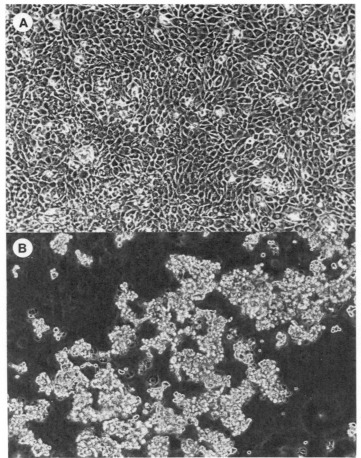

Figure 3. Cytotoxicity of "medical grade" O-rings. A medical grade O-ring was added to confluent monolayers of SHEC. After 20 hours the cultures were examined. A: SHEC, control; B: in the presence of the O-ring. The phase contrast micrograph was taken at a site distal to the O-ring, demonstrating the diffusible nature of the toxic compound(s). Original magnifications 100 x

inhibitor (PAI-1) levels [45,46]. To evaluate their adaptation to the prolonged regimen of cyclic deformations, HAMVECs were seeded in the "Flexercell" system onto flexible bottom plates, coated with various adhesive proteins. After "exercising" for up to 14 day at a frequency of 1 Hz, the cells were harvested periodically. The expression of PAI-1 and PGI_2 was normalized to age matched control cells, grown on identical substrates under static conditions.

When the cells were grown on silicone surfaces, either covalently modified with amine groups or coated with collagen type I, PGI_2 production was gradually enhanced, by the active environment, albeit to a moderate extent only. (**Figure 4a**). By contrast, cells growing on fibronectin, started to significantly increase their PGI_2 output after 6 days of "exercising". After 14 days the relative amount of PGI_2 production was more than 30-fold above that of the (static) control cells. Interestingly, plating the cells onto a more complex membrane, comprised of both fibronectin and collagen type I, synergistically enhanced the augmentation of PGI_2 production after 6 days of flexing.

Concomitantly with assaying for PGI_2, we also measured PAI-1 expression in the same endothelial cell cultures, plated onto fibronectin and collagen. As seen in **Figure 4b**, exposure of the cells to cyclic deformations caused a dramatic transient increase in PAI-1 production, suggesting that the initial exposure of the cells to mechanical stress enhances the procoagulant properties of the cell surfaces. However, PAI-1 levels decrease gradually within 6-8 days to baseline. Indeed, after 14 days of "exercising" the cells had lower PAI-1 expression than their matched controls, kept under static conditions.

Taken together, these data demonstrate that monolayers of HAMVECs will attain a flow-

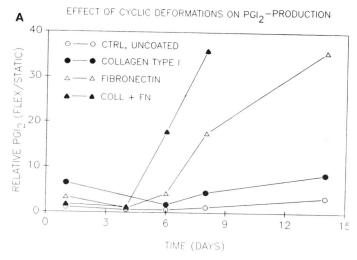

A EFFECT OF CYCLIC DEFORMATIONS ON PGI$_2$-PRODUCTION

RELATIVE PGI$_2$ (FLEX/STATIC)

TIME (DAYS)

○——○ CTRL, UNCOATED
●——● COLLAGEN TYPE I
△——△ FIBRONECTIN
▲——▲ COLL + FN

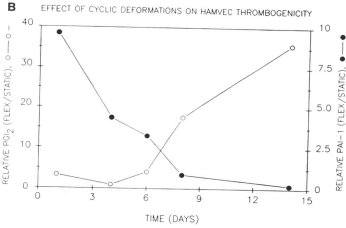

B EFFECT OF CYCLIC DEFORMATIONS ON HAMVEC THROMBOGENICITY

RELATIVE PGI$_2$ (FLEX/STATIC), ○——○

RELATIVE PAI-1 (FLEX/STATIC), ●——●

TIME (DAYS)

Figure 4. Effect of cyclic mechanicaldeformations on the production of modulators of fibrinolysis in human microvascular endothelial cells. A: Normalized prostacycline production of HAMVECS seeded onto different substrates. B: Comparison of the production of prostacycline and of tissue plasminogen activator inhibitor-type 1 in HAMVECS, seeded on collagen and fibronectin and flexed for 14 days. The data are presented after adjusting the PGI$_2$ and PAI-1 levels for protein concentrations/well and then normalizing to static conditions (details see text).

and flex- resistant configuration, provided they are given sufficient time (6 days) under static conditions. These 6 days might represent the time period required to synthesize extracellular matrix proteins and to deposit a functional basement membrane[47]. Furthermore, after growing for 6 days in static culture, monolayers of HAMVECs will adapt to cyclic deformations by assuming an anticoagulant configuration, much like their endothelial counterparts lining the inside of a beating heart. We hypothesize, that the assembly of a functional, complex basement membrane might be a pivotal prerequisite for conferring to cultured endothelial cell monolayers both durability and anticoagulant properties.

Endothelialization of skeletal muscle pouches: An alternate form of cardiac assist is by use of the patient's own skeletal muscle to aid the heart in its mechanical contraction[48]. An autologous skeletal muscle has many advantages over cardiac transplantation and artificial cardiac pumps [49]. Skeletal muscle tissue can be trained by electrical stimulation to acquire typical characteristics of cardiac muscle[50]. Over the past 20 years, the feasibility of using the autologous *latissimus dorsi muscle* (LDM) as a ventricular assist device has been explored in both animals and humans[51,52].

Several possibilities exists, to use a appropriately trained skeletal muscle as an autologous blood pump, in order to assist or to replace the failing heart. One of the options is to generate a muscle pouch underneath the *latissimus dorsi* and to use this pouch for cardiomyoplasty, or as an autologous cardiac contractile assist device[53]. However, the potential of thrombus formation on the blood contacting surfaces, which is comprised of fibrous material and/or mesothelial cells[54,55], have so far restricted the long term clinical usage of such pouches as ventricular replacements. Most recently, skeletal muscle ventricles have been lined with autologous tissue, derived from either pleura or endocardium[52]. However, the significant incident of thrombus formation in some of these ventricles, warrants the continued search for an optimal, thromboresistant surface, which is an endothelial lining[52].

We followed the rational developed for vascular grafts and artificial blood pumps (see above, and seeded the LDM pouches with autologous Canine Adipose Microvascular Endothelial Cells (CAMVECs). Endothelial cells were harvested from subcutaneous fat, essentially following the procedures used by Williams and colleagues[24,25].

At the time of harvesting the adipose tissue, several sterile, cylindrical silicone mandrels were inserted under the belly of the LDM in order to generate fibrous pouches of mesothelial origin. Four to six weeks after the first operation, the dogs were reoperated on. One pouch was left as an untreated control. The other pouches were opened, the mandrels removed and the pouches either filled with growth medium or sodded (seeded at high density) with CAMVECs. The cells were allowed to settle and to adhere to the pouch surfaces for 20 min. The mandrels were reinserted and the pouches sutured. Four to six weeks after the second operation, the dogs were sacrificed and the area of the pouches was carefully excised, and prepared for light and electron microscopic inspection. To identify the cells lining the pouch surface, some segments were examined by indirect immunofluorescence In addition, some pouches were incubated at 37°C for 4 hours with growth medium containing 10 μg/ml DiI-acetylated LDL, prior to fixation.

In initial experiments we verified that placement of silicon mandrels underneath the LDM resulted in the generation of mesothelial, fibrous muscle pouches. The gross appearance of such a non-endothelialized LDM pouch (either untreated, or filled with growth medium only) is smooth and glistening white. Upon fixation and staining with Hematoxylin and Eosin, light microscopic inspection revealed that the pouches abut onto the striated muscle and consist mainly of thickened fibrous connective tissue, which is lined by a collagenous surface, occasionally interspersed with a few cells. Previous studies by others had suggested, that this lining was cellular and comprised of mesothelial cells. [54,55]. The blood supply underneath the mesothelial pouch surfaces is well developed (**Figure 5a**). Following endothelialization, the pouch surfaces were lined by a well defined cellular monolayer (**Figure 5b**). The presence of a cellular monolayer on the endothelialized surfaces was also confirmed by scanning (**Figures 5c and 5d**) and by transmission electron microscopy(**Figures 5e and 5f**). Particular striking in the ultrastructural studies is the preponderance of fibrous (collagenous) strands on the untreated surfaces, which confirms earlier observations[54,55], and supports the findings by others, that these collagenous surfaces of mesothelial origin might be exceedingly thrombogenic[52].

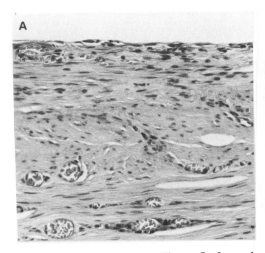 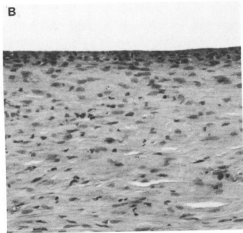

Figure 5. Legend on following page.

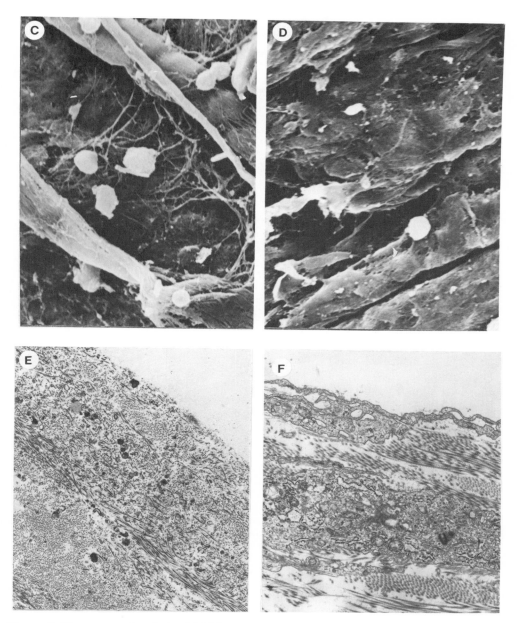

Figure 5. Ultrastructural analysis of LDM pouches by light and electron microscopy.
A and B:light micrographs of hematoxylin and eosin stained paraffin sections of an untreated (a) and an endothelialized pouch (b); **C and D**: non-endothelialized (c) and endothelialized (d) pouches; **E and F**:transmission electron micrographs of LDM pouches, transverse sections. (e) non-endothelialized control, note the fibrous, acellular tissue (f) endothelialized (note continuous thin endothelial cell on top of a mesothelial cell surface). Original magnifications: panels a,b: 200 x; panels c,d: 1,120 x, panels e,f: 6,000 x.

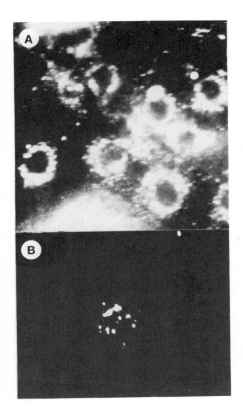

FIGURE 6. Identification of endothelial cells on the LDM pouch surfaces: Fluorescence micrographs of unfixed LDM pouches, incubated with diI-acetylated LDL, after image processing and background subtraction. A) endothelialized pouches; b) non-endothelialized control; original magnifications 400 x

To identify the nature of the cellular lining in the endothelialized pouches, we used several EC-specific markers. Shown in **Figure 6a** is a micrograph of an endothelialized pouch, that had been incubated for 4 hours with fluorescent DiI-acetylated LDL. The majority of the cells appear to have taken up this marker molecule, confirming the EC nature of the cellular lining. By contrast, virtually no fluorescence could be detected in non-endothelialized pouches (**Figure 6b**).

It is interesting to note, that the cells lining the endothelialized surfaces, appear to have attained some large vessel EC characteristics. For example, in tissue culture, CAMVECs do not show cobble stone morphology, and lose (due to in vitro dedifferentiation) staining for vWF. The cells lining the LDM pouches, expressed both attributes, namely cobble stone morphology and staining for vWF. At this point it is not clear, whether the cells lining the pouch surfaces are the same HAMVECs that were seeded, or their descendants. Alternately, endothelialization might have recruited the formation of a new endothelial cell lining.

Our results imply, that we might have solved the problem of how to endothelialize latissimus dorsi pouches[52]. Future studies will reveal, whether these endothelialized skeletal muscle ventricles can be used as thromboresistant autologous cardiac assist devices.

Concluding remarks: There are striking similarities in the experimental approaches to improve the hemocompatibility of different manmade blood conduits by endothelializing their blood contacting surfaces. This is the "good news", since one can adapt the successes and avoid the mistakes made in other systems. The "bad news" is, that for each type of cardiovascular prosthesis, special attention has to be paid to the individual hemodynamic constraints, which the endothelialized surfaces will be exposed to, upon implantation.

From the clinical point of view, successful endothelialization of synthetic, blood contacting surfaces would solve a host of complications, which to date have prevented the long term usage of manmade cardiovascular prostheses. From the vantage point of basic science, this avenue of research might reveal some clues about the role of hemodynamic factors in modulating the phenotypic heterogeneity of endothelial cell[56]. On the practical side, this work might eventually rekindle the quest for "factitious angiogenesis", i.e. generation of manmade blood conduits by extracorporeal reconstruction of the entire vessel wall from its parts, by coculturing e.g. ECs, pericytes, mesenchymal cells, smooth muscle cells and neurons[57,58].

Acknowledgements: This work was supported, in part, by grants in aid from the Milwaukee Heart Research Foundation (to P.I.L, C.W.C. and D.L.A) and by the American Heart Association, Wisconsin Affiliate (to P.I.L). We are grateful to Dr. B.R. Unsworth for critically reading and editing this manuscript, and to B. Haber for excellent secretarial assistance. Our work on the skeletal muscle pouches would have been impossible without the surgical skills of Drs. J.R. Edgerton and H. Gao. Our concept of endothelialization was inspired by the vision of Dr. R. Flemma, an eminent cardiovascular surgeon, who died, while on the waiting list for a cardiac transplant. We wish to dedicate this paper to his memory.

REFERENCES

1. Yeager, H. and Callow, A.D. Thrombin formation in small caliber grafts. *ASAIO Trans* 34:88-94, (1988).
2. Leonard, E.F., Turitto, V.T. and Vroman, L. *Blood in contact with natural and artificial surfaces*, N.Y.:New York Academy of Sciences, pp. 1-688, (1987).
3. Binns, R.L., Ku, D.N., Stewart, M.T., Ansley, J.P. and Coyle, K.A. Optimal graft diameter: effect of wall shear stress on vascular healing. *J Vasc Surg* 10:326-337, (1989).
4. Zarins, C.K., Zatina, M.A., Giddens, D.P., Ku, D.N. and Glagov, S. Shear stress regulation of artery lumen diameter in experimental atherogenesis. *J Vasc Surg* 5:413-420, (1987).
5. Kamiya, A. and Togawa, T. Adaptive regulation of wall shear stress to flow change in the canine carotid artery. *Am J Physiol* 239:H14-H21, (1980).
6. Gimbrone, M.A.,Jr. Vascular endothelium: nature's blood-compatible container. *Ann N Y Acad Sci* 516:5-11, (1987).
7. Stern, D.M., Handley, D.A. and Nawroth, P.P. Endothelium and the regulation of coagulation. In: *Endothelial cell biology in health and disease*, edited by Simionescu, N. and Simionescu, M. New York: Plenum Press, p. 275-306,(1988)
8. Ross, R. Endothelial injury and atherosclerosis. In: *Endothelial cell biology in health and disease*, edited by Simionescu, N. and Simionescu, M. N.Y.: Plenum Press, p. 371-384, (1988)
9. Popma, J.J. and Topol, E.J. Factors influencing restenosis after coronary angioplasty. *Am J Med* 88:16N-24N, (1990).
10. Mansfield, P.B., Wechezak, A.R. and Sauvage, L.R. Preventing thrombus on artificial vascular surfaces: true endothelial cell linings. *Trans Am Soc Artif Intern Organs* 21:264-272, (1975).
11. Herring, M. and Glover, J.L. *Endothelial seeding in vascular surgery*, Orlando:Grune & Straton, pp. 1-177, (1987).
12. Zilla, P.P., Fasol, R.D. and Deutsch, M. *Endothelialization of vascular grafts*, Basel, Switzerland:Karger, pp. 1-257, (1987).
13. Nerem, R.M. and Girard, P.R. Hemodynamic influences on vascular endothelial biology. *Toxic Path* 18:572-582, (1990).
14. Heimlinger, G., Geiger, R.V., Schreck, S. and Nerem, R.M. Effects of pulsatile flow on cultured vascular endothelial cell morphology. *J Biomech Eng* 113:123-131, (1991).
15. Lelkes, P.I. and Samet, M.M. Endothelialization of the luminal sac in artificial cardiac prostheses: a challenge for both biologists and engineers. *J Biomech Eng* 113:132-142, (1991).
16. Lampugnani, M.G., Resnati, M., Dejana, E. and Marchisio, P.C. The role of integrins in the maintenance of endothelial monolayer integrity. *J Cell Biol* 112:479-490, (1991).
17. Basson, C.T., Knowles, W.J., Bell, L., Albelda, S.M., Castronovo, V., Liotta, L.A. and Madri, J.A. Spatiotemporal segregation of endothelial cell integrin and nonintegrin extracellular matrix-binding proteins during adhesion events. *J Cell Biol* 110:789-801, (1990).
18. Kramer, R.H., Cheng, Y.-F. and Clyman, R. Human microvascular endothelial cells use ß1 and ß3 integrin receptor complexes to attach to laminin. *J.Biol.Chem.* 111:1233-1243, (1990).
19. Bowersox, J.C. and Sorgente, N. Differential effects of soluble and immobilized fibronectins on aortic endothelial cell proliferation and attachment. *In Vitro Cellular & Developmental Biology* 23:759-764, (1987).
20. Madri, J.A., Pratt, B.M. and Yannariello-Brown, J. Synthesis of extracellular matrix by endothelial cells. In: *Endothelial cell biology in health and disease*, edited by Simionescu, N. and Simionescu, M. N.Y.: Plenum Press, p. 167-188, (1988)

21. Williams, S.K., Jarrell, B.E., Rose, D.G., Pontell, J., Kapelan, B.A., Park, P.K. and Carter, T.L. Human microvessel endothelial cell isolation and vascular graft sodding in the operating room. *Ann Vasc Surg* 3:146-152, (1989).

22. Park, P.K., Jarrell, B.E., Williams, S.K., Carter, T.L., Rose, D.G., Martinez-Hernandez, A. and Carabasi, R.A.,III Thrombus-free, human endothelial surface in the midregion of a Dacron vascular graft in the splanchnic venous circuit--observations after nine months of implantation. *J Vasc Surg* 11:468-475, (1990).

23. Williams, S.K. Isolation and culture of microvessel and large-vessel endothelial cells: their use in transport and clinical studies. In: *Microvascular perfusion and transport in health and disease*, edited by McDonagh, P.F. Basel: Karger, p. 204-245, (1987).

24. Jarrell, B.E., Williams, S.K., Stokes, G., Hubbard, F.A., Carabasi, R.A., Koolpe, E., Greener, D., Pratt, K., Moritz, M.J., Radomski, J., et al. Use of freshly isolated capillary endothelial cells for the immediate establishment of a monolayer on a vascular graft at surgery. *Surgery* 100:392-399, (1986).

25. Jarrell, B.E., Williams, S.K., Solomon, L. and et al., Sodding of microvascular endothelial cells. *Ann Surg* 203:671-678, (1986).

26. Fasol, R., Zilla, P., Deutsch, M., Grimm, M., Fischlein, T. and Laufer, G. Human endothelial cell seeding: evaluation of its effectiveness by platelet parameters after one year. *J Vasc Surg* 9:432-436, (1989).

27. Grondin, C.M., Pomar, J.L., Hebert, Y., Bosch, X., Santos, J.M., Enjalbert, M. and Campeau, L. Reoperation in patients with patent atherosclerotic coronary vein grafts: a different approach to a different disease. *J Thorac Cardiovasc Surg* 87:379-385, (1984).

28. Dichek, D.A., Neville, R.F., Zwiebel, J.A., Freeman, S.M., Leon, M.B. and Anderson, W.F. Seeding of intravascular stents with genetically engineered endothelial cells. *Circulation* 80:1347-1353, (1989).

29. Callow, A.D. The vascular endothelial cell as a vehicle for gene therapy. *J Vasc Surg* 11:793-798, (1990).

30. Levinson, M.M., Smith, R.G., Cork, R.C., Gallo, J., Emery, R.W., Icenogle, T.B., Ott, R.A., Burns, G.L. and Copeland, J.G. Thromboembolic complications of the Jarvik-7 total artificial heart: case report. *Art Org* 10:236-244, (1986).

31. Moritz, A., Wolner, E. and Nose, Y. Clinical use of the artificial heart, indications and results. *Wien Klin Wochenschr* 18:161-167, (1988).

32. Okkema, A.Z., Grasel, T.G., Zdrahala, R.J., Cooper, S.L. and Solomon, D.D. Bulk, surface and blood-contacting properties of polyether urethanes modified with polyethylene oxide. *J.Biomat.Res.,Polymer Ed.* 1:81-100, (1989).

33. Schoen, F.J. Biomaterials science, medical devices, and artificial organs: synergistic interactions for the 1990s. *Trans Am Soc Artif Intern Organs* 37:44-48, (1991).

34. Zilla, P., Fasol, R., Grimm, M., Fischlein, T., Eberl, T., Preiss, P., Krupicka, O., von Oppell, U. and Deutsch, M. Growth properties of cultured human endothelial cells on differently coated artificial heart materials. *J Thorac Cardiovasc Surg* 101:671-680, (1991).

35. Bernhard, W.F. A fibrillar blood-prosthetic interface for both temporary and permanent ventricular assist devices: experimental and clinical observations. *Art Org* 13:255-271, (1989).

36. Ives, C.L., Eskin, S.G. and McIntire, L.V. Mechanical effects on endothelial cell morphology: in vitro assessment. *In Vitro Cell Dev Biol* 22:500-507, (1986).

37. Wechezak, A.R., Viggers, R.F. and Sauvage, L.R. Fibronectin and F-actin redistribution in cultured endothelial cells exposed to shear stress. *Lab Invest* 53:639-647, (1985).

38. Olesen, S.-P., Clapham, D.E. and Davies, P.F. Haemodynamic shear stress activates a K+ current in vascular endothelial cells. *Nature* 331:168-170, (1988).

39. Frangos, J.A., Eskin, S.G., McIntire, L.V. and Ives, C.L. Flow effects on prostacyclin production by cultured human endothelial cells. *Science* 227:1477-1479, (1985).

40. Lelkes, P.I. and Samet, M.M. Analysis of pulsatile flow and EC morphology in VAD-like cell culture chambers. *ASAIO* transactions, in press (1991).

41. Pratt, K.J., Williams, S.K. and Jarrell, B.E. Enhanced adherence of human adult endothelial cells to plasma discharge modified polyethylene terephthalate. *J Biomed Mat Res* 23:1131-1147, (1989).

42. Davies, P.F., Remuzzi, A., Gordon, E.J., Dewey, C.F.,Jr. and Gimbrone, M.A.,Jr. Turbulent fluid shear stress induces vascular endothelial cell turnover in vitro. *Proc.Natl.Acad.Sci.USA* 83:2114-2117, (1986).

350

43. Sumpio, B.E., Banes, A.J., Buckley, M. and Johnson, G.,Jr. Alterations in aortic endothelial cell morphology and cytoskeletal protein sythesis during cyclic tensional deformation. *J Vasc Surg* 7:130-138, (1988).

44. Sumpio, B.E., Banes, A.J., Link, G.W. and Iba, T. Modulation of endothelial cell phenotype by cyclic stretch: inhibition of collagen production. *J Surg Res* 48:415-420, (1990).

45. Etingin, O.R., Hajjar, D.P., Hakkar, K.A., Harpel, P.C. and Nachman, R.L. Lipoprotein (a) regulates plasminogen activator inhibitor-1 expression in endothelial cells: a potential mechanism in thrombogenesis. *J.Biol.Chem.* 266:2459-2465, (1991).

46. Stiko-Rahm, A., Wiman, B., Hamsten, A. and Nilsson, J. Secretion of plasminogen activator inhibitor-1 from cultured human umbilical vein endothelial cells is induced by very low density lipoprotein. *Arteriosclerosis* 10:1067-1073, (1990).

47. Franke, R.-P., Schnittler, H.J., Fuhrmann, R., Hopken, S., Sondermann, R., Mittermayer, Ch. and Drenckhahn, D. Human vascular endothelium in vitro: mass culture, growth factors, substrate, shear stress. In: *Endothelialization of vascular grafts*, edited by Zilla, P.P., Fasol, R.D. and Deutsch, M. Basel: Karger, p. 130-135, (1987)

48. Sola, O.M. Transplanted skeletal muscle for cardiac assist: historical basis and clinical implications. In: *Biomechanical cardiac assist: cardiomyoplasty and muscle-powered devices*, edited by Chiu, R.C.-J. Mount Kisco, N.Y.: Futura Pub. Co., p. 29-42, (1986)

49. Carpentier, A. and Chachques, J.C. The use of stimulated skeletal muscle to replace diseased human heart muscle. In: *Biomechanical cardiac assist: Cardiomyoplasty and muscle-powered devices*, edited by Chiu, R.C.-J. Mount Kisco, N.Y.: Futura Pub. Co., p. 85-102, (1986)

50. Acker, M.A., Mannion, J.D. and Stephenson, L.W. Methods of transforming skeletal muscle into a fatigue-resistant state: potential for cardiac assistance. In: *Biomechanical cardiac assist: cardiomyoplasty and muscle-powered devices*, edited by Chiu, R.C.-J. Mount Kisco, N.Y.: Futura Pub.Co., p. 19-28, (1986)

51. Stephenson, L.W., Macoviak, J.A., Armenti, F., Bitto, T., Mannion, J.D. and Acker, M.A. Skeletal muscle for potential correction of congenital heart defects. In: *Biomechanical cardiac assist: cardiomyoplasty and muscle-powered devices*, edited by Chiu, R.C.-J. Mount Kisco, N.Y.: Futura Pub. Co., p. 129-139, (1986).

52. Anderson, D.R., Pochettino, A., Hammond, R.L., Hohenhaus, E., Spanta, A.D., Bridges, C.R.,Jr., Lavine, S., Bhan, R.D., Colson, M. and Stephenson, L.W. Autogenously lined skeletal muscle ventricles in circulation: up to nine months' experience. *J Thorac Cardiovasc Surg* 101:661-670, (1991).

53. Stern, D.M., Handley, D.A. and Nawroth, P.P. Endothelium and the regulation of coagulation. In: *Endothelial cell biology in health and disease*, edited by Simionescu, N. and Simionescu, M. N.Y.: Plenum Press, p. 275-306, (1989)

54. Bernhard, W.F., Colo, N.A. and Szycher, M. Development of a nonthrombogenic collagenous blood prosthetic interface. *Ann Surg* 192:369-381, (1980).

55. Bernhard, W.F., Colo, N.A. and Szycher, M. Development of collagenous linings on permeable prosthetic surfaces. *J Thorac Cardiovasc Surg* 79:552-564, (1980).

56. Lelkes, P.I. New aspects of endothelial cell biology. *J.Cell.Biochem.* 45:242-244, (1991).

57. Jones, P.A. Construction of an artificial blood vessel wall from cultured endothelial and smooth muscle cells. *Proc.Natl.Acad.Sci.USA* 76:1882-1886, (1979).

58. Dewey, C.F. Effects of fluid flow on living vascular cells. *J Biomech Eng* 106:31-35, (1984).

MECHANISMS OF EMBRYONIC BLOOD VESSEL FORMATION

H. Drexler, H. Schurch, G. Breier, G. Drexler and W. Risau

MaxPlank Institut fur Psychiatrie, Abt. Neurochemie
Am Klopferspitz 18A, 8033 Martinstried, Germany

All blood vessels develop form angioblasts which differentiate in the early embryonic mesoderm of the area pellucida and area opaca. The mechanisms of angioblast induction are still unknown but the participation of the endoderm in this induction event seems to be a prerequisite (1). One model system to study the early development of the vascular system are embryonic stem cells, which in suspension culture spontaneously differentiate into angioblast-containing cystic embryoid bodies. This process apparently in an inherent program of embryonic stem cell differentiation (2).

After the in situ differentiation of angioblasts a primary capillary plexus is formed by the fusion of the growing endothelial cells. This process is described as vascularogenesis to distinguish it from a second mechanism, termed angiogenesis, by which new capillaries are formed through spouting from prexisting vessels (2).

The developing brain as well as the embryonic kidney are vascularized by such an angiogenic process. Attempts to characterize angiogenesis factors from these organs resulted in the purification of acidic and basic FGF (3, 4). However, hybridization studies using a chick specific single stranded DNA probe for aFGF indicate that the pattern of expression does not correlate with endothelial cell proliferation and angiogenesis (3). Also, the receptors for FGF do not seem to be present on embryonic endothelial cells at the time of brain vascularization (6).

Other factors than members of the heparin binding growth factor family may be responsible for the vascularization of the developing brain. The expression during brain development e.g. of vascular endothelial cell growth factor (VEGF), a secrered endothelial cell specific mitogen, correlates with endothelial cell proliferation (G. Breier: in preparation). Also, it is conceivable that factors which are not mitogenic but chemotastic for endothelial cells might be important to brain vascularization. During embryonic development angiogenesis occurs simultaneously with the regression of blood vessels. Localized regression can be observed during chondrogenesis in the embryonic limb bud (7). Locally acting factors controlling endothelial cell proliferation or endothelial cell death might be responsible for this morphogenetically important event (8).

1.   Flamme, Anat. Embryol. 180, 259-272 (1989)

2. W. Risau, H. Sariole, H.G. Zerwes, J. Sasse, P. Ekblom, R. Kemler, and T. Doetschman, Development 102, 471-478 (1988).
3. W. Risau and P. Ekblom, J. Cell Biol. 103, 1101-1107 (1986).
4. W. Risau, Proc. Natl. Acad. Sci., USA 83, 3855-3859 (1986).
5. H. Schnurch, W. Risau, Development 111, 1143-1154 (1991).
6. J. G. Heuer, C.S. von Bartheld, Y. Kinoshita, P.C. Evers and M. Bothwell, Neuron 5, 283-296 (1990).
7. R. Hallmann, R.N. Feinberg, C.H. Latker, J. Sasse and W. Risau Differentiation 34, 98-105 (1987).
8. M.S. Pepper, R. Montesano, L. Orci, J.D. Vassali, Biochem. Biophys. Res. Comm. 176, 633-638 (1991).

NEW TOOLS TO STUDY VASCULARISATION IN THE DEVELOPING EMBRYO

Luc Pardanaud and Francoise Dieterlen-Lièvre

Institut d´Embryologie cellulaire et moleculaire du CRNS
et du Collège de France,
49 bis av. de la Belle Gabrielle, 94130 Nogent s/Marne,
Paris, France

Our experiments in the avian embryo have uncovered the existence of
two modes of endothelial network formation, depending on the germ layer
associated with the mesoderm. When mesoderm is interacting with endoderm,
it gives rise to endothelial precursors. Such is the case when splanchnopleura from 2 to 20 pairs of somites embryos or internal organ rudiments
are transplanted from the quail in the chick or inversely. In contrast,
when mesoderm is in contact with ectoderm, no intrinsic endothelial precursors appear; thus somatopleura or external organ rudiments are colonized by extrinsic precursors. We have detected a unique pattern of expression of the c-ets1 oncogene in endothelial cells. c-ets1 is turned on in
endothelial cell nuclei of immature blood vessels and becomes inactive
when the vessel wall differentiates. This expression, studied by in situ
hybridization, appears as an interesting marker for emerging blood vessels
and would be more efficiently analyzed at the level of the protein. Thus
we are engaged in the preparation of a monoclonal antibody directed
against one of the splicing products of the ets1 locus (p68 c-ets1). The
use of this antibody will be combined with experimental approaches. We
are also attempting to obtain cloned endothelial cell lines from the
quail.

Pardanaud, L., Altmann, C., Dieterlen-Lièvre, F., and Buck, C., 1987,
    Vascularogenesis in the early quail blastodisc as studied with a
    monoclonal antibody recognizing endothelial cells, Development., 100:
    339-349.
Pardanaud, L., Yassine, F., and Dieterlen-Lièvre, F., 1989, Relationship
    between vasculogenesis, angiogenesis and hemopoiesis during avian
    ontogeny, Development., 105:473-485.
Vandenbunder, B., Pardanaud, L., Jaffredo, T., Mirabel, M.A., and Stehe-
    lin, D., 1989, In situ hybridization analysis of the expression of
    c-ets1, c-my b and c-my c during the development of blood cells and
    vascular endothelia in the chick embryo, Development., 107:265-274.

ASPECTS ON PEPTIDE GROWTH FACTORS AND THEIR RECEPTORS IN HUMAN BIOPSY

SPECIMENS

Hans-Arne Hansson

University of Goteborg, Department of Histology
P O Box 33031, S-400 33 Göteborg, Sweden

Human biopsy specimens were examined with regard to the occurence and distribution of the receptors (R) for FGF (type I and II), IGF-I and PDGF as well as that of several peptide growth factors. Immunohistochemical techniques were mainly used.

In the normal epidermis, type I FGF-R are recognized in the basal cells in the startum germinativum, while the IGF-I-R is localized to the more superficial keratinocytes in the startumperminativum. In addition type I FGF-R could be seen in vascular wall cells. Aberrant patterns were seen at and in acute and chronic ulcers.

In the nasal mucosa, the basal cells expressed type I FGF-R while only rare basal cells showed IGF-I-R. In healing nasal mucosa, there was a marked increase in the occurence of type I FGF-R and IGF-I-R as well as of IGF-I, IGF-II, TGFB and aFGF. Difference between cells in blood vessels were seen.

Myofibroblasts in contracted capsules around smooth-surfaced silicone breast implants showed high immunoreactivities for type I FGF-R and IGF-I-R. There was as well enrichment of IGF-II, TGFB and aFGF in these cells. However, in non contracted capsules the myofibroblasts usually lacked type I FGF-R but had mostly and IGF-I-R. A diverging pattern was seen in the myofibroblasts in the fairly chick capsules around textured implants, in which these cells mostly lacked type I FGF-R. A prominent feature was the impressive occurence of type I FGF-R in vascular wall cells.

SPROUT FORMATION AND FUSION IN TUMOR ANGIOGENESIS: MORPHOLOGICAL AND
MORPHOMETRIC STUDIES ON A HUMAN XENOGRAFTED MELANOMA

C. Van Ackern, M.A. Konerding, F. Steinberg, and C. Streffer

Institut fur Anatomie, Institut fur Medizinische Strahlenbiologie, Universitatklinikum Essen
Hufelandstrabe 55, 4300 Essen 1, Germany

We have studied the morphological characteristics of tumor angiogenesis using series of a xenografted human melanoma on nude mice. The
animals were sacrificed 8 to 79 days after transplantation with intervals
of 4 to 10 days. Light-, scanning- and transmission electron microscopic
specimens were prepared with the usual methods.

Structurally stabilised, destroyed and newly formed vessels are found
closely side by side at all times. In the case of tumor angiogenesis, at
least two different types of lumen formation can be distinguished as a
general principle and which occur during all stages of tumor growth. The
first is characterised by an intracellular formation of a vacuole giving
rise to a primitive intracellular lumen with connsecutive connection to
already established parent vessel. Here, more plump pseudopodia or the
eccentrically situated pericaryons of involved endothelial cells leading
to the formation of a fine slit veer out from the endothial structure. A
slit-like, intercellular lumen is thus already formed at the start of bud
formation and not, as usually postulated, after formation of a solid endothelial cell cord.

The total amount of sprouts decreases with age. In early stages, all
types of lumen formation can be observed whereas in later ones the intercellular mode prevails. Serial sections show that the newly formed
sprouts may fuse with established vessels as well as with other sprouts.
Usually the fusion of sprouts takes place at their tips.

After the fusion of sprouts new loops develop. Such early forms are
characterised by a clearly higher endothelium and cell organelle content.
Concerning the fate of these early forms, two aspects have to be considered: either the early vessels can differentiate into an established vessel with a typical thin endothelium or they degenerate with consecutive
endothelial cell death.

Summing up, our studies have shown that tumor angiogenesis clearly
differs in essential ways from other forms of secondary angiogenesis such
as wound healing or chronic inflammation.

ANGIOGENESIS AND INVASIVENESS: PHARMACOLOGICAL SUPPRESSION OF NEOPLASTIC SPREAD OF MALIGNANT CELL IN THE BRAIN

Steven Brem, Ana-Maria Tsanaclis, D. Zagzag and S. Gately

Neurosurgual Oncology, Northwestern Medical Faculty Found. Inc.
233 East Evie, Suite 500, Chicago, Ill. 60611, USA

The malignant phenotype is determined by at least three interdependent, coordinated cellular functions: 1) mitosis; 2) angiogenesis; and 3) neoplastic spread. Malignant cells can infiltrate along microvascular pathways[1-2] or directly penetrate the surrounding tissue. Although human brain tumors are among the most vascular[3], and malignant tumors display a high rate of cell turnover[4], the pathological feature that renders all forms of current treatment ultimately futile is the neoplastic infiltration of malignant cells in brain, and the main reason for tumor recurrence[5]. Despite the importance of neoplastic invasion, relatively little is known of its biology. Some clues, however, can be found in the study of angiogenesis, which depends upon similar mechanisms of: a)adhesion of the cell to the extracellular matrix; b)proteolytic breakdown of the matrix; and c)directed cellular migration[6-8]. Neovascularization itself has been viewed as merely one example of the more general process of tissue invasion[9]. Because of the link between angiogenesis and tissue invasion, we tested angiosuppressive therapy-copper depletion and penicillamine treatment in the rat 9L gliosarcoma, selected because of its pattern of extensive tumor invasion. Anticopper treatment prevented the infiltrative spread of gliosarcomatous cells into the surrounding neuropil[10]. Ultrastructured studies showed that pseudopodial protrusions, cytological organelles of locomotion, were absent in neoplastic cells at the tumor-brain interface in treated animals[10]. Because of the known affinity of basic fibroblast growth factor (bFGF) to copper ion[11], and the role of bFGF in neoplastic transformation[12] and cellular migration[13], we examined the malignant phenotype among two lines of bFGF-transfected 3T3 cells into the brains of athymic mice. The cell line lacking the bFGF-signal peptide (B-7) was as tumorigenic as the cells (6-1) transfected with signal peptide-bFGF[14]. The 6-1 cells, however, displayed an invasive phenotype, infiltrating the adjacent neuropil; in contrast, the B-7 cells aggragated together within the boundary of a thin membrane composed of moderately electron dense material, similar to basement membrane, forming a "capsule" between the neoplastic cells and the surrounding brain[14]. No difference in the microvascular density was detected in the brain tumors, although the 6-1 line was slightly more angiogenic in the rabbit cornea[14]. Experimental tumor systems can differentiate among the three components of the malignant phenotype: invasiveness, tumorigenity, and angiogenesis. The invasive phenotype itself can be modulated by genetic engineering as well as systemic metabolic and pharmacological therapy. Our finding suggest a biological role of copper in the neoplastic spread of brain tumor cells. An expanding knowledge of the structural, biochemical and genetic factors that regulate the invasion cascade will provide precise molecular targets for therapy designed to block neoplastic spread.

References

1. R. F. Nicosia, R. Tchao, J. Leighton, J. Cancer Res. 43:2159–2166, (1983).
2. D. Zagzag, S. Brem, F. Robert, Neovascularization and tumor growth in the brain, Am. J. Pathol.131:361–372 (1988).
3. S. Brem, R. S. Cotran, J. Folkman, J. Natl. Cancer Inst. 48:347–356 (1972).
4. A. M. Tsanaclis, F. Robert, J. Michaud, S. Brem, Can. J. Neurol. Sci. 18:12–17 (1991).
5. V. A. Levin, Semin Oncol. 2:57–61 (1975).
6. D. H. Ausprunk, J. Folkman, Microvasc. Res. 14:53–65 (1977).
7. L. Liotta, E. Schiffman, in: Important Advances in Oncology, Philadel. J. P. Lippincott, pp. 17–30 (1988).
8. D. Moscatelli, D. B. Rifkin, Biochim. Biophys. Acta 948:67–75 (1988).
9. R. Eisenstein, N. Sorgente, L. W. Soble, A. Miller, K. E. Kuettner, Am. J. Pathol. 73:765–774 (1973).
10. S. Brem, A. M. Tsanaclis, D. Zagzag, Neurosurgery 26:391–396 (1990).
11. Y. Shing, J. Biol. Chem. 263:9059–9062 (1988).
12. S. Rogeli, R. A. Weinberg, P. Fanning, M. Klagsbrun, Nature 331:173–175 (1988).
13. R. Tsuboi, Y. Sato, D. B. Rifkin, J. Cell Biol. 110:511–517 (1990).
14. S. Brem, S. Gately, A. M. Tsanaclis, M. Klagsbrun, Unpublished data.

ANGIOSUPPRESSION OF BRAIN TUMORS BY DEPLETION OF COPPER AND PENICILLAMINE

S. Brem, D. Zagzag, A. M. Tsanaclis, S. Gately, J. L. Gross

Neurosurgual Oncology, Northwestern Med. Faculty Found. Inc.
233 East Evie, Suite 500, Chicago, Ill. 60611, USA

Copper metabolism is altered in the neoplastic state, with increased intestinal absorption, serum and tissue concentration, and cellular uptake of the trace metal noted in human and experimental tumors[1-4]. Copper[5,6] affinity is also a property common to many known angiogenic molecules. Copper depletion and penicillamine treatment (CDPT) inhibits neovascularization in the cornea[7,8], and copper has been suggested as a constituent, a "cofactor", of angiogenesis[9]. To test this hypothesis, we depleted copper in the serum and tissues of VX2 rabbits[6], using a brain tumor model for studies of angiogenesis and the blood-brain barrier[10,11]. Normocupremic animals developed large, vascularized tumors. By contrast, small, circumscribed, relatively avascular tumors were found in the brains of rabbits copper-depleted and treated with penicillamine[6]. The CDPT rabbits showed a significant decrease in serum copper, copper staining of tumor cell nuclei, microvascular density, tumor volume, endothelial cell turnover, and an increase in the vascular permeability as well as peritumoral brain edema[6]. Copper depletion and penicillamine treatment also inhibited the intracerebral growth of the 9L gliosarcoma in F-344 rats[6]. These findings were recently extended to the rat C6 glioma. In two experiments, tumor growth was inhibited by 55% and 74% in a rat kidney capsule model of angiogenesis[12]. Similar results were obtained in rats by penicillamine treatment alone[12]. C6 glioma-induced vascularization was prevented by CDPT. The molecular mechanisms for angiosuppression may include inhibition of one or more of the following physiological processes: 1) growth factor binding, 2) collagen biosynthesis, and 3) protease activity. Taken together, these data demonstrate that angiosuppression represents a valid approach for growth inhibition of highly vascularized human brain tumors.

References

1. D. I. Cohen, B. Illowsky, M. C. Linder, Am. J. Physiol. 236:E309-315 (1979).
2. O. Miatto, M. Casaril, G. B. Gabrielli, Cancer 55:774-778 (1985).
3. S. Kobayashi, J. Sayato-Suzuki, Biochem. J. 249:69-75 (1988).
4. R. J. Coates, N. S. Weiss, J. R. Daling, Cancer Res.49:4353-4356 (1989).
5. J. Folkman, M. Klagsbrun, Science 235:442-447 (1987).
6. S. Brem, D. Zagzag, A. M. Tsanaclis, Am. J. Pathol. 137:1121-1142 (1990).
7. M. Ziche, J. Jones, P. M. Gullino, JNCI 69:475-482 (1982).
8. T. Matsubara, R. Saura, K. Hirohata, J. Clin. Invest. 83:158-167 (1989).
9. P. M. Gullino, Anticancer Res. 6:153-158 (1986).

10. D. Zagzag, S. Brem, F. Robert, <u>Am. J. Pathol.</u> 131:361-372 (1988).
11. D. Zagzag, M. Goldenberg, S. Brem, <u>AJNR</u> 10:529-534 (1989).
12. J. L. Gross, D. Hertel, W. F. Herblin, <u>Proc. Am. Assoc. Cancer Res.</u> 32:57 (abstr. #338) (1991).

REACTIVITY OF MoAb E-9 HUMAN TUMOUR, FOETAL AND NORMAL TISSUE VASCULATURE

J.M. Wang and S. Kumar

Christie Hospital
Manchester M20 9BX, England

Monoclonal antibody E-9 has been raised against proliferating human umbilical vein endothelial cells. In order to characterise it, we have compared its reactivity with a pan-endothelial cell antibody 5.6E (PECAM/ CD31) using a panel of normal, foetal, regenerating and tumour tissues. The results show that (1) E-9 antibody binds to the vasculature of tumours (61/61), foetal organs (7/7) and regenerating tissues (5/5), (2) E-9 antibody has limited reactivity in normal adult tissues (20/87) wherein usually only a very small percentage of blood vessel stained. All blood vessels in "normal" tonsils and cervix stain with E-9 antibody. (3) E-9 antibody binds to the outer layer of trophoblast in all 11 placentas studied, but surprisingly shows no reactivity to blood vessels in the placental villi. The results indicate that there is a qualitative and quantitative difference in the vasculature of normal tumour tissues.

These are early results therefore it is not possible to explain the paradoxical staining pattern. For example, blood vessels of foetal organs stain but those of placental villi do not. Is the antigenic difference recognized by E-9 related to the structure, function or histogenetic origin of blood vessels in these tissues? Similarly, why blood vessels in some normal organs (cervix, tonsil), are positive, whereas blood vessels of normal colon are consistently negative, cannot be explained at present. Could this be due to differences in the extracellular matrix or does it represent the influence of local microenvironmental factors? Once the exact nature of the antigen recognized by E-9 has been established, it might be possible to explain some of the observations.

CORRELATION OF BASEMENT MEMBRANE COLLAGEN BIOSYNTHESIS WITH ANGIOGENESIS IN THE CAM SYSTEM

N.E. Tsopanoglou, G.C. Haralabopoulos and M.E. Maragoudakis

University of Patras, Medical School, Dept. of Pharmacology
26500 Rio, Patras, Greece

Basement membrane (BM) collagen biosynthesis is known to increase greatly during angiogenesis in the CAM system. On the other hand, agents inhibiting BM biosynthesis have been shown to prevent angiogenesis (Maragoudakis et al., 1988).

In the present study an attempt was made to correlate angiogenesis induced by PMA and leukotrienes (LTB$_4$, LTC$_4$, LTD$_4$) with increased BM collagen formation in the CAM system in order to further establish whether this latter parameter can be used as an biochemical index of angiogenesis.

The tumor promoting agent PMA in agreement with results in the literature (Montesano and Orci, 1985, Morris et al., 1988) caused a strong angiogenic response in vivo (at doses ranging from 6 to 600 ng) which was accompanied by an increase in collagenous proteins and a less consistent increase in non collagenous proteins. In contrast PMA in vitro inhibited collagen biosynthesis under similar conditions.

Leukotriene C$_4$ (LTC$_4$), which is known to stimulate tube formation and migration in endothelial cell culture (Kanayasu et al., 1989) as well as LTD$_4$ also caused strong angiogenic responses in vivo (in doses ranging from 10^{-3} to 10^4 pg) which were followed by an increase in both collagenous and non collagenous proteins, although the increase in the former was much more substantial. LTB$_4$ had no effect on either of these processes. In vitro LTD$_4$ caused an increase in collagenous proteins biosynthesis while LTC$_4$ caused an inhibition.

These results suggest that BM collagen formation can indeed be used as a biochemical index of angiogenesis in vivo.

1. T. Kanayasu et al., Biochim. Biophys. Res. Commun. 159(2), 572-578 (1989).
2. M. E. Maragoudakis et al., J. Pharm. and Exper. Therap. 244, 729 (1988).
3. R. Montesano and L. Orci, Cell 42, 469-477 (1985).
4. P. B. Morris et al., Am. J. Physiol. 254, C318-322 (1988).

VASCULARIZATION AND CURABILITY OF CERTAIN HUMAN TUMORS, A PRACTICAL APPROACH

C.S. Delides, Efi Protopapa and L. Revesz

University of Crete, Dept. of Pathology[1] and Metaxas Cancer Institute of Pireaus[2]
Iraklion, Crete[1] and Botasi 51, 18537 Pereaus[2], Greece

The original observations of Kolstad on an association of vascularity in carcinomas of the uterine cervix with the response of the tumors to radiotherapy, were confirmed by studies performed with different methods in cervical, laryngeal and rectal carcinomas.

By adopting a morphometrical method to calculate a volume corrected vascularity index (VI) we found a correlation to exist between VI and survival time in nasopharyngeal and bladder carcinomas, but we failed to find such a relation in brain astrocytomas.

Variance analysis of VI in different regions of cervical carcinomas suggested a certain individual vascularization pattern in these tumors.

Deficient vascularization inhibits the supply of oxygen and other nutricients leading to the development of areas with radioresistant hypoxic cells. It appears that these cells are also less accesible to chemotherapeutic agents. In a preliminary study we calculated the VI in breast carcinomas treated preoperatively by chemotherapy and we found a significant greater VI in patients with long survival. This study is now in progress and the above results are based on a limited number of cases.

If further studies could confirm the predictive value of VI then by using it as complementary to other factors, a more individualized treatment can be applied since greatly varying results can be seen in tumors of the same classification, treated in similar way.

IMMUNOLOCALIZATION OF BASIC FIBROBLAST GROWTH FACTOR TO THE MICROVASCULATURE OF HUMAN BRAIN TUMORS

Ana-Maria Tsanaclis

Universidade de Sao Paulo, Faculdade de Medicina
Rua dos Franceses-470, Ed. Renan-271, 01329 Sao Paulo,
Brazil

A prominent feature of some tumors of the central nervous system is angiogenesis. In this process, basic fibroblastic growth factor (bFGF) plays a key role. Surgical specimens of 23 human brain tumors were examined, by immunocytochemistry with a highly purified murine monoclonal antibody to bFGF. Alternate sections were incubated with the Ki-67 monoclonal antibody to measure the growth fraction. Innumoreactivity to bFGF was observed in 20/23 (87%) of the neoplasms. All 18 malignant brain tumors, with elevated growth fractions, were immunoreactive as was a recurrent meningioma and an ependymoma. The three non-reactive tumors were benign, non-invasive, slow growing neomplasms, with a low index of proliferation, an acoustic schwannoma, a meningioma, and a cholesteatoma. Immunostaining to bFGF was found in the microvascular compartment in 18/23 (78%) of the tumors. Reactivity to bFGF was noted in the tumor cell nuclei in six (26%) and in the cytoplasm of two (9%) tumors. The immunostaing was not present after preabsortion of the antibody with pure human bFGF. In view of the role of bFGF in cell mitosis and invasion, the presence of bFGF predominantly within the tumor microvasculature indicates a cellular depot for a growth factor that mediates angiogenesis and tumorigenesis: bFGF provides a molecular target for adjuvant therapy of malignant CNS tumors by angiosuppression.

A POSSIBLE MECHANISM FOR THE ANTI-PROLIFERATION OF GOLD DRUGS ON ENDOTHELIAL CELLS INVOLVING INHIBITION OF PROTO-ONCOGENEPRODUCTS

T. Sahinoglu, M. Roncalli, D. Springall, J. M. Polak and
D. R. Blake

The London Inflammation Group, London Hospital Med. College
Turner Str., London E1 2AD, United Kingdom

The unusual persistence of the inflammatory process in rheumatoid arthritis has repeatedly been correlated with the continuous growth and proliferation of microvascular endothelial cells (EC), constituting synovial angiogenesis.

In addition to the ascribed role of reactive oxygen species (ROS) in the persistence of the inflammatory process in rheumatoid arthritis, low concentrations of these species are known to stimulate proliferation of fibroblasts. This proliferation has been correlated with the induced expression of a number of nuclear proto-oncogene products (c-Jun, c-Fos, and c-Myc) which are believed to be involved in signal transduction processes related to growth control.

One of the theories put forward to explain the underlying mechanism of the persistent synovitis in RA is hypoxia/reperfusioninjury. Increasing evidence now implicates the enzyme xanthine oxidase as a source of ROS in hypoxic reperfusion injury. Xanthine oxidase activity has been detected in rheumatoid synovial tissue and has recently been localised to microvascular EC in the synovium. We have previously reported modulation of the conversion of this enzyme to its ROS-generating form by the anti-arthritic gold drug, Myocrisin. Additionally, this drug has been shown to inhibit proliferation of cultured EC.

To investigate whether the anti-proliferative effects of Myocrisin are mediated through proto-oncogene expression in cultured human EC we have used immunocytochemistry employing antibodies to c-jun, c-fos and c-myc protein products.

A two-hour incubation of EC with Myocrisin resulted in a significant decrease in the levels of all three proto-oncogeneproducts and a corresponding decrease in proliferation.

We speculate that Myocrisin, by causing an increase in xanthine oxidase-derived ROS, induces a reduction in the expression of nuclear proto-oncogene products thereby influencing EC proliferation.

LOCALIZATION OF XANTHINE OXIDO-REDUCTASE TO MICROVESSEL ENDOTHELIUM IN THE SYNOVIUM

C. R. Stevens, T. Sahinoglu, R. Harrison and D. R. Blake

The London Intlammation Group, London Hospital Med. College
Turner Str., London E1 2AD, United Kingdom

There is no doubt that the inflamed rheumatoid synovium incorporates an oxygen radical generating system. The evidence for this is based upon studies which show that synovial fluid effusions contain lipid peroxidation products, oxidatively damaged IgG and products of oxidative carbohydrate degradation. In rheumatoid arthritis the synovial fluid and adjacent synovium are hypoxic, minimising the likelihood that radical generation is via the respiratory burst of activated inflammatory cells. An alternative mechanism involving ischaemic/reperfusion has been proposed. The unique topography of the component parts of the joint predispose the synovium to pressure induced fluctuations in the blood supply, creating the environment for ischaemic/reperfusioninjury.

We propose that the postischaemic reperfusion of the microvasculature generates reactive oxygen species via the enzyme xanthine oxidase. The non-pathological form of this enzyme oxidises hypoxanthine and xanthine to urate NAD^+ as an electron acceptor. Under ischaemic conditions, however, this enzyme can be converted to a form which catalyses the same reaction using O_2 as an electron acceptor resulting in superoxide anion generation. It has previously been shown that the rheumatoid synovium exhibits xanthine oxido-reductase (XOR) activity and here we show that the enzyme is ideally situated to affirm our hypothesis.

Anti-bovine milk xanthine oxidase (a-BMXO) was raised in rabbits immunised with purified XOR from fresh cow´s milk using standard methods. Serum antibody level was assessed by ELISA and test confirmed using immunoblotting and dot-immunobinding assays. Frozen blocks containing normal and rheumatoid synovium were sectioned (6 um) and stained using Vectastain Elite Rabbit IgG ABC kit. The method employs an ulabelled an ulabelled primary antibody (a-BMXO), followed by a biotinylated secondary antibody and then a preformed avidin and biotinylated horseradish peroxidase macromolecular complex. Peroxidase binding sites corresponding to enzyme localisation were revealed using H_2O_2 and DAB substrate. Controls included the omission of the primary antibody, inclusion of non-immune rabbit serum and a-BMXO which had been absorbed overnight with affinity purified BMXO from Biozyme, UK. In addition a-BMXO staining distribution was assessed in control human tissue. XOR antigen was found to be localised to both normal and rheumatoid synovium. In large vessels the endothelial staining was less intense and there appeared to be a synovium. In large vessels the endothelial staining was less intense and there appeared to be a diffuse positive staining in the smooth muscle. Staining of human gut and liver gave similar XOR distribution to that previously described.

We conclude that oxygen radical generation by the rheumatoid synovium is localised to the endothelium and is achieved via the xanthine oxidase system during post-ischaemic reperfusion of the microvasculature.

ATHEROSCLEROSIS AS AN ANGIOGENIC DISEASE: IMPLICATIONS FOR THE PATHOGENESIS AND THERAPY OF SYMPTOMATIC CAROTID ARTERIAL STENOSIS

Steven Brem

Neurosurgual Oncology, Northwestern Medical Faculty Found. Inc.
233 East Evie, Suite 500, Chicago, Ill. 60611, USA

Angiogenesis is a major event in the stepwise of the atherosclerotic plaque and its clinical complications[1-3]. New vessels arise from the outer medial and adventitial layers. The inner medial layer is relatively hypoxic[4]. The transition from a clinically innocuous fibrous plaque to a thickened, stenotic, symptomatic, complex plaque is marked by a series of characteristic changes in the microvasculature of the arterial wall[1]. The extent of neovascularization is linked to the severity of atherosclerosis[5]. Associated with proliferation of smooth muscle cells (SMC), a critical thickness of the arterial wall is exceeded, and neovascularization ensues[6]. The new capillaries are permeable and flagile,[7] and the pathological substrate for intraplaque hemorrhages[1], that, in turn, can precipitate plaque rupture, thrombosis, and embolism[8]. We tested the angiogenic activity of 278 carotid arterial plaque fragments from patients with cerebral ischemia who underwent carotid endarterectomy, using the rabbit corneal assay. Angiogenesis was stimulated by the atherosclerotic plaque, but control tissues- nonatherosclerotic human arterial tissue, boiled plaque, or normal rabbit carotid artery - failed to induce angiogenesis[9]. Histologic examination of the atherosclerotic plaques revealed that the cellular zones, composed mainly of SMC, were highly angiogenic, whereas the acellular zones composed of amorphic, necrotic, lipid-laden, calcific material were generally non-angiogenic[9]. An animal model that would allow for the quantitative study of neovascularization would be valuable to determine the cellular and biochemical indicators that regulate SMC-induced angiogenesis. For example, we observed that arterial wall implants placed in corneal micropockets elicit neovascularization and macrophage infiltration in hypercholesterolemic, but not in normolipemic rabbits[10]. The microenvironmental factors governing the production, release and degradation of angiogenic growth factors remain to be defined fot the atherosclerotic plaque, but candidate molecules include fibroblast growth factor[11-13], platelet-derived endothelial cell growth factor[14], tumor necrosis factor-alpha[15], and interleukin-1[16], and low-molecular-weight endothelial cell growth factor[15]. There are many analogies between the growth of hyperplastic SMC within the atherosclerotic plaque and the principles regulating the growth of neoplastic cells:[18] 1) the role of oncogenes[19] and growth factors;[11-14] 2) the transition from a relatively "avascular" medial wall to a "vascular" phase with capillary ingrowth[1]; and 3) microenvironmental control of stimulators and inhibitors leading to either cellular proliferation or plaque regression depending upon perturbations in the normal balances[9,12]. If the principle of angiogenesis-dependency, pivotal for the growth of solid tumors[10], is applied to the atherosclerotic plaque, then angiosuppression[21] may be a valid approach to arrest the plaque in a quiescent, non-vascularized, and clinically stable phase.

References

1. M.C. Winternitz, R. M. Thomas, P. M. LeCompte, in: The Biology of Arteriosclerosis, C. C. Thomas, Springfield, Ill, 1938.
2. A. C. Barger, R. Beeuwkes III, L. Lainey, 1985, New Engl. J. Med. 310: 175-177 (1985).
3. S. Brem, Can. Med. Assoc. J. 143:755 (1990).
4. T. Björnheden, G. Bondjers, Arteriosclerosis 7:238-247 (1987).
5. B. R. Karnat, S. J. Galli, A. C. Barger, Hum Pathol 18:1036-1042 (1987).
6. H. Wolinsky, S. Glagov, Circ. Res. 20:99-111 (1967).
7. M. Friedman, S. O. Byers, Br. J. Exp. Pathol. 43:363-372 (1962).
8. J. A. Fryer, P. C. Myers, M. Appelberg, J. Vasc. Surg. 6:341-349 (1987).
9. H. Alpern-Elran, N. Morog, F. Robert, J. Neurosurgery 70:942-945 (1989).
10. Y. Cornair, S. Brem, A. M. Tsanaclis, Unpublished data
11. M. Klagsbrun, E. R. Edelman, Arteriosclerosis 9:269-278 (1989).
12. V. Lindner, D. A. Lappi, A. Baird, Circ. Res. 68:106-113 (1991).
13. V. Lindner, M. A. Reidy, Proc. Natl. Acad. Sci, USA 88:3739-3743 (1991).
14. F. Ishikawa, K. Miyazono, U. Hellman, Nature 338:557-562 (1989).
15. S. J. Leibovich, P. J. Polverini, H. M. Shepard, Nature 329:630-632 (1987).
16. C. G. Gay, J. A. Winkles, Proc. Natl. Acad. Sci. USA 88:296-300 (1991).
17. G. A. Hoover, S. McCormick, N. Kalant, Arteriosclerosis 9:76-83 (1989).
18. W. A. Thomas, D. N. Kim in: Vol. 1 E. Rubin, I. Damjanov (eds), Williams and Wilkins, Baltimore, pp. 41-51 (1984).
19. A. Penn, S. J. Garte, L. Warren, Proc. Natl. Acad. Sci. USA 83:7951-7955 (1986).
20. J. Folkman, Anti-angiogenesis, Ann. Surg. 175:409-416 (1972).
21. S. S. Brem, D. Zagzag, A. M. Tsanaclis, S. Gately, E. Marie-Pierre, S. E. Brien, Am. J. Pathol. 137:1121-1142 (1990).

ANGIOGENESIS IN ARTHRITIC DISEASE: EFFECT OF HYALURONAN ON ENDOTHELIAL CELL MIGRATION

A. Sattar and S. Kumar

Dept. of Pathological Sciences, University of Manchester
Manchester M13 9PZ, England

Angiogenesis is a hallmark of rheumatoid arthritis but the mechanism of angiogenesis in this disease is poorly understood. Two events that always occur in angiogenesis are endothelial cell proliferation and migration. Hyaluronan (HA) is an important component of syhovial fluid. Both its concentration and size distribution varies in synovial fluid from normal and patients with rheumatoid arthritis. Depending on its molecular size, HA can stimulate or inhibit angiogenesis in vivo and endothelial cell proliferation in vitro. However, very little is known about the possible effect of HA size on endothelial cell migration. We have examined this using two different methods.

Firstly the length of migration into a filter matrix was studied using nitrocellulose filters and secondly the ability of cells to cross pores on a very thin polycarbonate filter. Both methods demonstrated that HA fragments (2-10 disaccharides) are able to promote migration of bovine endothelial cells. The fragments had no effect on the migration of bovine aortic smooth muscle cells or human lung fibroblasts. The data suggest that HA fragments specifically act on endothelial cells and may play an important part in the angiogenesis seen in rheumatoid

aFGF, EGF AND THEIR RECEPTORS IN THE NORMAL GASTRIC MUCOSA AND IN ULCER

HEALING

Hans-Arne Hansson and Elisabeth Norstrom

University of Göteborg, Department of Histology
P.O. Box 33031, S-400 33 Göteborg, Sweden

Mechanisms promoting the healing of gastric wounds are not known in detail, but reduction in gastric juice acidity or administration of epidermal growth factor (EGF) have been shown to be beneficial. The present work was aimed to investigate if there is any local formation of peptide growth factors in the normal gastric mucosa and, if so, if the corresponding receptors were demonstrable in the gastric mucosa as well as possible changes in their distribution during the healing of experimentally induced wounds. Immunohistochemical methods were mainly used.

Acidic fibroblasts growth factor (aFGF) is demonstrable in the cytoplasm of mucons neck cells in the normal rat oxyntic glands, and EGF in rare, basal chief cells. A fraction of the mucous neck cells express as well type IFGF receptors (FGF-R) and EGF receptors (EGF-R). One day after induction of a gastric ulcer in the rat stomach by applying acetic acid during a minute to the gastric serosa, there was a prominent increase of the aFGFimmunoreactive cells and concomitanly the proliferating oxytic gland epithelial cells showed an increasing number of cells expressing FGF-R and EGF-R. Numerous chief cells in the basal half of the oxyntic glands at the wound expressed EGF immunoreactivity. Maximal aFGF and EGF immunoreactivity was seen on day 3 to 7; the same was true for the FGF-R and EGF-R. The ulcers were usually reepithelialized in 2 weeks and grossly healed in 3 weeks. The poorly differentiated epithelial cells in the gastric mucosa scar showed still after three weeks strong immunoreactivity for FGF-R and EGF-R while EGF and in low frequency, aFGF-positive cells were mainly recognized in the basal parts of the oxyntic glands enveloping the scar. An intricate system in the gastric mucosa thus promote its maintenance and healing and involve locally formed trophic peptides, i.e. aFGF and EGF, and that epithelial stem cells appropriately express the corresponding receptors. The gland cells in the gastric mucosa scar did not attain their normal differentiation in 3 weeks but remained immature as reflected by their morphology and by their persisting expression of FGF-R and EGF-R, tentatively related to relative scarcity of aFGF forming cells. Vascular cells were fairly scanty in type I FGF-R.

DIABETES INDUCED VASCULAR CHANGES IN HUMAN INTERNAL MAMMARY ARTERY AND SAPHENOUS VEIN GRAFTS

C. Karasu, K. Ayrancioglu, H. Soncul, G. Ozansoy, V. Melih Altan
University of Ankara, Faculty of Pharmacy, Dept. of Pharmacology.
06100 Tandogan, Ankara, Turkey

Alterations in the reactivity of blood vessels to neurotransmitters and circulating hormones have been suggested to cause or contribute to some of cardiovascular complications, such as hypertension, that are associated with diabetes mellitus. On the other hand, macrovascular disease is a significant cause of morbitidity and mortality in diabetic patients.

In this study, cumulative contractile response to noradrenaline and endotheline (ET-1) were found to be enhanced in internal mammary arteries and saphenous veins obtained from noninsulin-dependent diabetic patients undergoing coronary artery bypass surgery compared to those from nondiabetic controls.

Endothelium-dependent relaxations produced by acetylcholine in internal mammary artery and saphenous vein rings precontracted with noradrenaline were significantly attenuated from noninsulin-dependent diabetic patients.

Alterations in vascular responsiveness may have a role in promoting vasospasm and vascular occlusion contribute to increased vascular resistance in diabetic hypertension. Furthermore, enhanced contractions to noradrenaline and ET-1 in diabetic saphenous vein and internal mammary artery may be important for graft function and patency in diabetic patients undergoing coronary bypass grafting.

THROMBIN IMMOBILIZED TO ECM PROMOTES VASCULAR SMOOTH MUSCLE CELL PROLIFERATION VIA A NON-ENZYMATIC DOMAIN

Benezra M., Vlodavsky and R. Bar-Shavit

Department of Oncology, Hadassah University Medical Center
Jerusalem, 91120 Israel

Thrombin (EC 21.3.4) a serine protease with a central role in hemostasis is ubiqiutous to injury sites where the endothelial layer has been disturbed. We have demonstrated that thrombin can be immobilized onto the subendothelial extracellular matrix (ECM) through a short anchorage binding site ("loop B" domain) leaving most of the molecules free for further interaction with the surrounding milieu. Indeed, while immobilized onto the ECM, thrombin was found capable of stimulating effectively vascular smooth muscle proliferation. This growth stimulation was completely independent of the proteolytic pocket since esterolytically inactive diisopropyl phosphofluoride-conjugated thrombin (DIP-a-thrombin) induced ^{3}H-thymidine incorporation and increased cell number in growth arrested SMCs. This stands in contrast to the mitogenic effect in human skin fibroblast (HF) where only the enzymatically active thrombin was capable of inducing cell proliferation. In addition, a variety of thrombin species that were chemically modified to alter thrombin procoagulant or esterolytic functions stimulated ^{3}H-thymidine in SMCs. Likewise, DIP-a-thrombin induced the rapid and transient expression of the c-fos proto-oncogene in these cells. Binding of ^{125}I-a-thrombin to SMC cultures revealed the presence of specific cell surface receptors and cross-linking studies of ^{125}I-a-thrombin to SMCs showed a specific 55KDa cell surface protein. These observations lend support to the possibility that thrombin once present within the arterials wall can be firmly bound to the ECM providing a localized and persistent growth stimulation of SMCs. Therefore, thrombin may be directly involved in the progression of the atherosclerotic plaque formation.

RADIATION INDUCES CHANGES IN GLYCOSAMINOGLYCANS OF ENDOTHELIAL CELLS

D. Pye, S. Kumar, J. Gallagher, and R. Hunter

Christie Hospital
Manchester M20 9BX, England

Proteoglycans (PGs) are macromolecules found widely distributed in animal tissues principally in the extracellular matrix (ECM) and cell surface memebranes. A PG is a glycoprotein consisting of a protein core, linked to which are one or more sulphated polysaccharide chains, glycosaminoglycans (GAGs). We have observed that X-irradiation of bovine aortic endothelial cells in culture result in either changes in GAG production levels or molecular organization or both. A variety of techniques including chromatography and electrophoresis can be employed to elucidate structural changes in the GAGs, which may occur on irradiation of BAEC. In particular one type of GAG heparan sulphate, is under study and shows promise as a biological indicator of radiation induced damage to endothelial cells during radiotherapy.

In a related study 2-D gel electrophoretic mapping of endothelial cell proteins has demonstrated marked differences in irradiated compared with unirradiated endothelial cells. Isolation of such radiation induced proteins may eventually allow development of antibodies and quantitative assay capable of measuring them in body fluids of cancer patients receiving radiotherapy.

ADHESION MOLECULES, THEIR ANTIBODIES AND AN EXAMPLE OF A THERAPEUTIC APPLICATION WITH ANTI-LFA1

André Van Agthoven

Immunotech,
13288 Marseille Cedex 9, France

Recently a strong interest has arised in adhesion molecules in the different fields of research.
- Platelet research with possible applications of thrombosis, haemostasis and cardiac disease.
- T cell immunology including immunosuppression, immune stimulation and cytotoxicity.
- Inflammatory disease including interaction of leucocytes with the endothelial wall. This latter phenomenon also having impact on auto-immune disease.
- Lymphocyte homing including the study of homing receptors.
- Cancerology studying the process of metastasis.

The adhesion molecules can be subdivided in three families: the integrins, the members of the immunoglobulins superfamily and the LECCAMS. Also molecules having unique sequence patterns have been identified. The member of families of adhesion molecules might be extended in the future.

Most of the adhesion molecules have been cloned and the sequence is known. Monoclonal antibodies against these structures have been raised and there is a vivid discussion in the possibilities of these reagents in diagnostics and therapy.

Since 1986, an antibody against the integrin molecule LFA1 has been used in an experimental therapeutical protocol in France concerning bone marrow transplantation of patients having partial immune deficiency diseases. The results of this therapeutical trial support the idea of the use of anti-adhesion molecule monoclonal antibodies in therapy.

QUANTITATIVE METHODS TO EVALUATE ANGIOGENESIS AND ANGIOSUPPRESSION: A RAPID IMMUNOCYTOCHEMICAL CYTOKINETIC BRAIN TUMOR ASSAY

S. Brem, S. Brien

Neurosurgual Oncology, Northwestern Med. Faculty Found. Inc.
233 East Evie, Suite 500, Chicago, Ill. 60611, USA

More than 20 years ago, Judah Folkman envisioned a time when "anti-angiogenesis" would be applied clinically to treatment of human cancer, but one fundamental problem remained: how could the growth of microvasculature, and its inhibition, be scientifically (i.e. quantitatively) assessed, and avoid the subjective and confusing terms of "highly vascular" or "relatively avascular" terms popular in the older literature[1]. Working with Ramzi Cotran, we developed a "MAGS" (microscopic angiogenesis grading system) method to quantitate angiogenesis in histologic sections, based on 3 features of endothelial cell (EC) growth: a) cytology, b) EC hyperplasia, and d) microvascular density (MVD)[2]. The MAGS method has proved useful to quantify tumor angiogenesis in a variety of human[3-5] and experimental[6,7] tumor systems. Because the intense hyperplastic and dedifferentiated forms of the capillary endothelium appear unique to the microvessels of human brain tumors[2], from a practical perspective, microvascular density has proved to be the most valuable scale to assess extracerebral tumors. Vascular density is among the factors that are decisive for the growth rate of tumors[8]. For example, histolofic quantitation of MVD was shown to correlate with the risk of distant metastasis in patients with invasive breast carcinoma[9]. We recently applied a monoclonal antibody to bromodeoxyuridine that provides a rapid, reproducible, nontoxic, immunohistochemical method to measure cellular kinetics of intracerebral tumor angiogenesis[10]. The rabbit brain tumor model of angiogenesis consists of tumor and endothelial cell populations with high proliferative rates that demonstrate the close interdependence between microvascular and neoplastic growths as well as topographic gradients, heterogeneity, and regional microdomains of cell proliferation[7]. The labelling index of endothelial cell was 25.8% at the tumor periphery, compared to 1.7% in the tumor center (P<0.001). Concomitant with an increased turnover of neoplastic cells at the tumor periphery, the tumor LI was 26.6% with a LI of 7.7% in the center (P<0.01). Furthermore, labeled tumor cells tended to be organized around proliferating capillaries, with less DNA synthesis further form the nearest blood vessel. The established normal microvessels of the brain, e.g. in the opposite tumor-free hemisphere, were mitotically inactive with a LI of <0.001%. Quantitation of vascular cytokinetics should be useful in further studies of the pathophysiology of brain tumor angiogenesis and the development of pharmacolgoical approaches directed toward the microvasculature. For example, using this method, dexamethasone has been shown to be selectively suppress EC proliferation, validating its place as an angiostatic[12], or angiosuppressive, steroid.

References

1. J. Folkman, Personal Communication
2. S. Brem, R. Cotran, J. Folkman, J. Natl. Cancer Inst. 48:347-356 (1972).
3. J. Vafidis, J. E. Meats, H. Reid, in: Brain Oncology, Chatel M., Darcel F., Pecker J. (eds), Boston, Martinus Nijhoff, pp. 149-152 (1987).
4. N. Kochi, E. Tani, T. Morimura, Acta Neuropathol. 59:119-126 (1983).
5. A. Srivistava, P. Laidler, R. P. Davies, Am. J. Pathol. 133:419-423 (1988).
6. B. R. Deane, P. L. Lantos, J. Neurol. Sci. 49:67-77 (1981).
7. D. Zagzag, S. Brem, F. Robert, Am. J. Pathol. 131:361-372 (1988).
8. E. K. Rofstad, Cell Tissue Kinet 17:91-101 (1984).
9. N. Weidner, J. P. Semple, W. R. Welch, J. Folkman, New Engl. J. Med. 324:1-8 (1991).
10. S. Brien, S. Brem, Neurosurgery 25:715-719 (1989).
11. S. Gately, A. M. Tsanaclis, S. Brem, FASEB J. 4:A487 (Abst. #1284) (1990).
12. J. Folkman, D. E. Ingber, Ann. Surg. 206:374-383 (1987).

Participants of the NATO Advanced Studies Institute "Angiogenesis in Health and Diseases" held at Porto Hydra, Greece, during 16 - 27 June, 1991. The organizing committee included Pietro Gullino (ASI Co-Director), Peter Lelkes (ASI Co-Director), Michael Maragoudakis (ASI Director and Chairman), Juliana Denekamp, Olga Hudlicka and Shank Kumar.

PARTICIPANTS

ALPER, R.
University City Science Center
Connective Tissue Res. Institute
3624 Market Street, PA 19104, U.S.A.

ANTONIADES, H.
Harvard School of Publich Health
Cancer Biology, 655 Huntington Ave.,
Boston, Mass. 02115, U.S.A.

AUERBACH, R.
Laboratory of Development, Biology,
Zoology, Research Building, Univ.
of Wisconsin, 1117 West Johnson Str.
Madison, WI53706, U.S.A.

BASTAKI, M.
The Gray Laboratory, P.O. Box 100,
Mount Vernon Hospital, Northwoood,
Middlesex, HA6 2JR, UNITED KINGDOM

BEBOU-TSOLAS, L.
University of Ioannina, Medical Sch.
Department of Biochemistry, Ioannina
GREECE

BENEZRA, M.
Department of Radiation and Clinical
Oncology, Hadassah University Hospi-
tal, P.O. Box 12000 Jerusalem,
ISRAEL

BINDER, K.
Department of Pharmacology, Dr. K.
Thomae GmbH, Birkendorfer Stabe 65,
Postfach 1755, D-7950 Biberach,
GERMANY

BREM, H.
Department of Surgery, Harvard Medical School, The Children´s Hospital,
300 Longwood Ave., Boston, Mass.
02115, USA

BREM, S.
Neurosurgual Oncology, North Western
Medical Faculty Foundation Inc.
Streetville Center, 233 East Evie,
Suite 500, Chicago, Ill. 60611,
U.S.A.

BROWN, R.
Department of Experimental Pathology
Institute of Orthopaedics, R.N.O.H.
Brockley Hill, Stanmore, Middlesex
HA7 4LP, UNITED KINGDOM

CHEN, C.T.
Institut für Arterios Kleroseforschung, University Münster, Dogmak str. 3, Münster, GERMANY

CHRISTENSSON, P. I.
KABI Pharmacia Therapeutics AB, P.O. Box 941, S-25109 Helsingborg, SWEDEN

DEJANA, E.
Instituto di Ricerche Pharmacologiche Mario Negri, Via Eritrea 62, 20157 Milano, ITALY

DELIDES, G. S.
University of Crete, Medical School, Department of Pathology, Crete, GREECE

DENECAMP, J.
The Gray Laboratory, P.O Box 100, Mount Vernon Hospital, Northwood, Middlesex, HA6 2JR, UNITED KINGDOM

DIETERLEN-LIEVRE, F.
Institut D´Embryologie Cellulaire et Moleculaire UMR 009, 49 bis Avenue de la Gabriello, 94736 Nogent-Sur-Marne Cedex, Paris, FRANCE

DREXLER, H.
Max Planck Institut für Psychiatre Abt. Neurochemie, Am Klopferspritz 18a, D-8033 Martinsried, München, GERMANY

FAN, T. P. D.
University of Cambridge, Tennis Court Road, Cambridge CB2 1QJ, UNITED KINGDOM

GARNER, A.
Bioscience Department, ICI Pharmac. Mereside, Arderley Park, Macclesfield Chesire SK10 4TG, UNITED KINGDOM

GRANT, D.
Section of Dev. Biology and Anomalies National Inst. of Dental Research Bldg. 30, Room 414, Bethesda, MD20892, U.S.A.

GUERIN, C.
Brain Tumor Service, Johns Hopkins Hospital Neurosurgery, 600 North Wolfe Str./Meyer 7-113, Baltimore, Maryland 21205, U.S.A.

GULLINO, P.
Dipartimento di Scienze Biomediche e Oncologia Umana, Sezione di Anatomia e Istologia Patologia, Via Santena 7, 10126 Torino, ITALY

HAMOT, D.
Institut d´Embryologie Cellulaire et Moleculaire UMR 009, 49 bis Avenue de la Gabriello, 94736 Nogent-Sur-Marne Cedex, Paris, FRANCE

HANSSON, H. A.
University of Göteborg, Department of Histology, Institute of Neurobiology P. O. Box 33031, S-400 33 Götebörg SWEDEN

HANSSON, M.	University of Göteborg, Department of Histology, Institute of Neurobiology P.O. Box 33031, S-400 33 Göteborg, SWEDEN
HARALABOPOULOS, G.	University of Patras Medical School, Department of Pharmacology, 26500 Rio, Patras, GREECE
HAUDENSCHILD, C.	Mallory Institute of Pathology, Cardiovascular Research Laboratory, 784 Massachusetts Ave., Boston, MA 02118, U.S.A.
HUDLICKA, O.	University of Birmingham, Medical School, Department of Physiology, Vincent Drive, Birmingham B15 2TJ UNITED KINGDOM
ITHAKISSIOS, D.	University of Patras, School of Pharmacy, Department of Pharmaceutical Technology, 26500 Rio Patras, GREECE
KARAKIULAKIS, G.	Aristotle University of Thessaloniki Department of Pharmaceutical Sciences Laboratory of Pharmacology, 54006 Thessaloniki, GREECE
KARASU, C.	University of Ankara, Faculty of Pharmacy, Department of Pharmacology, 06100 Tandogan, Ankara, TURKEY
KARG, C.	Dr. K. Thomae GmbH, Department of Pharmacology, Birkendorfer Stabe 65 Postfach 1755, D-7950 Biberach, GERMANY
KEFALIDES, N.	University of Pennsylvania, Department of Medicine, University City Science Center, 3624 Market Street, Philadelphia Pennsylvania 19104, U.S.A.
KONERDING, M. A.	University of Essen, Institute of Anatomy, Hufelandstr. 55, 4300 Essen I GERMANY
KUMAR, S.	Clinical Research Laboratories, Christie Hospital, Manchester M20 9BX UNITED KINGDOM
LELKES, P. I.	Laboratory of Cell Biology, University of Wisconsin, Medical School, Department of Medicine, Sinai Samaritan Medical Center, Mount Sinai Campus, 950 North Twelfth Str., P.O. Box 342, Milwaukee, WI 153201-0342, U.S.A.
LITTBRAND, B.	University of Umea, Department of Oncology, S-901 85 Umea, SWEDEN

MALYSKA, E.	Agricultural-Pedagogical University ul. 3 Maja 54 (Rector) p. 111, Siedlce 08-110, POLAND
MANOLOPOULOS, E.	University of Patras Medical School, Department of Pharmacology, 26500 Rio Patras, GREECE
MARAGOUDAKIS, M. E.	University of Patras Medical School, Department of Pharmacology, 26500 Rio Patras, GREECE
McLAUGHLIN, B.	Wolfson Angiogenesis Unit, Department of Rheumatology, Clinical Science Bldg. Hope Hospital, Salford M6 8HD, UNITED KINGDOM
MOREIRA, C.	University of Lisboa, Faculty of Medicine, Institute of Biochemistry, Egas Moniz Avenue, 1699 Lisboa Cedex PORTUGAL
OZANSOY, G.	University of Ankara, Faculty of Pharmacy, Department of Pharmacology, 06100 Tandogan, Ankara, TURKEY
PAPACONSTANTINOU, E.	Kantosspital Basel, Univeritätskliniken Department Forshung, Hebelstrasse 20, CH-4031 Basel, SWITZERLAND
PAPADIMITRIOU, E.	University of Patras Medical School Department of Pharmacology, 26500 Rio Patras, GREECE
PAPAIOANNOU, S.	University of Patras, Department of Pharmaceutical Sciences, Laboratory of Pharmacology, 26500 Rio Patras, GREECE
PARDANAUD, L.	Institut d´Embryologie Cellulaire et Moleculaire, UMR 009, 49 bis Avenue de la Belle Gambrielle, 94736 Nogent-sur-Marne cedex, Paris, FRANCE
POLVERINI, P.	Northwestern University Medical and Dental Schools, Ward Memorial Bldg. B 179, 303 East Chicago Ave., Chicago Ill. 60611-3008, U.S.A.
PRESTA, M.	Universita degli Studi di Brescia, Facolta di Medicina e Chirgurgia Dipartimento di Scienze, Biomediche e Biotecnologie, Via Valsabbina 19, I-25123 Brescia, ITALY
PROTOPAPA, E.	Anticancer Institute of Pireaus "Metaxa", Botassi 51, Pireaus, GREECE
PYE, D.	Christie Hospital, Clinical Research Laboratories, Manchester M20 9BX, UNITED KINGDOM

SAHINOGLOU, T.	University of London, London Hospital Medical College, Bone and Joint Res. Unit, Turner Str., London E1 2AD, UNITED KINGDOM
SARMONIKA, M.	University of Patras Medical School, Department of Pharmacology, 26500 Rio Patras, Greece
Sattar, A.	Christie Hospital, Clinical Research Laboratories, Manchester M20 9BX, UNITED KINGDOM
SHARMA, H. S.	Max-Planck Institute, Department of Experimental Cardiology, Benekestrasse 2, D-6350 Bad Nauheim, GERMANY
STEVENS, C.	London Hospital Medical College, University of London, Bone and Joint Research Unit, Turner Str., London E1 2AD, UNITED KINGDOM
TOUMBIS, C.	Internal Medicine and Experimental Pathology, Roswell Park Elm and Carlton Streets, Buffalo, NY 14263, U.S.A.
TOUMBIS, E.	Mavromateon 10, 10682 Athens, GREECE
TSANACLIS, A. M. C.	Rua dos Franceses-470, Ed. Renan 271, 01329 Sao Paolo, SP, BRAZIL
TSIGANOS, C.	University of Patras, Department of Organic Chemistry, Biochemistry and Natural Products, 26500 Rio, Patras, GREECE
TSOLAS, O.	University of Ioannina Medical School, Department of Biochemistry, Ioannina, GREECE
TSOPANOGLOU, N.	University of Patras Medical School, Department of Pharmacology, 26500 Rio, Patras, GREECE
VAN ACKERN, C.	University of Essen, Institute of Anatomy, Hufelandstr. 55, 4300 Essen 1 GERMANY
VAN AGTHHOVEN, A.	Immunotech, Luminy Case 915, 13288 Marseille Cedex 9, FRANCE
VAN HISBERGH, V. W. M.	TNO Gaubius Institute, P.O. Box 430, 2300 AK Leiden, THE NETHERLANDS
WANG, J.	Christie Hospital, Clinical Research Laboratories, Manchester M20 9BX, UNITED KINGDOM
WEGRZYN, M.	Agricultural Pedagogical University ul. 3 Maja 54 (Rector) p.111, Siedelce 08-110, POLAND

WEISS, J. B.   Wolfson Angiogenesis Unit, Department
                                      of Rheumatology, Clinical Science Bldg.
                                      Hope Hospital, Salford M6 8HD, UNITED
                                      KINGDOM

WILLIAMSON-GRANT, S.   Section of Development Biology and
                                      Anomalies, National Institute of
                                      Dental Research, Bldg. 30, Room 414,
                                      Bethesda, MD 20892, U.S.A.

ZAGRIS, N.   University of Patras, Department of
                                      Biology, Laboratory of Physiology,
                                      26500 Rio, Patras, Greece

INDEX